Have you been to our website?

For code downloads, print and e-book bundles, extensive samples from all books, special deals, and our blog, please visit us at:

www.rheinwerk-computing.com

Rheinwerk Computing

The Rheinwerk Computing series offers new and established professionals comprehensive guidance to enrich their skillsets and enhance their career prospects. Our publications are written by the leading experts in their fields. Each book is detailed and hands-on to help readers develop essential, practical skills that they can apply to their daily work.

Explore more of the Rheinwerk Computing library!

Kofler, Gebeshuber, Kloep, Neugebauer, Zingsheim, Hackner, Widl, Aigner, Kania, Scheible, Wübbeling

Hacking and Security: The Comprehensive Guide to Penetration Testing and Cybersecurity

2023, 1141 pages, paperback and e-book
www.rheinwerk-computing.com/5696

Tobias Fertig, Andreas Schütz

Blockchain: The Comprehensive Guide to Blockchain Development, Ethereum, Solidity, and Smart Contracts

2024, 654 pages, paperback and e-book
www.rheinwerk-computing.com/5800

Veit Steinkamp

Python for Engineering and Scientific Computing

2024, 511 pages, paperback and e-book
www.rheinwerk-computing.com/5852

Michael Kofler

Scripting: Automation with Bash, PowerShell, and Python

2024, 470 pages, paperback and e-book
www.rheinwerk-computing.com/5851

Michael Kofler

Linux: The Comprehensive Guide

2024, 1178 pages, paperback and e-book
www.rheinwerk-computing.com/5779

www.rheinwerk-computing.com

Sandip Dholakia

Modern Cryptography

The Practical Guide

Editor Megan Fuerst
Acquisitions Editor Hareem Shafi
Copyeditor Julie McNamee
Cover Design Graham Geary
Photo Credit iStockphoto: 1486820020/© MicroStockHub, 468900078/© D3Damon
Layout Design Vera Brauner
Production Hannah Lane
Typesetting SatzPro, Germany
Printed and bound in Canada, on paper from sustainable sources

ISBN 978-1-4932-2565-1
© 2024 by Rheinwerk Publishing, Inc., Boston (MA)
1st edition 2024

Library of Congress Cataloging-in-Publication Control Number: 2024034055

All rights reserved. Neither this publication nor any part of it may be copied or reproduced in any form or by any means or translated into another language, without the prior consent of Rheinwerk Publishing, 2 Heritage Drive, Suite 305, Quincy, MA 02171.

Rheinwerk Publishing makes no warranties or representations with respect to the content hereof and specifically disclaims any implied warranties of merchantability or fitness for any particular purpose. Rheinwerk Publishing assumes no responsibility for any errors that may appear in this publication.

"Rheinwerk Publishing", "Rheinwerk Computing", and the Rheinwerk Publishing and Rheinwerk Computing logos are registered trademarks of Rheinwerk Verlag GmbH, Bonn, Germany.

All products mentioned in this book are registered or unregistered trademarks of their respective companies.

Contents at a Glance

PART I Foundations of Modern Cryptography

1	Fundamentals of Cryptography	27
2	Symmetric Cryptography	79
3	Asymmetric Cryptography	113
4	Cryptography Services	147

PART II Modern Cryptography in Practice

5	Storage Security: Data Encryption at Rest	177
6	Web Security: Data Encryption in Transit	199
7	Cloud and Connected Device Cryptography	237
8	Cryptography in Cryptocurrency	271
9	Cryptography and Artificial Intelligence	297
10	Post-Quantum Cryptography	317
11	Homomorphic Encryption	345
12	Cryptography Attacks and Ransomware	367
13	Cryptography Standards and Resources	389
14	Future Trends and Concluding Remarks	399

Contents

Foreword 15
Acknowledgments 19
Preface 21

PART I   Foundations of Modern Cryptography

1   Fundamentals of Cryptography   27

1.1   History of Cryptography 27
1.1.1   Historical Development of Cryptography 28
1.1.2   Why Is Cryptography Required? 33
1.2   Introduction to Cryptography 34
1.2.1   Definitions 34
1.2.2   Classical versus Modern Cryptography 36
1.2.3   Encoding, Hashing, and Encryption 38
1.2.4   Security Concepts 40
1.2.5   Types of Modern Cryptography 48
1.2.6   Steganography 51
1.2.7   Additional Techniques 53
1.2.8   Confusion and Diffusion 55
1.2.9   Refresher on Mathematics 56
1.3   Primer on Ciphers 57
1.3.1   Substitution Cipher 57
1.3.2   Transposition Ciphers 60
1.3.3   Stream Ciphers 62
1.3.4   Block Ciphers 70
1.4   Summary 78

2   Symmetric Cryptography   79

2.1   Primer on Symmetric Cryptography 79
2.2   Symmetric Key Algorithms 83
2.2.1   Data Encryption Standard 83

7

	2.2.2	Triple DES and DES Variants	91
	2.2.3	International Data Encryption Algorithm	93
	2.2.4	Advanced Encryption Standard	95
	2.2.5	AES Finalists	107
2.3	Summary		110

3 Asymmetric Cryptography   113

3.1	Primer on Asymmetric Cryptography		114
	3.1.1	Properties of Asymmetric Cryptography	115
	3.1.2	Introductory Mathematics	119
3.2	Asymmetric Cryptography Algorithms		124
	3.2.1	RSA Algorithm	124
	3.2.2	Diffie–Hellman–Merkle Key Exchange Algorithm	130
	3.2.3	Elgamal Cryptosystem	136
	3.2.4	Elliptic Curve Cryptography	139
3.3	Summary		144

4 Cryptography Services   147

4.1	Hash Functions and Algorithms		148
	4.1.1	Primer on Cryptographic Hash Functions	148
	4.1.2	Message Digest Algorithms	153
	4.1.3	SHA Family of Algorithms	157
	4.1.4	SHA-3 Algorithm	161
4.2	Message Authentication Codes		168
4.3	Digital Signature		169
	4.3.1	Primer on Digital Signatures	169
	4.3.2	Digital Signature Standard	170
4.4	Merkle Trees		172
4.5	Summary		173

PART II Modern Cryptography in Practice

5 Storage Security: Data Encryption at Rest — 177

5.1	Primer on Data Security	178
	5.1.1 Understanding the Data to Be Protected	178
	5.1.2 Understanding Data Security	181
5.2	Data-at-Rest Encryption Methods	183
	5.2.1 Disk Encryption	184
	5.2.2 Volume Encryption	186
	5.2.3 File Encryption	188
	5.2.4 Database Encryption	190
	5.2.5 Application Encryption	194
5.3	Summary	196

6 Web Security: Data Encryption in Transit — 199

6.1	Primer on Web Security	200
6.2	Web Security Protocols	209
	6.2.1 Implementing the TLS Protocol	209
	6.2.2 Implementing VPNs Using IPS	222
6.3	Securing Web-Based Applications	227
	6.3.1 Securing Email Communication	227
	6.3.2 Securing Streaming and Downloading	230
6.4	Public Key Infrastructure	232
6.5	Summary	236

7 Cloud and Connected Device Cryptography — 237

7.1	Primer on Cloud Cryptography	238
	7.1.1 Securing Infrastructure in the Cloud	240
	7.1.2 Securing Data in the Cloud	243
	7.1.3 Securing Applications in the Cloud	247

Contents

7.2		**Encryption Key Management**	**248**
	7.2.1	Cloud Service Provider-Managed Key	251
	7.2.2	Customer-Managed Key	252
	7.2.3	Bring Your Own Key	253
7.3		**Cryptography as a Service by Major Cloud Service Providers**	**254**
	7.3.1	Amazon Web Services	255
	7.3.2	Microsoft Azure	257
	7.3.3	Google Cloud Platform	258
	7.3.4	Cryptography Services by Other Cloud Service Providers	259
7.4		**Lightweight Cryptography and the Internet of Things**	**259**
	7.4.1	The IoT Concept	260
	7.4.2	Risks and Attacks Associated with IoT	261
	7.4.3	Securing Connected Devices with Lightweight Cryptography	263
	7.4.4	Cryptography in Cars	266
7.5		**Summary**	**268**

8   Cryptography in Cryptocurrency   271

8.1		**Primer on Cryptocurrency**	**272**
	8.1.1	History of Money	272
	8.1.2	Introduction to Cryptocurrency	275
	8.1.3	Primer on Blockchain	278
8.2		**The "Crypto" in Cryptocurrency**	**282**
	8.2.1	Cryptographic Transactions	282
	8.2.2	Cryptography Algorithms Used in Cryptocurrency	287
	8.2.3	Cryptocurrency Wallets	290
8.3		**Outlook: Cautiously Optimistic**	**293**
8.4		**Summary**	**296**

9   Cryptography and Artificial Intelligence   297

9.1		**Primer on AI**	**298**
9.2		**AI for Cryptography**	**303**
	9.2.1	Role of AI in Cryptography	303
	9.2.2	AI Algorithms for Cryptography	305

9.3	**Cryptography for AI**	309
	9.3.1 Security Risks of AI	310
	9.3.2 Securing AI Models with Cryptography	313
9.4	**Best Practices and Ethical Use of AI**	315
9.5	**Summary**	316

10 Post-Quantum Cryptography   317

10.1	**Primer on Quantum Computing**	317
	10.1.1 History of Quantum Mechanics	318
	10.1.2 Quantum Computing 101	319
	10.1.3 Quantum Computing Technologies	323
10.2	**Quantum Computing and Cryptography**	325
	10.2.1 The Risk of Quantum Computing	325
	10.2.2 Quantum Cryptography	328
	10.2.3 Quantum-Resistant Cryptography Algorithms	331
10.3	**Preparing for Post-Quantum Cryptography**	336
	10.3.1 NIST: Initiative on Post-Quantum Cryptography	336
	10.3.2 Preparing and Implementing Post-Quantum Cryptography	338
10.4	**Future of Cryptography in Post-Quantum Computing**	341
10.5	**Summary**	344

11 Homomorphic Encryption   345

11.1	**Primer on Homomorphic Encryption**	346
	11.1.1 History of Homomorphic Encryption	346
	11.1.2 Understanding Homomorphic Encryption	348
	11.1.3 Types of Homomorphic Encryption	351
	11.1.4 Homomorphic Encryption Using a Symmetric Key	353
11.2	**Practical Applications of Homomorphic Encryption**	354
	11.2.1 Healthcare	354
	11.2.2 Electronic Voting System	357
	11.2.3 Artificial Intelligence and Other Use Cases	360

11.3	Advantages, Challenges, and the Future	361
	11.3.1 Advantages	361
	11.3.2 Challenges	362
	11.3.3 Future of Homomorphic Encryption	364
11.4	Summary	365

12 Cryptography Attacks and Ransomware   367

12.1	Primer on Cryptography Attacks	368
	12.1.1 Cryptanalytic Attacks	368
	12.1.2 Man-in-the-Middle Attacks	374
	12.1.3 Other Cryptography Attacks	377
	12.1.4 Cryptography Attacks in a Nutshell	378
12.2	Ransomware Attacks	379
	12.2.1 Anatomy of Ransomware	380
	12.2.2 Ransomware as a Service	384
	12.2.3 Protecting Against Ransomware Attacks	386
12.3	Summary	387

13 Cryptography Standards and Resources   389

13.1	Government Standards for Cryptography	389
	13.1.1 National Institute of Standards and Technology	390
	13.1.2 European Standards and Regulations for Cryptography	391
13.2	Other Standards for Cryptography	392
13.3	Best Practices and Other Resources	392
	13.3.1 Best Practices for Cryptography	393
	13.3.2 Other Resources	394
13.4	Further Reading	396
13.5	Summary	397

14 Future Trends and Concluding Remarks — 399

14.1 Future Trends in Cryptography — 399
14.2 Concluding Remarks — 401
14.3 Summary — 403

Appendices — 405

A Bibliography — 405
B The Author — 415

Index — 417

Foreword

Cryptography is the backbone of security in our digital world, and it continues to grow in importance as services, capabilities, and our lives become ever more digital. Cryptography increases in importance daily as we see new reports about cyberattacks. Hospitals, retail stores, businesses of all sizes, governments—all are under a constant threat of attack to exfiltrate data, disrupt critical systems, or other nefarious purposes.

For the last few decades, vendors have provided all the implementations of cryptography that organizations use. No one expected the level of complexity of cryptographic subsystems, and organizations became unaware of the strengths and weaknesses of their overall environments. The time has come for every organization to own and manage their cryptographic systems (although they will mostly still be provided by vendors and open-source libraries). It's becoming essential that cryptographic assets are regarded as high-priority items, and that teams are dedicated to understanding, managing, and updating these implementations. This book provides a valuable step for security professionals to understand the practical issues related to using cryptography in our digital world.

Modern cryptography is strong against the capabilities of current computers, if we assume the implementations are current and free of vulnerabilities. Data breaches, for example, haven't been successful because they break cryptography; rather, successful attackers mostly obtain digital identities and masquerade as authorized entities to gain access to data and systems. Consequently, we operate on the well-founded belief that the cryptography used to protect our digital lives is, in fact, very secure. One can easily see that the practical implementations of cryptographic systems have been the weak link in our systems. Understanding the importance and the subtle nature of implementing these systems has proven to be of utmost importance.

A new threat—quantum computing—is challenging that belief. Cryptographers have known for decades (in fact, well before the development of quantum computers began) that a quantum computer of sufficient power would be able to break some of the most important cryptographic algorithms we depend on today. Significant advances in quantum computing are being announced every few months, putting pressure on organizations to prepare for their arrival. This will force many organizations to understand, modify, and manage their cryptographic infrastructure, which again will require security professionals to gain more knowledge.

To counter the threat of quantum computing, organizations worldwide face challenges of moving from a small set of well-known, traditional cryptographic algorithms to a larger set of new algorithms specifically developed to withstand quantum computing. These new algorithms are known as post-quantum cryptographic algorithms.

Foreword

It's difficult to think of any area of an organization's digital landscape that isn't touched by cryptography. Every user login and machine authentication, such as a web server identifying itself to a browser, requires cryptography. Securing data in transit and at rest requires cryptography. With such a widespread use of cryptography today, the transition to post-quantum cryptography will take time. Organizations will only succeed in that transition by first understanding where and how they use cryptography. Each organization owns the overall security posture of its own digital environment, although the organization may not have a complete understanding of their cryptographic assets.

Organizations need a comprehensive cryptography inventory. The inventory provides insight into the algorithms, keys, protocols, and software libraries in use today and where they are used. Underscoring its fundamental importance, US federal government agencies are currently mandated to generate a full cryptography inventory. Similarly, organizations processing credit card data must produce a cryptography inventory by March 2025 to meet industry compliance standards.

Only with a comprehensive cryptography inventory can organizations truly understand their cryptography landscape and prioritize the areas to tackle first in their journey to post-quantum readiness.

The US National Institute of Standards and Technology (NIST) is currently standardizing a family of new algorithms capable of withstanding the threats of quantum computing. These new algorithms present significant differences compared to the traditional algorithms they are intended to replace. There is uncertainty over which post-quantum algorithms will stand the test of time and, of course, we should expect more post-quantum algorithms to be developed and standardized. Consequently, unlike the past, software and hardware security solutions will need to offer customers a set of cryptographic algorithms that they can easily select from and subsequently change their selection for specific applications.

When considered at the scale of large organizations, selecting and later modifying the selection of cryptographic algorithms for applications needs to be manageable. This means that organizations need to be able to change the cryptographic algorithms used by applications in a timely, policy-driven manner without requiring changes to the applications themselves. Simply put, this is the definition of agile cryptography, often referred to as *crypto-agility*.

There is no stopping progress in the digital age, and there is no uncertainty about cryptography's importance to that continuous progress. A post-quantum future enabled by crypto-agility will be a better, more manageable place for organizations to benefit from cryptography's essential capabilities.

The most important step for organizations to handle the changing landscape of cryptography is to understand what they have, communicate with their vendors, and improve the level of understanding of their staff in the area of cryptography. In this book, author Sandip Dholakia will provide the basic knowledge needed in this space.

Modern Cryptography: The Practical Guide provides an excellent overview of most of today's applications, methods, and issues related to cryptography. You'll benefit from reading the book in the order presented, but the topics are fairly independent, so reading about any specific topic will be beneficial as well. The level of detail is sufficient for security engineers who are working with or building systems that include cryptography. The book assumes general knowledge of the systems involved. I found the history of the use of cryptography, including methods and applications, to be quite informative.

Dr. Taher Elgamal
General Partner, Evolution Equity Partners

Acknowledgments

A book of this magnitude and complexity would not be possible without the help and support of many individuals. Writing a book on cryptography has long been an ambition of mine, but I never found the courage to start. This book would not have come to life without the persistent encouragement of Hareem Shafi from Rheinwerk Publishing. Thank you, Hareem, for giving me the opportunity to write about what I love!

A very special thank you to my editor, Megan Fuerst! Your patience with my late submissions and flexibility with my schedule were invaluable. Your detailed guidance and tireless support throughout the process were simply amazing. To sum it up in four words: *you are the best*!

Reviewing an initial unedited draft and providing constructive feedback is no small feat, but my friend and colleague Jay Thoden Van Valzen volunteered without hesitation. Thank you, Jay, for your time, insights, and unwavering support. I also want to thank Jack Malay for helping create several figures for this book, Sohan Bhat for helping with examples of software implementations, Tapan Pandya for partially reviewing the initial draft, Pritam Pal for verifying the mathematics, and Joe Campbell for encouraging me to write about cryptography and discussing the initial concept of the book.

I extend my heartfelt gratitude to the SAP management team for their encouragement and support in my writing journey. My thanks also go to the staff of Rheinwerk Publishing and everyone who directly or indirectly contributed to bringing this concept to life.

I am deeply honored that Dr. Taher Elgamal, *an inventor of cryptographic algorithms and the father of SSL*, agreed to write the foreword for this book. Thank you once again, Dr. Elgamal.

Lastly, I want to express my deepest appreciation to my family for their patience and understanding as I wrote in the evenings, late at night, on weekends, and even during vacations. And, of course, thank you to my loyal cheerleader, Simba, who was always by my side as I wrote.

Preface

This book aims to provide you with a comprehensive understanding of modern cryptography and its practical applications, ranging from lightweight cryptography to post-quantum cryptography and from cryptocurrency to homomorphic encryption. After reading this book, as a security professional, you'll be equipped to apply your newfound knowledge to protect data and information, create and implement cryptography guidelines and policies, and develop a robust cryptography program for your organization.

My journey with cryptography began more than 20 years ago when I developed a widely used anti-theft subsystem. As a math and number addict, this project fueled my passion for mathematics and cryptography. Growing up, I loved solving problems, proving geometric theorems, and studying calculus. This led me to pursue a graduate thesis in Partial Differential Equations (PDE) for an engineering application. Although, like most cryptographers, I did not choose a career as a mathematician, working on cryptography reignited my passion for mathematics. Since my first encounter with cryptography, like many security professionals, I've always focused on the practical applications of cryptography. Throughout my career, I've successfully implemented cryptography algorithms to secure data and applications at some of the great corporations of our time.

What inspires me most about cryptography is how it has rapidly proliferated everywhere. Every time we enter a password, cryptography gets to work. Every time we start a car, cryptography performs its magic. Cryptography keeps us secure whenever we log in to the bank or purchase something from a retailer. We use cryptography almost everywhere but don't even realize it.

If cryptography is now ubiquitous in our daily lives, why do most security professionals know little about it? There are two possible reasons:

- Cryptography works very seamlessly; we don't even notice it's working for us while starting a car or ordering a purse from Amazon.
- Most security professionals avoid opening a can of worms because of the complexity and difficulty of mathematics involved in cryptography. They often leave cryptography to mathematicians to innovate new primitives and to programmers to implement them correctly (fingers crossed!).

Who This Book Is For

This book is written for security practitioners who want to deepen their understanding about cryptography and implement it to secure their business-critical resources. You don't need to be a programmer or math major to grasp the concepts in this book. I've minimized the use of mathematics while explaining the workings and implementation of cryptography primitives. However, I assumed that readers have a basic understanding and some experience or education in the field of information technology. The ideal audience for the book is any security engineer, architect, or practitioner, but it's useful to students and beginners also. The book can also be used as a textbook or supplement/reference to a cryptography course.

Why Another Book on Cryptography?

There are two primary reasons for writing this book. First, when I began working in cryptography, I had to read many books, papers, and articles to understand various aspects of cryptography. Most of the books I found were written by academics, and some were even used as a textbook. Other books were heavy on mathematics. They are all great resources, and I learned a lot from them, but most of them lack comprehensive, practical guidance for implementing cryptography. It took a lot of time and effort to compile every detail from various books and sources that were needed to understand and then implement cryptography primitives.

Many years later, I asked myself, if I were to start working with cryptography today, what would I like in a book? How can one book deliver everything security practitioners need to know and apply? And that is precisely what I've tried to answer in this book! This book contains everything you need to understand and use cryptography. It's not an outcome of any research or byproduct of any graduate course. Instead, it's a compilation of my notes over the years, designed to be a practical resource for security practitioners.

Second, in the 1990s, cryptography wasn't taught to us in school. My knowledge in the field of cryptography is largely thanks to the wonderful community of cryptographers. I learned everything from published papers, webinars, and books. I benefited a lot from the cryptography community; however, I never actively contributed to the community. As a practitioner, my focus has always been on implementation and not research. I always thought being a practitioner should not stop me from contributing. This book is my humble attempt to give something back to the community by connecting inventors with implementors!

How This Book Is Organized

This book starts with the basics of cryptography and slowly progresses to advanced topics and applications. I recommend reading the chapters in order, but it's not required. The chapters are structured so that experienced readers can read them in any order based on their interests or needs.

The book is divided into two parts. **Part I**, which consists of the first four chapters, discusses the foundational concepts of modern cryptography, walking readers through the principles, methods, and algorithms to protect data.

Chapter 1 briefly outlines the history and development of cryptography, basic security principles, and fundamental cryptography concepts and compares classical and modern cryptography. The chapter concludes with a detailed discussion of various ciphers, including block cipher modes. **Chapter 2** and **Chapter 3** cover symmetric and asymmetric cryptography algorithms, respectively—the foundational concepts of cryptography. **Chapter 4** focuses on cryptography services. These services are contributions of modern cryptography and are kind of an extension of cryptography as we know it.

Part II examines modern cryptography in practice, showing readers how the basic concepts discussed in Part I are applied to today's security challenges. Part II also explores some advanced topics and new trends in cryptography.

Chapter 5 and **Chapter 6** discuss the most basic applications of cryptography—data encryption at rest and in transit.

Lightweight cryptography has attracted attention recently. These primitives are developed for Internet of Things (IoT) devices and other resource-constrained applications. The increasing use of IoT devices and cloud computing in people's professional and personal lives has opened up more avenues for hackers. **Chapter 7** explains how cloud computing and IoT devices can be secured using cryptography.

Cryptocurrency is the topic that everyone wants to talk about. **Chapter 8** dissects cryptocurrency and blockchain technology to explain how cryptography is used in cryptocurrency. The use of AI with cryptography is a game changer. **Chapter 9** surveys various AI algorithms and how they can be used with cryptography. You'll also learn about how cryptography can help AI.

Will quantum computers crack conventional cryptography? Are we prepared for it? **Chapter 10** walks you through the fundamentals of quantum mechanics, quantum cryptography, and various quantum-resistant algorithms. The chapter wraps up with a strategy to migrate and implement post-quantum cryptography that security professionals can use as a blueprint for their migration projects.

The holy grail of cryptography, homomorphic encryption, is reviewed in **Chapter 11**. It's an evolving encryption method intended for data in use. It allows systems to work with encrypted data without decrypting it first. The chapter covers types of homomorphic encryption, the application of homomorphic algorithms, and the future of this area of cryptography.

The use of cryptography in a cyberattack is a new dimension. **Chapter 12** focuses on cryptanalysis, cryptography attacks, and ransomware, as well as outlines cryptanalytic attacks, the anatomy of ransomware, and strategies for protecting your organization from it. **Chapter 13** presents various resources, best practices, and standards to learn more and keep yourself up to date with the technology.

What is on the road ahead for cryptography? In **Chapter 14**, we wrap up the discussion with the future trends and outlook of the technology and industry.

Conclusion

Cryptography has always fascinated me. I've enjoyed learning about it, working with it, and presenting it. Writing about cryptography was no different—it was truly an amazing experience. I hope you enjoy reading this book as much as I enjoyed writing it. Hopefully, this book will be helpful to you, and it will ignite a fire within you to learn more about cryptography. Don't make this book your destination. Cryptography is an ever-evolving field, and learning and working with cryptography is a journey. I sincerely hope you continue the journey and cherish it as much as I do.

PART I
Foundations of Modern Cryptography

Chapter 1
Fundamentals of Cryptography

Cryptography has been around in one form or another for a long time—probably much longer than you would imagine. Interestingly, it has been used throughout history and by almost all cultures around the world. What motivated people to use cryptography thousands of years ago, and what meaningful purpose is it serving for society in the 21st century? This chapter introduces the fundamentals of cryptography.

What is cryptography, and why do we need it? What basic terms and definitions do you need to know to understand and use cryptography? This chapter answers these and other similar questions. The chapter introduces the fundamental concepts of cryptography for those who are new to the subject and serves as a quick refresher to those who already know about the topic.

> **Chapter Highlights**
> - Overview of the history and necessity of cryptography
> - Learn core concepts of cryptography
> - Deep dive into classical and modern ciphers

This chapter is divided into three sections. Section 1.1 briefly outlines the history of cryptography. Section 1.2 focuses on the core concepts of cryptography, such as definitions, security concepts, classical cryptography, and so on. The section concludes by introducing modern cryptography, the hero of our story, and various types of modern cryptography. Section 1.3 provides a primer on ciphers. It dives into various classical and modern ciphers—with an emphasis on stream ciphers and block ciphers.

1.1 History of Cryptography

Most children have used Pig Latin (or a variation) while growing up. I was no different. I used and enjoyed Pig Latin all the time. Pig Latin, as defined by the Oxford dictionary, is "An invented or modified version of a language; spec. a systematically altered form of English used as a sort of code, esp. by children." This is an elementary form of a cipher used primarily in spoken language, but it's enough to confuse children if they don't know the deciphering key—moving a consonant to the end of the word.

1   Fundamentals of Cryptography

Although I didn't know it then, this was my first unofficial exposure to cryptography.

My personal history with cryptography started in school with Pig Latin, but the history of cryptography goes back much further. In the following sections, we'll examine the past and discover why cryptography has been and still is a necessary tool.

1.1.1   Historical Development of Cryptography

Cryptography has been used in one form or another since the inception of communication. Throughout history, various civilizations have used different methods to hide messages for one reason or another. Let's trace the use of cryptography throughout history.

Early Cryptography

The first known use of cryptography was in Egypt in 1900 B.C. Egyptians used carved hieroglyphs as an alternate form of communication [Kahn, 1973]. Although the purpose of these alternate symbolic writings is unclear (there are several theories), creating an alternate script for communication to hide the message was born almost 4,000 years ago.

Around 400 BC, in ancient India, Kama Sutra [Burton, 1991] mentioned using an alternate form of writing to communicate between lovers. *Kama Sutra* is an ancient Indian book about the principles of love written by a Brahmin scholar, Vatsyayana, in which she discusses 64 arts that women must know to have successful love from their lover. The list includes some common tasks for that period, such as cleaning, cooking, and so on. Some less obvious arts, such as carpentry, also made the list. Surprisingly, the 45th art on the list is an "alternate form of writing or communicating." The book describes using the Vatsyayana cipher [Singh, 2000]. Vatsyayana proposed pairing the two letters from the alphabet and substituting them for each other [CEMC, n.d.]—for example, pairing the alphabet with a displacement of 5 in which A pairs with E, B pairs with F, and so on. In cipher text, A is replaced by E, and E is replaced with A. Similarly, B is replaced by F, and F is replaced by B. Table 1.1 shows an example of the Vatsyayana cipher with a pairing by 13.

AN	BO	CP	DQ	ER	FS	GT	HU	IV	JW	KX	LY	MZ
Plain text: THE BEST CRYPTOGRAPHY BOOK													
Cipher text: GUR ORFG PELCGBTENCUL OBBX													

Table 1.1 Example of Vatsyayana's Substitution Cipher

It's a straightforward concept, and it worked for that time. Just imagine how many pairing combinations you can have with 26 letters!

1.1 History of Cryptography

The Polybius Square, named after the Greek historian Polybius, was a Greek contribution to cryptography. The Polybius Square cipher converts letters to numbers based on a 5 × 5 square or grid. The columns and rows are numbered from 1 to 5. The corresponding row and column numbers represent each letter in the square. Table 1.2 shows an example of the Polybius Square cipher [Salomon, 2003]. In this example, T is 44, H is 23, and E is 15. THE is represented as 44 23 15.

	1	2	3	4	5
1	A	B	C	D	E
2	F	G	H	I	J
3	K	L	M	N	O
4	P	R	S	T	U
5	V	W	X	Y	ZQ

Plain text: THE BEST CRYPTOGRAPHY BOOK

Cipher text: 44 23 15   12 15 43 44   13 42 54 41 44 35 22 11 41 23 54   12 35 35 31

Table 1.2 Polybius Square Cipher with an Example

The advantage of the Polybius Square is its flexibility. The letters and numbers can be rearranged in any order, and it works as long as the sender and receiver have the same chart. Even if the chart is compromised once, users can rearrange the letters and numbers and reuse them, but the message will remain hidden. It's easier to crack the Polybius Square cipher once the eavesdropper knows the concept.

Egyptians, Indians, and Greeks, from different parts of the world and during different periods in history, used alternate forms of communication to hide messages. Clearly, the need for secret communication was always in demand, and people tried various ways to achieve it.

Julius Caesar was the first person to use cryptography in real-world situations. Caesar used a systematic method of encrypting and decrypting a message during wartime to communicate with his war generals or Cicero in Rome. He always feared that his messengers could be captured en route to generals or, even worse, that messengers could play a double agent or part of an enemy. This could put his war strategies at risk, which prompted him to send encrypted messages. Enemies must know the key to decrypt the message; otherwise, the intercepted message is useless to enemies. Julius Caesar used a shift 3 technique to encrypt the message—in simple terms, it replaces every letter with a third letter in the alphabet, like A substituted with C and B with E [Singh, 2000].

The Caesar cipher is a substitution, monoalphabetic cipher because it uses a fixed substitution of one letter for the entire message. This type of cipher is easy to crack with some minor effort.

Almost 1,600 years after the Caesar cipher, in the 16th century, Blaise de Vigenère introduced another cipher [Salomon, 2003] that was more difficult to crack than the Caesar cipher. The Vigenère cipher was a little more complex because it used a *keyword* to encrypt and decrypt the message. This was probably the first recorded use of an encryption key. A fixed table with a shifted alphabet is used every time. However, the keyword is changed with every message. A new keyword with every message makes the Vigenère cipher difficult for an attacker to decipher because the frequency analysis technique can't be used easily with this cipher. For this complexity, the Vigenère cipher remained unbreakable until the 19th century. You'll learn more about the Caesar cipher and Vigenère cipher in Section 1.3.

As we've seen briefly, cryptography made good progress between 1900 BC and the 19th century. However, due to two world wars, its development has accelerated since the dawn of the 20th century.

World Wars Fueled the Development of Cryptography

There are striking similarities between the history of aviation and cryptography. Amateurs or hobbyists developed both. Although not intended, both technologies were first used during the First World War and only after the Second World War, they tasted commercial success.

The telegram became an essential tool for business by the late 1800s. Most major cities in the West were connected by telegram before the start of the 20th century. Telegrams became the main form of communication during the First World War because of their speed. However, the problem was the secrecy. There was no way to keep the information secure from others while communicating over telegrams. This is when the use of encrypted messages gained momentum. The Germans used a book of code, which was developed with a numeric number for every word in the dictionary, to encrypt the messages over telegrams. The book also included a book number at the beginning. That helped identify which book to use in decoding the message. Table 1.3 shows an example of a code book.

5199 – Book number	93 – apple	526 – book	6846 – the
27 – an	357 – best	2520 – cryptography	9468 – zoo

Table 1.3 Example of a Code Book

Now, let's use this code book to encrypt our example message:

- **Plain text:** The Best Cryptography Book
- **Cipher text:** 5199 6846 357 2520 526

In this example cipher text, the first number, 5199, indicates the book number. The rest are just direct substitutions of numerical values for each word. This is a difficult cipher to crack unless you get your hands on book number 5199. However, this method of encrypting has many disadvantages:

- The code books were printed, making it very difficult to correct, change, or add new code.
- The code book must be safely sent to the message receiver. This needs to happen only once, but both sender and receiver must have the same book.
- Safeguarding the code book is crucial.
- Words in the code book and their corresponding numerals are in ascending order. That gives some clues to any cryptanalyst. For example, the word corresponding to code 357 may start with a or b (book, in our example), and the word corresponding to code 9468 must be high in the alphabet (zoo, in our example).

The case study of the Zimmermann Telegram is a very popular example of how Germans used code books in the First World War and how it was cracked. Encrypting messages picked up steam during and after the war in the early part of the 20th century. Military around the world started hiring people with some knowledge or skills in developing and deciphering secret messages. Soon after the end of the First World War, electromechanical rotor-based devices were invented to encrypt messages. By the early 1920s, a mechanical rotor-based machine, Enigma, was available to encrypt messages. However, the machine was a commercial failure and had very limited use until the late 1920s. Enigma saw success after the German military started using it. By the 1930s, the Germans had a robust technology to encrypt the messages. Enigma used a series of rotors—between three and six—to implement a substitution cipher. The rotor setting was the key. The only way to decipher the message is to use a similar machine with the same key–rotor settings. Allied forces invested heavily in cracking the Enigma ciphers with little success. Eventually, the Polish military had initial success in rebuilding the Enigma. The technology was transferred to British forces, and, in 1940, the Enigma code was cracked [Singh, 2000].

Commercial Success

The military dominated cryptography until the end of the Second World War. After the war, going back to our comparison with aviation, air travel became more commercial, and so did cryptography (and the technology as a whole).

IBM was way ahead in developing computers and other digital machines. In 1966, IBM decided to offer data protection and security to its users. IBM started a crypto group led by Horst Feistel. Feistel developed a block cipher, Lucifer. In 1973, Feistel published a paper outlining the importance of cryptography in data security. However, the industry started emphasizing the importance of cryptography only in the late 1990s or even early 2000s. Feistel was almost 30 years ahead of his time! Lucifer has historic importance as this was the first nongovernment, nonmilitary cipher.

In 1977, the US National Institute and Standard Technology (NIST) released a modified version of Lucifer as a Data Encryption Standard (DES). Lucifer provided an initial foundation for DES development, which was later replaced by the Advanced Encryption Standard (AES). We examine DES, AES, and many other cryptographic algorithms in subsequent chapters. By the mid-1980s, modern cryptography became an essential tool to protect our data and information.

Rise of Hackers

Caesar and others thought of protecting the messages because they feared someone stealing or eavesdropping on their secrets. Hackers and adversaries were always ready to crack any secret code. The science and history of cracking secrets, cryptanalysis, is as interesting as cryptography itself.

For older encryption methods, such as the Caesar cipher where a basic substitution cipher is used, the process of cracking was largely dependent on the frequency analysis. *Frequency analysis* is a basic or common-sense approach to code breaking based on the fact that, in most languages, some letters are used more commonly than others. Cryptanalysts use this concept to match a few common words, and then the rest becomes easy. For example, in the English alphabet, "e" is the most frequently used letter, followed by "t" while "q" and "z" are the least used letters. The goal for the cryptanalyst is to find the three-letter word "THE" to start the process.

The Babington Plot [Singh, 2000] is an example of how cryptanalysis was used to crack the code. In the 16th century, Anthony Babington was the prime conspirator in the assassination and replacement of Queen Elizabeth I of England, a protestant, with her Catholic cousin, Mary, Queen of Scots. Babington used the nomenclator substitution cipher to communicate with Mary. In this variation of the substitution cipher, the letters are replaced with other symbols or codes. The chart of the used symbols or codes is shared with the receiver of the message. The letter was decoded by Queen Elizabeth's spymaster Sir Francis Walsingham and his cryptanalyst Thomas Phelippes. You guessed it right; the frequency analysis method was used to crack the code. For example, if "Symbol4" appeared most in the sentence and "Symbol7" was the second most frequent, then Philippes needed to find a three-letter word and try to fit symbol4 and symbol7 to see if it matches with "THE" anywhere, and the process begins. (Obviously, real-world cracking isn't this simple. The examples here are overly simplified to explain the concept.)

Like cryptography, the frequency analysis method to crack the code is also very old. In the ninth century, Arab scholar Al-Kindi developed probably the first frequency analysis technique. One key assumption in this method is that the cryptanalyst knows the language used in the communication. The Vigenère cipher was uncrackable for a long time but was eventually cracked by a variation of this method three centuries later. In that case, the concept was to find a repetitive letter at a certain frequency that will help you find the keyword length. Then, you use frequency analysis to crack the code.

> **For the History Buffs**
>
> The old and long history of cryptography and cryptanalysis is very interesting. This section has hopefully given you a good overview of the historical development of the science of secrecy. It's neither possible nor practical to cover the entire history in detail in this book.
>
> If you're into the history of cryptography, we highly recommend reading *The Code-Breakers* [Kahn, 1973] and *The Code Book* [Singh, 2000]. *The CodeBreakers* is very detailed with more in-depth mathematical information. It's primarily for people who really want to go deep into the history and cryptography. *The Code Book* takes a more entertaining, story-driven approach to the history of cryptography. The highlights of *The Code Book* are also available as a four-part documentary on YouTube.

1.1.2 Why Is Cryptography Required?

By now, based on the history, we have a very clear understanding of why cryptography is required. People have used cryptography for many reasons—from hiding a message from an enemy to protecting personal data and information. If people minded their own business and there were no criminals and wars in the world, we really wouldn't need cryptography. The following are some of the common reasons modern cryptography is necessary:

- **We need to protect assets or valuable messages.**
  Let's consider a bank vault. Anyone can get a vault in the bank by paying a small fee. But the vault is of no use if you don't have any valuables to keep in it. Once you have some valuables to keep in the vault, the value of the vault becomes much higher than the bank fees for the vault. Cryptography is very similar to this vault analogy. The primary purpose of cryptography is to protect data and information. If we have some important information that must be protected, then the use of cryptography is justified.

- **Nothing is secure.**
  There is no such thing as a 100% secure system. No matter how robust the security controls are, there is always a chance that an attacker will find a weakness in some control somewhere to get their hands on the data. However, if data is encrypted, then hackers won't get any information. If all controls fail, but the data is still encrypted, there is no real loss.

- **Security is as strong as the weakest link.**
  The security of any system is as robust as the weakest link. This applies to any security control within the system. Weakness or vulnerability in any one control is enough for the attacker to penetrate the environment. At this point, it doesn't matter how careful you are with designing and implementing the security control; it

isn't going to help because of the weakest link. And the weakest link is people. People will often leave workstations unattended or unlocked. People tend to catch the bait and click on the phishing links. People will write down passwords on a sticky note under the desk. Then there are other types of people with malicious intent and insider threats—like escalation of privilege or exfiltration of data. In both intentional and unintentional situations, encryption protects our data. If the stored data is encrypted, stealing the data or stolen data won't help.

- **Developers don't think like hackers.**
  When software engineers develop software, they think like engineers, and wear creative hats. They don't think like hackers. As a result, they leave some loopholes in the software, like backdoors. This works in favor of hackers. Cryptography is necessary for this situation. Even if the hacker uses a backdoor or other vulnerabilities left by developers to gain access to the system, the encryption at rest will protect data from exposure.

These are commonsense reasons and not technical reasons why cryptography is required. Usually, business requirements drive the decision of which data needs to be encrypted at rest and which data needs to be encrypted in transit or which backup is encrypted. Based on the risk priority and expense, the business decides what types of security controls are needed and what layers of controls should be implemented. However, there are some good technical reasons for using cryptography, such as confidentiality, integrity, authentication, and nonrepudiation. These are discussed in detail in the next section.

1.2 Introduction to Cryptography

This section introduces the fundamentals of cryptography such as definitions, types, security concepts, and so on. The section will prepare readers for the remaining chapters of the book and also be a good refresher for readers who have some experience.

1.2.1 Definitions

Like every trade, cryptography has its own terminology. Various terms introduced in this section are industry standard and used throughout this book. We'll define them here to enhance your understanding:

- **Cryptology**
  Cryptology is the science of writing and understanding the hidden messages. The word *cryptology* means "knowledge of hidden." *Crypto* is derived from the Greek word *krypto*, which means "hidden or secret." The suffix *-logy* means "knowledge of."

- **Cryptography**
  Cryptography is the science of concealing a message or information. The Greek word

krypto means hidden, and the suffix *-graphy* means "to write." Cryptography creates a message that is unreadable to everyone except for the intended recipient.

- **Cryptanalysis**
Cryptanalysis, on the other hand, is the process of decrypting a message by an unauthorized person without using the official encryption key. In other words, cryptanalysis is the process of cracking the code or eavesdropping on the message. The person who engages in this activity is known as a cryptanalyst.

> **How Does Cryptanalysis Differ from Cryptography?**
> Cryptology includes both cryptography and cryptanalysis. Cryptography is further divided into two techniques: encryption and decryption. Cryptanalysis also decrypts the message but, unlike cryptography, it deciphers the message without using the authorized decryption key.

- **Cipher**
This is a specific algorithm used in cryptography to encrypt and decrypt messages.
- **Plain text**
Plain text is a message written in natural language that is readable by anyone. The definitions you're reading right now are in plain text.
- **Cipher text**
The cipher text is the text generated after encrypting the plain text. The output of the encryption process is the cipher text. Cipher text can't be read (understood) without decrypting it.
- **Encryption**
This is the process of converting plain text into cipher text, which is usually done using an encryption algorithm and an encryption key.
- **Decryption**
Decrypting is a process of converting cipher text back into plain text using an algorithm and the authorized key. Don't confuse this with cryptanalysis, which deciphers the message without an authorized key or algorithm.
- **Encryption key**
The encryption key is a piece of encryption/decryption algorithm that is known only to the sender and receiver (and other authorized persons) of the information. The rest of the algorithm can be made public. The algorithm isn't complete and can't function as intended without this missing (key) information.
- **Keystream**
A keystream is a stream or sequence of random (usually pseudorandom) bits that is used with a stream cipher as an encryption key. The keystream is also used with block cipher modes.

- **Nonrepudiation**
    Nonrepudiation is the process of providing the sender's identity to the recipient without the sender being able to deny the authenticity of the signature.

Understanding these definitions helps build a strong foundation for cryptography.

1.2.2 Classical versus Modern Cryptography

The automobile industry has undergone huge advancements in the past 120 years. Despite these dramatic changes, we don't identify gasoline vehicles as classic automobiles or electric vehicles as modern automobiles. But we do make that distinction in the case of cryptography—and rightly so—as we'll discuss in the following sections.

Classical Cryptography

Cryptography strategies and methods used to convert plain text into cipher text without the use of technology are known as *classical cryptography*. Cryptography is commonly defined as "an art of writing hidden messages"—and classical cryptography was indeed an art. There were no defined methods or standard techniques to create a cipher text. People needed special interests or hobbies in creating cipher text. The more creative a person is, the stronger the cipher message. Classical cryptography was primarily dependent on substitution ciphers. Every ciphering technique from Vatsyayana, Caesar, and Vigenère to the German Enigma machine, used one or another form of substitution technique.

Classical cryptography had several limitations:

- Classical cryptography was heavily dependent on substitution ciphers, which were very vulnerable to frequency analysis. Most codes were cracked relatively easily. The Vigenère cipher was uncrackable for almost three centuries, but eventually the cipher was cracked. Even the Zimmerman Telegram and Enigma cipher texts were broken.
- Successful decoding of the message relied on the fact that the substitution mechanism, the code book, or other methods, was shared before the communication took place. The Vigenère cipher was the first one to use some sort of key or keyword in the encryption process. Every cipher before that was largely dependent on the secrecy of the method (the algorithm itself). Once the method or the code book was intercepted, there was no secrecy.
- Another limitation of classical cryptography was its inability to scale. Classical cryptography wasn't designed for large amounts of data. Most of the processes and computations were performed manually in classical cryptography. Manual processing limits the amount of data that can be handled, which restricts the scalability.

Modern Cryptography

After World War II, technology started to influence every part of our lives, and cryptography couldn't escape. The use of the latest technology and computing power changed the tide from classical cryptography to modern cryptography. In modern cryptography, messages aren't sent on horses, and keys aren't exchanged via pigeons. The use of computing power allowed modern cryptography to use advanced mathematical algorithms to secure not only communication but also data and information at all stages of life.

Modern cryptography took off after the release of DES by NIST in 1977. That was really a turning point in the history of cryptography.

One of the major differences between classical cryptography and modern cryptography is its impact on our lives—classical cryptography was limited to military use while modern cryptography is everywhere—from banking to shopping and from working to vacationing. Almost anything you do now, you have to experience the magic of modern cryptography—and without even realizing it. It may be time to rename *modern cryptography* to *everyday cryptography*!

Although the dictionary definition of cryptography still defines it as "an art of writing hidden messages," modern cryptography isn't an art anymore. It's an exact science. It uses complex mathematics and is very difficult, if not impossible, to crack.

Modern or everyday cryptography has several advantages:

- One of the prime advantages of modern cryptography is its seamless integration. Users never even realize that when you start a car, you encrypt and decrypt codes multiple times or when you enter a password, you use cryptography.
- The speed at which modern cryptography works is the second advantage. Thanks to the latest technology and computational horsepower available to us, even the most complex algorithm runs in a microsecond.
- Modern cryptography protects data in communication, in storage, in applications, in the cloud, and in Internet of Things (IoT) devices. The reach is mind-boggling.
- Modern cryptography offers a very secure key exchange mechanism using key exchange algorithms, which allow for the exchanging of keys without the sender and receiver physically meeting each other.
- Modern cryptography not only offers confidentiality of the data but also offers integrity, authenticity, and nonrepudiation.
- Modern cryptography thrives on public key infrastructure (PKI). Generation and verification of digital certificates are vital to the success of modern cryptography.

1 Fundamentals of Cryptography

Like every coin, modern cryptography also has another side. The challenges of modern cryptography are as follows:

- Modern cryptography is heavily dependent on computational power, and, as a result, it can also be easily cracked with that power. Quantum computing can potentially create risk if we don't upgrade our encryption algorithms. We've covered the topic of post-quantum cryptography in detail in Chapter 10.
- Hackers can use this seamless modern cryptography and generate various attacks, such as ransomware.

> **What Is a Cryptography Algorithm?**
>
> Modern cryptography involves extensive mathematical computations. An *algorithm* is a process of executing mathematical computations in predefined steps. The software and hardware designed to perform these computations are executed with great computational power.

Modern cryptography, unlike classical cryptography, is heavily dependent on the secrecy of the encryption key. In classical cryptography, it was necessary and very important to keep the algorithm secret. If the attacker knew the algorithm or the method of encryption, the attacker could easily decrypt the message. The concept was known as security through obscurity (STO). The encryption cipher or algorithm was shared with only those who needed to encrypt or decrypt messages. For example, Germans needed to guard the code book (refer to the Zimmerman Telegram in Section 1.1.1) and the design of the Enigma machine. However, in modern cryptography, only the encryption key needs to be kept confidential.

> **Kerchoff's Principle**
>
> What we just learned is known as *Kerchoff's principle* [van Tilborg, 2005]. Kerchoff stated that the cryptosystem should be secure even if everything about a cryptosystem is known except the encryption key. Kerchoff was the first to introduce this concept in the late 19th century.

Classical cryptography laid a foundation, but modern cryptography made it possible to secure our world.

1.2.3 Encoding, Hashing, and Encryption

Although it's somewhat confusing and used interchangeably at times, there is a clear difference between encoding, hashing, and encryption. Before learning more about encryption, in this section, we'll learn how encryption differs from encoding and hashing.

Encoding

Encoding is the process of converting plain text messages into different formats, usually for transmission purposes. The purpose of encoding isn't to hide the message from an attacker or any eavesdropper. In encoding, each alphabet is assigned a special character or sequence of characters. The encoding chart is openly shared with everyone. Anyone who receives the encoded message can convert it back to plain text—as long as they have the right tools and technology.

Encoding was necessary for the use of telegraphy because it's possible to transmit electrical signals, but there is no way to transmit alphabets using electrical signals. Encoding made it possible to transmit alphabets using electrical signals. This wasn't the first use of encoding, but its use in telegraphy was probably the first experience of encoding for the mass population. Generation X is probably the last generation to see telegraphy in action, and most Generation Y people have only heard of telegraphy, but it played a vital role in communication in the 19th century and early part of the 20th century.

The telegraph used Morse code, which is one of the early encoding systems that has been used for a long time. Morse code is named after its inventor, Samuel Morse, who was also one of the inventors of telegraphy. In Morse code, the alphabets are defined in dots and dashes. The dots and dashes are defined by different signal durations. The longer durations are dashes, and short durations are dots. There is no distinction between the uppercase and lowercase alphabet in Morse code. The chart of Morse code—the conversion from alphabets to dots and dashes—is publicly available.

In the computer and technology industry, the American Standard Code for Information Interchange (ASCII) is used for encoding. As the name implies, ASCII is used to communicate on the internet, and ASCII converts the plain text to bits that the computers and networks understand. ASCII encoding includes uppercase and lowercase alphabets, 0 to 9 numerals, and punctuation.

Unicode is the latest encoding mechanism, formally introduced in 1991. Unicode is a universally acceptable encoding mechanism used right now regardless of platforms, programming languages, or devices. Unicode is the encoding behind text messages and even emojis. Unicode is accepted by all software providers and is used across all operating systems; various devices, including smartphones; and across the internet. Unicode is made up of more than 100,000 characteristics, including all Latin and many non-Latin languages.

Over time, the term *encoding* was also used for other, but conceptually similar, purposes. It's also used for analog to digital conversion, as well as for compressing audio and video files.

Whether it was Morse code in the early 20th century or Unicode now, if the code isn't converted back into human-readable text, it becomes useless to people. The process of decoding is converting codes back to plain text.

Hashing

The hashing algorithm takes in a message of any length and generates a fixed string of data known as a hash value, hash code, hash digest, or just hash. The hash value changes even if a single bit in the original message is changed.

Hash is also known as a *one-way function* (and sometimes even *one-way encryption*) simply because the receiver can't "unhash" it. However, don't be misled by the use of word *encryption* in the term. The message is transmitted in plain text only. Hashing doesn't provide any secrecy to the message itself.

Hashing is performed in a fixed length of data. The message of any length is divided into small chunks of data, and, via the sender, each chunk is passed through the hash function to generate a value. This hash value is attached to the message. The receiver then runs the entire message through the same hash function and generates the hash string. The hash generated by the receiver should be exactly the same as the string sent by the sender with the message. If the strings aren't identical, the message is tempered.

Message digest (MD) and secure hash algorithm (SHA) are examples of commonly used hashing algorithms. We'll learn more about hashing in Chapter 4.

Encryption

We'll only briefly touch on the concept of encryption here as most chapters in this book directly or indirectly discuss encryption. Encryption is the process of converting plain text into cipher text or unreadable message. Encryption is primarily used to hide data or messages. Decryption is the process of converting cipher text back to plain text.

> **A Common Interview Question**
>
> A very common interview question for an entry-level security job such as an information security analyst or security engineer is the following: What is the difference between encoding, hashing, and encryption, and how do they help the security triad?
>
> Review this section and the next carefully for an answer to this question.

1.2.4 Security Concepts

Cryptography is a very important part of the overall security of data and information. In this section, we'll discuss some basic security concepts that will help you understand what role cryptography plays in an overall data security strategy. This is a refresher for readers who are somewhat familiar with information security concepts, but feel free to skip the section if you're familiar with these concepts.

Security Triad

The primary goal of security is to protect the confidentiality, integrity, and availability of data. As shown in Figure 1.1, these properties of security are represented as the *security triad*. Security controls are evaluated based on how well they protect these properties for the asset. In this section, we'll evaluate these properties and explain how cryptography is tied to each.

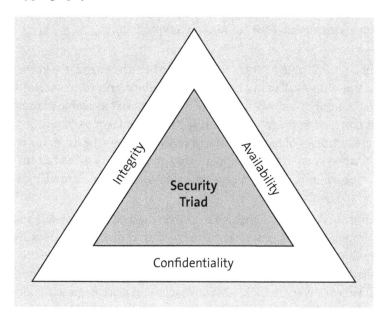

Figure 1.1 Security Triad

Let's take a closer look at the three sides of the security triad:

- **Confidentiality**

  Confidentiality is one of the most important factors in protecting data and information. In fact, confidentiality of data and information is the primary concern for most people and businesses. That confidentiality is compromised if anyone has unauthorized access to the data while data is in storage, in use, or in transit. Confidentiality is also compromised if the data is leaked or shared without proper permission or used for unintended purposes. Sometimes confidentiality is also compromised unintentionally by users. This happens when users are susceptible to phishing emails or other social engineering attacks.

  Various security controls can be implemented to protect confidentiality such as robust access control, extensive and continuous employee/user training, robust data labeling, classification, and handling procedures. The last but very important control is, of course, cryptography.

1 Fundamentals of Cryptography

- **Integrity**
  Data and information integrity is another important property of the security triad. If integrity is compromised, data isn't reliable. Integrity could be compromised by data modification by unauthorized users, unintentional modification of data by authorized users, or modification with malicious intent (insider threat).

  Various security controls can be implemented to protect the integrity of data such as stricter access controls, activity logging and monitoring, setting up baseline values, and verifications. Last but not least is, of course, cryptography.

- **Availability**
  If data or a system isn't available to authorized users when they need it, all other controls aren't very useful. Availability constitutes the third arm of the security triad. Availability is compromised when the data or system isn't available to users when need to use, store, or transmit the data. Distributed denial of service (DDoS) and ransomware are examples of technical attacks on availability. One major difference of availability when compared to confidentiality and integrity is the fact that availability can also be compromised by nontechnical factors, such as natural disasters, utility failures, and hardware failures.

  Various security controls can be implemented to protect the availability of data and systems, such as implementing network layer firewalls and monitoring network traffic while using the intrusion detection system (IDS) and intrusion prevention system (IPS). Regularly backing up data to a different location and having a business continuity and disaster recovery (BC/DR) plan and facility in place helps with nontechnical attacks. The hero of our story, cryptography, is also one of the important security controls in protecting availability.

Encoding, Hashing, and Encryption

Let's revisit these three key terms in the context of the security triad. Encoding, hashing, and encryption support the security triad in the following ways:

- **Encoding**
  Encoding/decoding loosely represents data *availability* in the security triad. It makes data available from one format to another.

- **Hashing**
  In the security triad, hashing represents the *integrity* of data because a change in the message changes the hash value, which indicates that the integrity has been compromised.

- **Encryption**
  Encryption represents *confidentiality* in the security triad. Encryption provides confidentiality by preventing the unauthorized exposure of data.

The Shift-Left Approach and DevSecOps

One of the important concepts in security is to implement security as early in the development cycle as possible. This concept is known as the *shift-left approach*. The automation of development and operation phases has made these tasks seamless, and the complete development and operations cycle is known as DevOps. However, because of the shift-left approach, security is embedded into the entire process, and DevOps is now referred to as DevSecOps.

Figure 1.2 shows the secure development and operational lifecycle (SDOL). The first introduction of security should be during the requirement gathering phase. While generating business requirements, security requirements should also be documented. Once the requirements are generated and the architecture is developed, threat modeling must be performed. Threat modeling plays out the possible attack scenarios and how the architecture or design will withstand the possible attacks. This is the time to modify the architecture and make appropriate changes in the design.

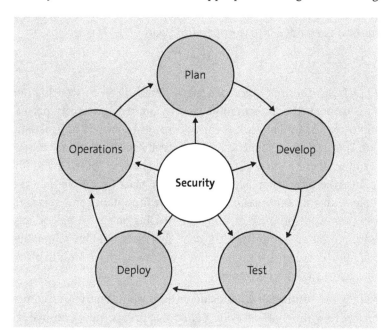

Figure 1.2 Secure Development and Operational Lifecycle

During the development phase, manual inspection or review of the code and static scans are performed. The process of running static scans is also known as static application security testing (SAST). SAST is used to discover errors and other coding vulnerabilities early in the DevSecOps cycle. Once the application is developed, dynamic application security testing (DAST) is performed. DAST is run on the application URL. Usually, DAST is performed in the preproduction or test environment. DAST is performed at a later stage in the DevSecOps cycle; and, as a result, it takes a little more time, resources, and money to fix vulnerabilities found in DAST.

The last two security tests in the cycle are penetration testing and red team testing. Penetration testing includes automated vulnerability testing as well as manual testing to simulate the attack. The penetration testing can be divided into three types: black box, white box, and gray box testing. Penetration testing is performed with a specific scope. Red teaming, on the other hand, is the process where the tester tries to mimic the real-world attack scenario. There is no predefined scope nor any time limit. The tester can use any technique available and try to find a weakness and exploit it. The difference between penetration testing versus red teaming is very similar to a cook versus a chef. The cook has a recipe, and prepares the food based on the prescribed recipe only. On the other hand, the chef can prepare and serve a variety of dishes. Chefs can even create their own dishes and see how people respond.

Dynamic vulnerability scanning, penetration testing, and red teaming are usually performed at regular intervals even after the application is deployed in production. That is why these security tests are extended into the operations part of the DevSecOps.

Defense-in-Depth

The *defense-in-depth* or *layered security* approach plays a critical role in securing our data and information. Through this information security approach, security professionals ensure multiple, redundant layers of security controls to protect critical data, assets, or applications. The idea isn't to depend on one security control. If a hacker penetrates through the outer layer of the security control, the hacker doesn't have full access to the data because there are other layers or controls in place to complement the first layer. For example, if an attacker breaks in through the firewall or gains access to the environment using some phishing attack, and if we're using only a firewall or only using an access control mechanism to protect the data, then we would be vulnerable, and our data or information could be up for grabs. But if we have both controls in place, then we increase the security of our environment.

But what if all controls fail or the attacker is an insider? This is where encryption comes to the rescue. No matter how much an attacker penetrates or how much an insider aggregates the privileges, encryption protects the data. The attacker won't be able to get any meaningful information. As shown in Figure 1.3, in the *defense-in-depth* strategy, security controls on data are part of the last layer. In summary, the layers contain the following:

- **Operational security**
  - Audit and compliance
  - Business continuity and disaster recovery

- Patch and vulnerability management
- Logging, monitoring, and incident response
- Security policies and standards
- Physical security
- Data loss prevention

- **Network and perimeter security**
  - Network segmentation
  - Virtual private network (VPN)
  - Virtual private cloud (VPC)
  - Network address translation (NAT)
  - Network firewall
  - Wireless security
  - Honeypot
  - Intrusion detection system (IDS)/intrusion prevention system (IPS)

- **Host security**
  - Host-based firewall
  - Endpoint detection and response
  - Virus and malware scans
  - OS and hardware management
  - Host-based detection (HDS)

- **Application security**
  - Web application firewall (WAF)
  - Web application penetration test (WAPT)
  - Secure software development lifecycle (SDLC)
  - ID and access management
  - Static and dynamic scans
  - Audit logs

- **Data security**
  - Data masking
  - Need-to-know
  - Encryption At rest/in transit
  - Secure data deletion

1 Fundamentals of Cryptography

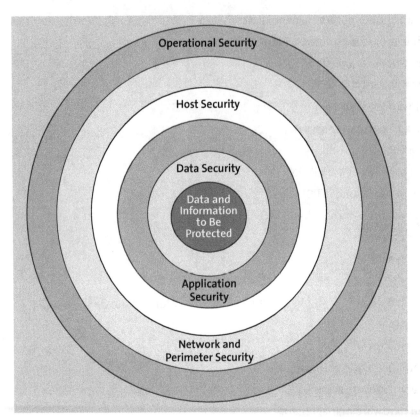

Figure 1.3 Defense-in-Depth Strategy with Encryption Control at the Core

Security Services and Cryptography

The main purpose of cryptography is to meet one or more of these security principles: confidentiality, integrity, authenticity, and nonrepudiation. These principles are known as *security services*, and, as you can see, they greatly reflect the security triad concepts we discussed previously. Confidentiality is the original and main purpose of cryptography; however, integrity, authenticity, and nonrepudiation are also used in cryptography applications to enhance data and message protection. Let's take a closer look at how cryptography enables each of these security services:

- **Confidentiality**

    Cryptography was invented for one purpose—secrecy (i.e., confidentiality). As we've seen in Section 1.1, the primary use of cryptography over the centuries has been to hide messages, and that is no different now. To date, the confidentiality principle of security is one of the most used security principles in cryptography.

    Attackers have access to very sophisticated tools to steal or expose data, and today encryption is used more than ever. Encryption provides confidentiality to the data at rest, data in transit, and data in use. Encrypted data provides a very high degree of confidentiality because, once encrypted, only people with authorized encryption keys can decrypt and access the data. This makes it very difficult for any unauthorized users or attackers to access the data.

Confidentiality can be compromised by insider threats, data leaks, data breaches, eavesdropping, and man-in-the-middle attacks, but various forms of encryption can protect data from these attacks.

- **Integrity**

  Cryptography is also used to protect the integrity of data and information. Data integrity can be compromised intentionally or unintentionally. It's compromised when attackers intercept the data in transmission and change the value or infect the storage device with a malicious virus. It can also be compromised unintentionally when a user accidentally deletes a part of the data or changes the value unknowingly. Device malfunction is another means by which data can be compromised unintentionally.

  Cryptography helps verify data integrity in transit and in storage. Cryptography uses hash algorithms (introduced previously) to generate a fixed size value or digest for the message or the data. This hash value is also known as a hash digest or just hash. The hash digest is attached to the message. The receiver uses the same hash algorithm to re-create the hash digest. The generated digest will be the same as the receiver if the value isn't altered. We discuss hashing in detail later in Chapter 4.

- **Authenticity**

  A digital signature uses asymmetric cryptography to prove the authenticity of a sender. The sender generates a hash digest and encrypts the digest with a private key. The encrypted hash digest is appended to the message. The receiver uses the sender's public key to decrypt the digest of the message. Once decrypted, the receiver runs the message through the same hash function and compares the value of the hash digest. The process is the same as occurs for integrity. However, using a public key of the sender confirms that the message was sent (or signed) by the sender. This basically confirms the authenticity of the sender.

- **Nonrepudiation**

  Nonrepudiation is the process of confirming the sender and preventing the sender from denying that they sent the message. The nonrepudiation process uses the digital signature for authenticity.

Remember the Purpose

It's important to note that using cryptography for integrity, authenticity, and nonrepudiation doesn't provide secrecy or confidentiality. The message or data itself isn't encrypted; only the hash digest is encrypted while using the digital signature. However, to keep the message confidential, it's possible to encrypt the data or message. Keep in mind that it's a separate process, and integrity, authenticity, and nonrepudiation don't provide confidentiality.

1.2.5 Types of Modern Cryptography

As we've discussed in Section 1.2.1, cryptology is divided into two broad categories: cryptography and cryptanalysis. Now that we understand the key terms, background, and underlying security concepts, we'll dive deeper into the types of modern cryptography in this section, as shown in Figure 1.4.

Cryptanalysis can be divided into three categories, depending on the purpose or usage:

- **Government or military**
  Cryptanalysis is heavily used by governments and militaries around the world to enhance the techniques used and for espionage purposes.
- **Research and academic**
  The use of cryptanalysis by researchers and the academic world falls into the second category. The goal of this category is to improve cryptography techniques for society's benefit.
- **Hackers and attackers**
  Hackers and attackers use cryptanalysis for personal gain or malicious intent. The cryptanalysis techniques used by these categories are very similar. The difference may be the availability of the resources.

Cryptography is divided into three types based on the type of encryption keys used:

- Symmetric cryptography
- Asymmetric cryptography
- Keyless cryptography

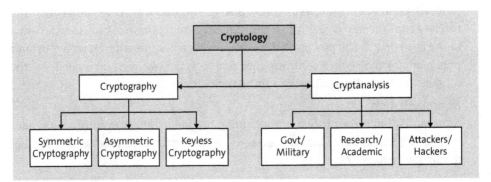

Figure 1.4 Types of Modern Cryptography

We'll discuss these three types in more detail in the following sections.

Symmetric Cryptography

Symmetry is defined in Webster's Dictionary as "correspondence in size, shape, and relative position of parts on opposite sides of a dividing line or median plane or about a center or axis." Cryptography that uses the same key for encrypting and decrypting (on both sides of the "axis") is known as *symmetric cryptography*. Symmetric cryptography is based on the *shared key* concept, which means that the same encryption key is used for encryption and decryption. The same key is shared and used by both actions or at both ends.

Figure 1.5 shows the concept of symmetric cryptography. The symmetric algorithm takes plain text (P) and symmetric encryption key (Ks). The output is cipher text (C). It's important to note that this communication can happen on public links such as the internet or Wi-Fi, but still, the message is secure. Mathematically, we can write this as cipher text C is a function of symmetric key Ks and plain text P:

$C = f(Ks,P)$

The decryption of the cipher text is simply reversing the process. When the cipher text C and symmetric key Ks are fed to the decryption algorithm, it generates the plain text P. Mathematically speaking, the plain text P is a function of cipher text C and symmetric key Ks:

$P = f(Ks,C)$

If a hacker or eavesdropper intercepts the communication, the attacker will only get the encrypted text. Even if they know the encryption algorithm, the attacker won't be able to decrypt the message. The only way to decrypt the message is to use the correct symmetric encryption key (Ks). As long as the encryption key is secure, the cipher text is safe.

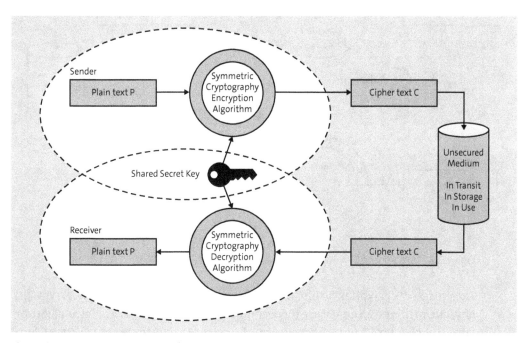

Figure 1.5 Symmetric Cryptography

1 Fundamentals of Cryptography

Symmetric cryptography works perfectly as long as the secret key can be shared easily between the sender and receiver because they both use the same key. However, the biggest risk lies in sharing the key. What if the key is compromised while sharing? If that happens, there won't be any confidentiality. We'll discuss symmetric cryptography in detail in Chapter 2.

Asymmetric Cryptography

The concept of encryption and decryption is the same here as in the case of symmetric cryptography but with a small twist. *Asymmetric cryptography* uses two keys: a private key and a public key. If one key is used for encryption, the other key must be used for decryption, and vice versa. You can't encrypt and decrypt with the same key.

Figure 1.6 shows the concept of asymmetric cryptography. Here, every user has two keys (the public key and the private key). The public key is shared with others, while the private key is confidential and must be kept secure. The sender encrypts the plain text message P using the public key Ke to generate the cipher text C. Mathematically speaking, the cipher text C is a function of plain text P and encryption Ke:

$C = f(Ke, P)$

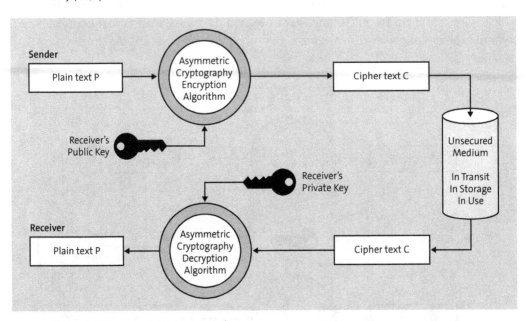

Figure 1.6 Asymmetric Cryptography

The process of decryption is similar to what we saw in the case of symmetric cryptography, but, in this case, the key use for decryption is different. The asymmetric algorithm takes in cipher text C and the private key of receiver K_d to recreate the plain text P:

$P = f(K_d, C)$

1.2 Introduction to Cryptography

The goal of asymmetric cryptography is to solve the problem of sharing the secret key in symmetric cryptography. In asymmetric cryptography, users can share the public key with almost anyone. The important thing to note here is that the sender must encrypt a message with the receiver's public key. That way, only the receiver can decrypt the message with the corresponding private key. We'll discuss asymmetric cryptography in detail in Chapter 3.

Keyless Cryptography

Keyless cryptography is also known as hashing, hash algorithms, or one-way functions (or one-way encryption). We introduced hash algorithms in Section 1.2.3, but here, as shown in Figure 1.7, we'll briefly recap the core concepts.

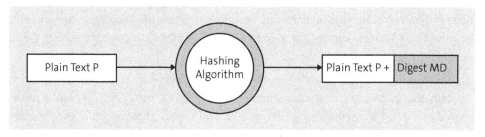

Figure 1.7 Keyless Cryptography

Remember, the hash function takes the plain text and generates a fixed value string or a digest. The digest or the hash value is unique to the message, and the digest changes if the message is changed. Once generated, neither the content nor the length of the digest can be reversed or changed, which is why it's known as one-way encryption. In addition, because there is no way to decrypt or reverse the hash value or digest, there is no need for any key—hence the term keyless cryptography. It's important to note that keyless cryptography is primarily used for data integrity and not confidentiality. The message itself isn't encrypted using a hash algorithm. However, we can combine encryption and hashing to ensure integrity and confidentiality. We'll discuss keyless encryption and hashing in detail in Chapter 4.

1.2.6 Steganography

Cryptography protects the messages by making them unreadable. However, anyone who comes across the message would know that the sender is trying to hide something. This probably piques the curiosity of attackers, hackers, and others who don't have access to this data. But what if we completely hide the message? If we hide the message, then anyone except the sender, receiver, and authorized users wouldn't know that there is a message. The process of hiding the message itself is known as *steganography*, which comes from the Greek words *steganos*, meaning covered or concealed, and *graphy*, meaning writing. Steganography means covered or concealed writing.

1 Fundamentals of Cryptography

Like cryptography, steganography also has a very old history. Over the centuries, people have used various forms and means to hide messages from enemies, spies, and other unauthorized users. In the 16th century, Italian scientist Giovanni Porta used steganography writing on a hard-boiled egg to pass messages. Porta wrote on the shell with his solution, hard-boiled the egg, and the writing disappeared from the outer shell and was unreadable until the shell was removed to see the message that had transferred to the egg.

Even before this, kings used steganography to send messages. One technique was to shave the head of the messenger and write the message on their head. The messenger is held captive until their hair regrows. Once the message is covered, the messenger will be sent to deliver the message. The recipient of the message shaves the head of the messenger to recover the message. Obviously, these messages weren't time sensitive! Furthermore, writing with invisible ink between the readable lines has always been a popular way to conceal messages. Germans used invisible ink during the First World War.

Another popular method to hide the message is the microdot technique. In microdot, the message is reduced to a very small size—typically 1 mm diameter—and represented as a dot. The message was first written and then a photograph of the message was taken. This image is then reduced to a dot. The dot is then used as a period or title of i or j in the normal readable message. Although there is a debate about the first use of microdot, Germans definitely used it heavily during the Second World War and made it a popular choice for steganography.

Steganography is very easy to use now with all the technology at our disposal. It's very easy to hide a message within a digital image or photoshop the message onto something else. One popular way to hide the message is to make it a part of the sound waves. The message is converted to sound waves and then included as a part of an audio file in the form of white noise. Similarly, the message can be hidden in a video file. Almost anything we can digitally modify can be used to hide a message.

The primary use of steganography is to hide the message or other sensitive information. But people can use this technique for a malicious purpose. Steganography can be used to hide a malicious virus or malware within a useful file or web link. As soon as the user opens a file or clicks the link, the malicious program is executed.

One of the downsides of using steganography is its vulnerability to reveal everything at once if the attacker finds out the message is hidden. Once the attacker suspects that the image has some message hidden, he can do some investigation and discover the hidden message. Once the message is discovered, he would get the information because the message is in plain text. One way around this vulnerability is to use cryptography in conjunction with steganography. That way, even if an attacker finds out there is a hidden message in the image or in an audio file and can extract the message, the message is in cipher text. The attacker will have to go through an extra step to decipher the encrypted message.

> **Steganography Tools**
>
> A simple Google search reveals that many tools are available to hide messages easily using various media. Some of the tools are even free! Some open source and free steganography tools are listed here:
>
> - **Steghide**
>   This is a command-line tool that allows you to hide your message in an image or an audio file.
> - **Crypture**
>   This command-line tool allows you to hide the message within a BMP file.
> - **OpenStego**
>   This tool also allows you to hide messages in images and encrypt data.
>
> These are just some examples. However, many more tools are available to hide messages in an image or in a different type of file.

1.2.7 Additional Techniques

In this section, we'll briefly describe the cryptographic techniques that are no longer used in modern cryptography or aren't the primary focus of this book but are important to know about from a conceptual point of view.

One-Time Pad

The key is the only thing that needs to be kept secret in the cryptosystem. It's very crucial to make the key secure or strong enough to withstand adversarial attacks. One way to make the key stronger, harder to guess, and difficult to break is to make it very long. The *one-time pad* is known as the holy grail of the cryptosystem. It's considered very secure and almost unbreakable because of the following factors:

- The key is as long as the message itself.
- The key is randomly generated.
- The key is used only once (hence the name "one-time").

Like most cryptosystems, the root of one-time pad goes back to war espionage. One-time pad is an improvement to the Vernam cipher, which was developed by Gilbert Vernam. In the Vernam cipher, a number is assigned to each letter in the alphabet. Table 1.4 shows an example.

1 Fundamentals of Cryptography

A	B	C	D	E	F	G	H	I	J	K	L	M
0	1	2	3	4	5	6	7	8	9	10	11	12
N	O	P	Q	R	S	T	U	V	W	X	Y	Z
13	14	15	16	17	18	19	20	21	22	23	24	25

Table 1.4 Vernam Cipher

For example:

- Plain text = CRYPTO
- Key = CIPHER

Using Table 1.4, you get the following:

- CRYPTO = 2 17 24 15 19 14
- CIPHER = 2 8 15 7 4 17

Or, consider another example where *Plain text + Key* = *4 25 39 22 23 31*. After looping around numbers that are larger than 26, this would equal *4 25 13 22 23 5*, which results in the cipher text E Z N W Y F. In the real world, modular addition using exclusive OR (XOR) is used for the implementation.

The problem with the Vernam cipher was that the key was repeated, making the system vulnerable. The one-time pad improved the Vernam cipher by restricting the use of the key to only one time.

The one-time pad offers perfect secrecy if the key is as long as the message itself, is generated randomly, and is used only one time. However, in practical implementations, this is very difficult to achieve for several reasons. First, when the message is very long, it takes a significant amount of time to perform the XOR operation. Second, it's difficult to share the key, and the risk of compromising the key is extremely high when the key is very long. Third, it's practically unmanageable when a high volume of messages is exchanged every day.

Although one-time pad remains unbreakable and has earned a place in history for perfect secrecy, the use of public key cryptography and symmetric cryptography are popular choices in modern cryptography.

Zero-Knowledge Proof

Zero-knowledge proof (ZKP) is a very useful and powerful concept in cryptography. ZKP allows a prover to convince the verifier that he knows the information without revealing the information to the verifier. The concept is simple but very important in cryptography because it offers the perfect balance between security and privacy. Some practical examples of ZKP in modern cryptography are as follows:

- **Identity and access management**
  ZKP is used in various scenarios in identity and access management (IAM). For example, in authentication, users can verify to authenticate to the system or use password-less authentication. The concept can also be used for authorization in access control.

- **Digital signature**
  We'll learn more about digital signatures in Chapter 4, but for now, it's important to note that the concept of ZKP helps private keyholders prove that they have sent the information without sharing the key itself.

- **Other usages**
  ZKP is also used with secure communication, voting machines, and blockchain.

ZKP is becoming very popular for its ability to offer security without compromising privacy, but it has its share of challenges too.

Resources to Learn about ZKP

The complete discussion is beyond the scope of this book but you can refer to these resources to learn more about the concept and implementation:

- "Zero Knowledge Proof: An Introductory Guide": *http://s-prs.co/v585600*
- "Zero-Knowledge Proofs: How It Works & Use Cases in 2024": *http://s-prs.co/v585601*
- "Demonstrate How Zero-Knowledge Proofs Work without Using Math": *http://s-prs.co/v585602*

1.2.8 Confusion and Diffusion

The principles of *confusion* and *diffusion* were first defined by Claude Shannon. In the simplest terms, confusion refers to when the relationship between plain text and cipher text is so complicated that the attacker can't determine the key (or plain text) from the cipher text. The purpose of the confusion principle is to obscure the relationship between the cipher text and the key.

Diffusion is a single change in the plain text resulting in multiple changes in the cipher text. The purpose of diffusion is to hide any apparent relationship between the plain text and cipher text.

The principle of confusion is used in stream cipher (Section 1.3.3), while confusion and diffusion are both used in block cipher (Section 1.3.4). In Chapter 2, we'll learn how confusion and diffusion are applied to symmetric algorithms.

1 Fundamentals of Cryptography

1.2.9 Refresher on Mathematics

As promised in the preface of the book, we'll explain the concept of cryptography with minimal use of mathematics or programming. However, even as a practitioner, you need to understand some basic concepts of mathematics. In this section, we'll quickly review modular arithmetic.

Addition, subtraction, multiplication, and division are the fundamental building blocks of mathematics. Modular arithmetic takes the concept of division to the next level. In traditional division problems, the answer, usually, is the quotient. In modulo arithmetic, the answer is the remainder. For example, take a traditional math problem, 27 divided by 12. The traditional answer is 2 remainder 3. However, modulo arithmetic focuses on the remainder, which is 3. The same problem is written as follows:

27 = 3 mod 12

Instead of 27, if 24 is divided by 12, then the remainder in this problem is 0. And the remainder keeps increasing by 1, as we go up from 24: at 25, the remainder is 1, at 26 the remainder is 2, and all the way up to 35. When 35 is divided by 12, the remainder is 11. However, this cycle resets at 36, and the remainder resets to 0 at 36. The cycle of the remainder starts again and resets at 48. For this reason, it's also known as a *wrap-around* problem. This is best described with an example of the clock. If you start at midnight, 3 pm is 15:00. However, because we're talking about wrapping around at 12, the answer is 3 pm or the remainder is 3. For 3 am the next day (which is 27 hours from the previous midnight), it will be 3 again. Generalizing this equation, it's written as follows:

x = y mod n

Here, *n* is known as the modulus, and *y* is the remainder when the number *x* is divided by modulus *n*.

Now let's take this concept one step deeper to congruence. The numbers are considered congruent when two numbers share the same remainder with the same modulus *n*. Let's go back to our example with a modulus of 12. In this example, 15, 27, 39, and so on are congruent because they have the common remainder 3 when divided by 12. The concept is represented as follows, where *a* and *b* are congruent numbers and *n* is a modulus:

$a \cong b \ (mod \ n)$

Unlike other mathematical concepts, modular arithmetic works on finite sets. Finite sets simply mean the numbers in the set or group end, and the counting starts over—the concept of wrap-around that we discussed previously. Why is this modular arithmetic useful in cryptography? Cryptography uses modular arithmetic in encryption/decryption algorithms. For example, the substitution cipher uses mod 26, and the XOR bitwise operation uses mod 2. We'll learn more about the ciphers and their operations in the next section.

> **Disquisitiones Arithmeticae**
>
> Carl Gauss is credited with modular arithmetic. He wrote the *Disquisitiones Arithmeticae* (or *Arithmetical Investigations*) [Internet Archive, n.d.] when he was only 21 and published his work on number theory in 1801. It's believed that this was one of the last mathematical publications in Latin. Gauss was exceptionally talented and made significant contributions to mathematics [Britannica, n.d.].

1.3 Primer on Ciphers

In this section, we'll start scratching the surface of the inner workings of cryptography by learning more about ciphers. A *cipher*, also known as an encryption algorithm, is the functional mechanism of how encryption is performed. Our focus will be on stream ciphers and block ciphers, their advantages, and disadvantages, and examples of each cipher. But first, we'll introduce other types of ciphers in this section such as the substitution cipher, transposition cipher, running key cipher, and so on due to their historical importance.

1.3.1 Substitution Cipher

A substitution cipher is probably the simplest and the oldest form of encryption algorithm. In this technique, the plain text alphabet is simply substituted by another alphabet. Because it's substituted by one bit at a time, it's also a type of stream cipher. We'll learn about stream ciphers in Section 1.3.3.

The Caesar cipher we learned about in Section 1.1 is probably the best and most basic example of the substitution cipher. As shown in Table 1.5, A is replaced by D, B is replaced by E, and so on.

A	B	C	D	E	F	G	H	I	J	K	L	M	N	O	P	Q	R	S	T	U	V	W	X	Y	Z
D	E	F	G	H	I	J	K	L	M	N	O	P	Q	R	S	T	U	V	W	X	Y	Z	A	B	C

Plain text: THE BEST CRYPTOGRAPHY BOOK

Cipher text: WKH EHVW FUBSWRJUDSKB ERRN

Table 1.5 Example of the Caesar Cipher

We saw that the alphabets are shifted by three letters, which is an example of a shift 3 or rotate 3 algorithm. However, the cipher can be made a little complicated by substituting symbols and other variations. As long as the sender and receiver know the algorithm, any variation of the substitution can be implemented.

1 Fundamentals of Cryptography

This type of substitution cipher is known as a *monoalphabetic cipher* because it substitutes only one alphabet at a time. In most cases, the substitution is fixed. As we saw in the Caesar cipher, the substitution was one alphabet and with a shift by 3. The substitution cipher with monoalphabetic cipher is easy to crack with frequency analysis.

Remember our discussion of modular arithmetic in Section 1.2.9? I promised that this would come back to haunt you as long as you study or work with cryptography, and now is a great example. The Caesar cipher (with English language implementation) is mod 26. That is because the English alphabet is a finite set of 26 characters. The alphabet resets to A every time after reaching Z. If we take a shift of 3, then X is substituted with an A.

Polyalphabetic ciphers, on the other hand, can make the substitution cipher a little more difficult as it uses multiple alphabets for substitution in the same message. The Vigenère cipher, which we introduced in Section 1.1.1, is an example of a polyalphabetic cipher. Table 1.6 shows the Vigenère cipher substitution arrangement.

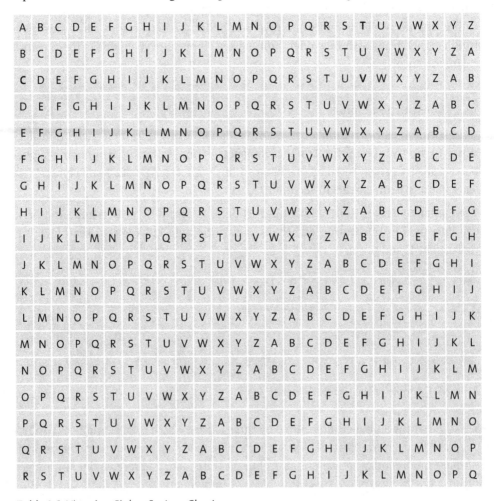

Table 1.6 Vigenère Cipher System Chart

1.3 Primer on Ciphers

Table 1.6 Vigenère Cipher System Chart (Cont.)

The Vigenère cipher uses the same chart every time to encrypt and/or decrypt the message. The Vigenère cipher works on the keyword, which should be changing for every message. For example, the keyword is CRYPTO. Now let's go back to our example—The Best Cryptography Book. Table 1.7 shows an example of the Vigenère cipher with the keyword *Crypto*.

Plain Text	THE BEST CRYPTOGRAPHY BOOK
Keyword	CRY PTOC RYPTOCRYPTOC RYPT
Cipher Text	VYC QXGV TPNIHQXPPIVA SMDD

Table 1.7 Example of the Vigenère Cipher

The best way to understand how the Vigenère cipher works is to follow these steps:

1. Write down the plain text. Our example is The Best Cryptography Book.
2. Write down a keyword under each letter and repeat it for the entire message. We used the keyword Crypto and wrote it down under the entire message.
3. To encrypt the message, first select a letter from the plain text in the top row. In our example, it's **T**.
4. Next, take the corresponding keyword, and select it from the first column. For example, **C** is for the first letter of the keyword crypto.
5. The intersection of this **T** column and **C** row is the cipher letter for **T**, which is **V**.

Unlike monoalphabetic ciphers, in the Vigenère cipher, the same letter O gets different cipher texts. In the example, the word *Book* has two Os but one is encrypted as M and the other is D. While the last K in the word *Book* is also D. This makes it difficult to decipher by an attacker. It's due to this complexity that the Vigenère cipher remained unbreakable until the 19th century.

1 Fundamentals of Cryptography

The substitution ciphers were popular because of the simplicity of implementation; however, the process of deciphering the message was equally easy. The substitution cipher with a polyalphabetic cipher can be cracked with period analysis. The *period analysis* technique is based on the size of the key and its repetition. To make the cryptanalysis difficult, other variations such as homophonic and polygraphic substitution ciphers were used. In a homophonic cipher, the plain text alphabet can be substituted with more than one alphabet or symbol, and in a polygraphic cipher, the substitution is done with a pair of alphabets instead of a one-to-one substitution.

Another disadvantage is the difficulty in sharing the algorithm or the key. The receiver should know the algorithm or the key before the message is sent; otherwise, there is a high risk of compromising the secrecy of the algorithm or the key in transfer.

1.3.2 Transposition Ciphers

The word *transpose* means to change positions with each other. The transposition cipher just does that by changing the position of the letter in the plain text, as opposed to the substitution cipher, which replaces the letters with other letters or symbols.

Although the transposition cipher is very similar to scrambling and unscrambling word games for children, if used with a twist, it could be somewhat difficult to break. For example, take a five-letter word, HELLO, and transpose the alphabet to LEOHL. There are 120 possible (5!) ways to rearrange the word HELLO. If we make that a six-letter word, DISNEY, and scramble it to SYDNEI — there are 720 possible (6!) ways to spell these six alphabets, and only one is spelled Disney. The point is that although this technique looks simple, it could be difficult to break with a longer message. The cipher can be made difficult to crack by making the message long, adding a key to transpose alphabets in an unpredictable way, and changing this system very often.

Let's consider an example where our key is CRYPTO. Next, we assign value to each alphabet based on alphabetic order: CRYPTO 1 4 6 3 5 2

Our plain text message is as follows: THIS IS THE BEST CRYPTOGRAPHY BOOK

As shown in Table 1.8, rearrange the plain text based on the key size of 6.

C	R	Y	P	T	O
1	4	6	3	5	2
T	H	I	S	I	S
T	H	E	B	E	S
T	C	R	Y	P	T
O	G	R	A	P	H
Y	B	O	O	K	x

Table 1.8 Example of a Transposition Cipher

Write this down as a sentence, forming a word for each column:

TTTOY HHCGB IERRO SBYAO IEPPK SSTHX

Now, reorder the words (columns) using our key for each column—146352:

TTTOY SSTHX SBYAO HHCGB IEPPK IERRO

Now, this becomes even more difficult to crack.

The transposition cipher was difficult to break when encryption and cryptanalysis were performed manually. But now, with the computing power that we have, this technique doesn't stand a chance.

There are other examples of transposition ciphers, such as Rail Fence, Scytale, Route, and Columnar transposition, which we'll briefly discuss before moving on:

- **Rail fence cipher**
  This the simplest form of transposition cipher. The idea is to scramble the letters of the plain text in an unrecognizable or unpredictable fashion. Table 1.9 shows an example of the rail fence cipher using the text THIS IS THE BEST CRYPTOGRAPHY BOOK. Our plain text is arranged in a shifted manner, and then we've created cipher text by listing the letters by the row and in groups of five letters. The last group has only three letters. We can create different groups or add padding values like X and Y at the end.

T		T		T			O		Y		Y
	H		S	H		S	C		T	G		H	B		X
	I	I		E	E			R	P		R	P		O	K
	S				B				Y			A		O	

Table 1.9 Rail Fence Cipher

The cipher text would read like this:

TTTOY YHSHS CTGHB XIIEE RPRPO KSBYAO.

- **Scytale cipher**
  In this cypher, a tape is wrapped about a rod or a cylindrical object. The message is written on the tape while it's wrapped around. The tape is unwrapped and sent with the message to the receiver. Any eavesdropper or even receiver won't understand the message unless they have a rod or a cylindrical shape of the same diameter. Once the tape is wrapped around the rod, the message becomes readable.

  The Scytale is a transposition cipher because it transposes the letters of the message when the tape or ribbon is unwound from the rod. The "key" to encryption and decryption is the size of the rod. If the incorrect size is used, the message could be decrypted incorrectly. This is very easy to use and was a popular choice historically.

- **Route cipher**
  This cipher follows a particular route, which is the key for encryption and decryption. As shown in Table 1.10, the message is written in a grid formation, and a specific route is given.

T	B	T	Y	O	A	Y	O
H	E	C	P	G	P	B	K
E	S	R	T	R	H	O	X

Table 1.10 Route Transposition Cipher

For example, the sender and receiver decide on a route to be "starting bottom right, counterclockwise." This means the cipher text will read like this:

XKOYA OYTB THESR TRHOB PGPCE

The route is the encryption key for route ciphers. Route ciphers can have complicated routes for larger messages. If the message is large, it could be very difficult for an attacker to guess the possible route. However, this advantage can also go against route ciphers. If the message is large and the route isn't properly chosen, then an attacker could guess it. Alternatively, if the route isn't well defined and shared, then the legitimate receiver may not be able to decrypt it properly.

- **Columnar cipher**
  In this cipher, the cipher text is generated column by column. The plain text is divided into a block of rows with fixed widths. Columns in random order are picked to create the cipher text. The width of the rows and the order of the columns are decided based on the keyword, and the keyword is usually converted into a number to set the order.

Substitution and transposition ciphers are very important in understanding the concept but aren't used anymore in modern cryptography. These classical ciphers are either not feasible to use with computers (e.g., Scytale) or can be easily cracked with modern cryptography.

1.3.3 Stream Ciphers

Stream ciphers are a very important part of the cryptosystem. Stream ciphers encrypt 1 bit or byte at a time (8 bits = 1 byte). Although the concept of encrypting 1 bit or byte at a time is very old, the stream cipher is very attractive to certain applications. In this section, we'll discuss how stream cipher works, the advantages and challenges of stream cipher, and some popular stream ciphers.

1.3 Primer on Ciphers

Interestingly, the stream cipher concept is based on the one-time pad, which is based on Vernam ciphers. We learned about the one-time pad and the Vernam cipher in Section 1.2.7.

> **Vernam Cipher**
>
> More than 100 years ago, Gilbert Vernam was the first to use plain text and key as the two feeds to generate the Vernam cipher text. The electromechanical machine took these two inputs to generate the cipher text. Stream ciphers also mix plain text with a keystream to generate the cipher text but use an XOR operation. Vernam co-invented the one-time pad, which is basically an improved version of the Vernam cipher.

Stream ciphers are classified into two types: synchronous stream ciphers and asynchronous stream ciphers. We'll discuss these main types and some popular stream ciphers in the next sections.

Synchronous versus Asynchronous

In *synchronous stream ciphers*, as shown in Figure 1.8, the keystream is dependent only on the key. The keystream is completely independent of the plain text or the cipher text. The keystream is generated using an initialized value, usually known as the *seed value*.

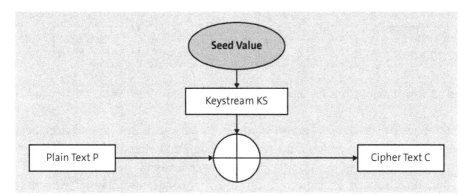

Figure 1.8 Synchronous Stream Ciphers

The keystream bit performs the XOR operation with the plain text bits and generates the cipher text. At the receiving end, the XOR operation is repeated by the keystream with the cipher text to generate the plain text.

Synchronous stream ciphers provide two main benefits:

- **No error propagation**
  First, it doesn't propagate the error, meaning if one bit is corrupted in encryption, only one bit is impacted at the receiving end. There is no impact of that error on other bits.

1 Fundamentals of Cryptography

- **Speed**
  The second advantage is speed. Because the keystream generation doesn't depend on any other factor than the seed value, it's fast.

One potential issue with synchronous stream ciphers is their ability to synchronize between encryption and decryption. Special care must be taken to ensure that they are synchronized; otherwise, the decrypted message at the receiving end may not make sense.

Synchronous stream ciphers can be implemented using output feedback (OFB) mode. OFB is one of the modes of operation for block ciphers. We'll learn about OFB in Section 1.3.4.

Asynchronous stream ciphers, on the other hand, depend on the cipher text. It takes a bit from cipher text and mixes it with the key to generate the new keystream bit. Figure 1.9 shows the basic implementation of the asynchronous stream cipher.

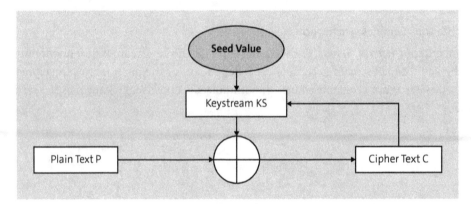

Figure 1.9 Asynchronous Stream Ciphers

Asynchronous stream ciphers are also referred to as self-synchronizing stream ciphers. The encryption and decryption functions of asynchronous stream ciphers are always in sync because they use the previous cipher text to generate the next keystream. However, this also introduces error propagation. An error in one bit will stay in the message and corrupt the message until the errored bit is passed through the keystream.

Asynchronous stream ciphers can be implemented using cipher feedback (CFB) mode, which is one of the modes of operation for block ciphers. We'll learn about CFB in Section 1.3.4.

It's important to note that stream ciphers use XOR operations to add the key to the plain text to encrypt and add the key to the cipher text to decrypt the data. A simple XOR operation performs the magic.

Implementation and Security

Now let's revisit our favorite topic—modular arithmetic. If we divide any number by 2, the possible remainder is either 0 or 1. That means mod 2 actually works like a Boolean function.

Let's analyze our stream cipher operation bit by bit with mod 2. Table 1.11 shows a comparison of mod 2 arithmetic with XOR addition. The first column shows the plain text bit P, and the second column shows the keystream bit KS. The third column shows the addition of a plain text bit with a keystream bit. The fourth column shows the mod 2 operation, $n = r \bmod 2$. The remainders, **r**, are shown in bold. The fourth column shows the equivalent XOR operation (indicated by the symbol $\oplus$). Again, the output of XOR, **r**, is in bold.

P	KS	n = P + KS	n = r mod 2, r = remainder	P $\oplus$ KS = r
0	0	0 + 0 = 0	0 / 2 = 0 with remainder **0**	0 $\oplus$ 0 = **0**
0	1	0 + 1 = 1	1 / 2 = 0 with remainder **1**	0 $\oplus$ 1 = **1**
1	0	1 + 0 = 1	1 / 2 = 0 with remainder **1**	1 $\oplus$ 0 = **1**
1	1	1 + 1 = 2	2 / 2 = 1 with remainder **0**	1 $\oplus$ 1 = **0**

Table 1.11 Mod 2 and XOR Comparison

> **Why Use XOR with Cryptography?**
>
> The obvious question at this stage is why use XOR with cryptography and not any other Boolean function (e.g., an AND or OR gate)? The answer lies in mathematics. It's all about probability and randomness. We want the cipher text to be hard to decrypt by guessing the probability or predicting. The XOR shows the property:
>
> - Plain text $\oplus$ Keystream = Plain text $\oplus$ 0 = Plain text
> - Plain text $\oplus$ Keystream = Plain text $\oplus$ 1 = Inverted plain text
>
> Based on Table 1.11, there is an exact 50% chance that the cipher text will be 1 or 0. This randomness makes the XOR operation a lucrative option for cryptography.

It's clear that stream ciphers are very simple to implement. However, security is completely dependent on the keystream. If an attacker or malicious user is able to guess or predict the keystream, then the stream cipher becomes useless.

To avoid this situation, the keystream must be generated using random numbers or values. This is why random numbers and their generation are very important in cryptography. Informally, we can define random numbers as numbers that aren't repeatable and predictable. There are three broad categories of random number generators:

- **Real or true random number generator (TRNG)**
  Random numbers are true or real random numbers if they aren't reproducible. In other words, the random numbers are truly random if we can't repeat the sequence. However, the TRNGs aren't useful in cryptography because we have to encrypt and decrypt the data using the same key. If we can't repeat the random keystream at the receiving end, we can't use the TRNG for the keystream.

> **Examples of TRNG**
> Some physical examples of TRNG are flipping a coin, rolling dice, noise and vibrations, and so on. All of these examples generate truly random sequences. If we roll a die four times, we may get 5, 2, 6, 1. However, there is no guarantee that we'll get the same sequence next time. Tracking mouse movements, measuring the keyboard clicking speed, gauge thermal noise, and so on are examples of digitally generated TRNG.

- **Pseudorandom number generator (PRNG)**
  The output of TRNG is nonreproducible. We want randomness, but we need something that we can control and reproduce. That is why we digitally generate random numbers, which we call pseudorandom numbers. Pseudorandom numbers are man-made random numbers. They aren't truly random but look almost random. They aren't truly random because they can be reproduced—but we want to reproduce them at the receiving end and that is why we created PRNGs.

  Another property of a truly random number is that the sequence or number shouldn't repeat. Digitally generated pseudorandom numbers have a finite period, and they will repeat, but the latest technology makes it possible to keep this period very long, and the number won't repeat for many cycles.

  PRNG is used for software testing, video games, and many other applications but isn't suitable for cryptography. The trouble is that if we can reproduce the sequence, an attacker will be able to reproduce it too. If that happens, there is no benefit of using encryption. That is why we want to generate secure random numbers, which is discussed next.

- **Cryptographically secure pseudorandom number generator (CSPRNG)**
  Another key property of randomness is unpredictability. No one should be able to guess what the next number will be. This property and the inability to reproduce the sequence are the core requirements for using any random number generator in cryptography. CSPRNG must possess these two qualities. One way to achieve this is to add a seed value (basically a key for PRNG to make it CSPRNG). Adding the seed value that is known to the sender and receiver only, the PRNG becomes CSPRNG. The goal of CSPRNG is to generate random numbers that any attacker or malicious user can't reproduce or predict even if they know the algorithm, but the authorized users

can reproduce because they know the seed values. If we achieve this, then we have a secure PRNG. You can also generate the keystream using hardware.

One technically efficient way to generate a stream of random bits is using linear feedback shift registers (LFSRs). Shift registers are logical circuits consisting of flip-flops. If the input from the shift registers is a linear function of the previous state, a series of LFSRs is connected to generate a random bit stream. The LFSRs are synchronized on a clock cycle and XORed with a previous bit. The use of LFSRs is technically very efficient but isn't very secure. There are a few ways to improve the security:

- **Connect the LFSRs in a nonlinear fashion**
  In this implementation, the LFSRs are connected in parallel, and the output is computed using a Boolean function. This becomes a little more difficult to crack compared to the linear connection.
- **Use the clock-controlled implementation**
  In this variation, the LFSRs aren't synchronized with a regular clock cycle but the trigger is based on the combination of events.
- **Filter the LFSR outputs**
  In this implementation, it's critical to keep the filtering parameters secret.

Advantages and Disadvantages

Stream ciphers offer the following advantages:

- One of the primary advantages of stream ciphers is the speed. Because the encryption and decryption happen 1 bit at a time (or byte), it can produce output almost instantly. Historically, this was a big advantage because microprocessors weren't very fast. For the most part, the speed was the decision factor in using the stream cipher.
- Lightweight and very low complexity is another advantage of the stream cipher. The cipher is very easy to implement in terms of software and hardware. This makes implementation not only easy but also inexpensive.
- Lastly, stream ciphers are an attractive choice for mobile and other devices where the computing power is relatively limited because of their low complexity to implement and speed of encrypting and decrypting.

Stream ciphers have one major disadvantage—they can easily be vulnerable to various attacks, even without any sophisticated tools or knowledge. For example, in the case of a synchronous cipher, even having the same seed values twice can help the attacker crack the cipher. In addition, the low complexity or simplicity of this cipher goes against its security because the simplicity makes it easy for attackers to understand the cipher and use it for malicious intent or attack.

Additional Stream Ciphers

To conclude our primer on stream ciphers, this section will cover a few popular examples. These examples will help showcase how stream ciphers are used in practical scenarios.

Our first example is Rivest Cipher version 4 (RC4), which was developed by Ron Rivest in 1987 for RSA Data Security Inc. RC4 is a symmetric encryption stream cipher. It's believed that internally within RSA Data Security, the work of developing a stream cipher by Ron Rivest was nicknamed as "Ron's Code" and hence the cipher is called RC. RC4 is a proprietary cipher. In 1994, the RC4 algorithm was published anonymously and later circulated all over.

RC4 uses the basic stream cipher principle, as shown in Figure 1.10: it generates cipher text by XORing the plain text with the keystream. The critical part of the algorithm is its key generation mechanism or key scheduling algorithm. RC4 uses the OFB mode to generate the keystream (for more on OFB, see our discussion in Section 1.3.4). At this stage, we only need to know that a 256-byte block is used to generate the keystream. The key generation algorithm is made up of 256-byte permutations and two 8-bit pointers. RC4 is a variable key size cipher because the key size can vary between 1 and 2,048 bits.

RC4 is a symmetric stream cipher, which means it uses the same key to decrypt the cipher text. The mechanism is simply reversed for decryption. Decryption in RC4 can be explained by simply changing the direction of the arrows in Figure 1.10, which shows the cipher text and plain text.

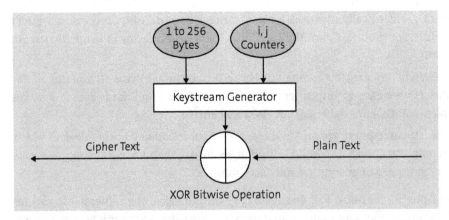

Figure 1.10 RC4 Cipher

Let's cover some of the advantages and disadvantages of RC4:

- **Primary advantages of RC4**
  RC4 is extremely simple to implement, which makes it very popular among coders and software engineers. This also minimizes implementation errors. RC4 is lightweight and has very low complexity.

- **Disadvantages of RC4**

    RC4 is considered vulnerable, and attacks have been published. Researchers have demonstrated that RC4 generates biased output, and bias in the bytes can be detected by observing only 256 bytes. However, RC4's defense is that it's secure if implemented correctly. RC4 can be vulnerable if the keystream isn't generated properly,

RC4 is used with many applications. Its lightweight makes it a popular choice with cell phones and wireless applications. It's used with Secure Socket Layer (SSL) and Wired Equivalent Privacy (WEP). Both SSL and WEP are considered vulnerable and are replaced with Transport Layer Security 1.3 (TLS 1.3) and Wi-Fi Protected Access 2 (WPA2), respectively. At one point, RC4 was also used in Microsoft Edge and Internet Explorer browsers.

> **Cypherpunks**
>
> The Cypherpunks mailing list was formed in the early 1990s by Eric Hughes, Timothy May, and John Gilmore. In the mailing list, they discussed cryptography, privacy, security, and many other concerns, and the list grew quickly to almost 700 subscribers. Julian Assange (WikiLeaks), Marc Andreessen (Netscape), Derek Atkins (RSA), Bruce Schneier (*schneier.com*), and many other notable people were part of the mailing list. On Friday, September 9th, 1994, an anonymous email was sent to the Cypherpunks mailing list with the subject: "RC4 source code." Several rightful owners confirm the validity of the code [Cypherpunk, 1994].

A5/1 is another popular stream cipher. A5/1 is used for over-the-air communication with the Global System for Mobile Communications (GSM) cellular technology. A5/1 was first published in 1987 and was kept confidential. The cipher was reverse engineered in the 1990s.

A5/1 is a symmetric key synchronous stream cipher. The cipher uses a 64-bit symmetric secret key, meaning the same key is used for encryption and decryption. The 64-bit key produces a 228-bit keystream to be used for encryption. The voice communication is performed in frames or bursts. Each frame is mixed with the keystream to generate the cipher text. The mixing operation is performed using the bitwise XOR operation. A5/1 uses three LFSRs with irregular clocking. The LFSRs have different bit lengths. The outputs of these LFSRs are XORed.

A5/1 is potentially vulnerable to various attacks, such as known-plaintext attacks, brute force attacks, and correlation attacks. There are some published attacks on A5. However, some attacks require lots of precomputing processing and lots of time and memory. Other attacks were based on guessing the contents of the LFSRs.

Let's look at one final popular stream cipher, Trivium. Trivium is a stream cipher that is popular for its simplicity. It's a hardware-based synchronous stream cipher. Trivium uses three shift registers of different lengths—93, 84, and 111, with a total length of 288. These are linear shift registers, the outputs of which are logically combined and fed back to create the nonlinear behavior.

Trivium uses an 80-bit key and an initialization vector (IV). We'll learn more about IVs in Section 1.3.4, but for the Trivium discussion, we need to know that an IV is added to provide some randomness in keystream generation. The 80 bits of the IV and key are loaded into the first two registers, and the remaining bits of these registers are set to 0. The last register is all set to 0 except the three most significant bits (MSBs); that is, bits 109, 110, and 111 are set to 1. Trivium generates 2^{64} bit keystream.

Trivium is attractive to many lightweight applications, such as embedded systems, IoT, wireless devices, and so on because of its simplicity and efficiency. There are already known attacks on Trivium. Trivium is somewhat attractive to various attacks, such as guess and determine attacks, algebraic attacks, resynchronization attacks, and more [De Canniere, 2006, 2008].

1.3.4 Block Ciphers

The block cipher is another flavor of cipher that is used in modern cryptography. Block ciphers are a type of symmetric cryptography, and they use the same key to encrypt and decrypt the data. Block ciphers can encrypt data in a block or chunk instead of one bit at a time. About 80% to 90% of cryptography applications use block ciphers. Encrypting blocks of data is the obvious use of the block cipher but it's also used with stream ciphers to generate a keystream, with a PRNG, with message authentication code (MAC), and with hash functions. (You'll learn more about MAC and hash functions in Chapter 4.) In this section, we'll introduce the following modes of block ciphers:

- Electronic code book (ECB) mode
- Cipher block chaining (CBC) mode
- Output feedback (OFB) mode
- Cipher feedback (CFB) mode
- Counter mode (CTR)
- Galois/Counter mode (GCM)

Why do we need so many modes of operations for block ciphers? Because the block cipher encrypts data in blocks of 64 or 128 bits, we need to have a mechanism to encrypt each block, construct, and transmit the message. At the receiving end, we need to decrypt each block and reconstruct the message in plain text. These modes of operations provide various ways to achieve this mechanism.

Figure 1.4 is expanded to add stream ciphers and block ciphers in Figure 1.11.

1.3 Primer on Ciphers

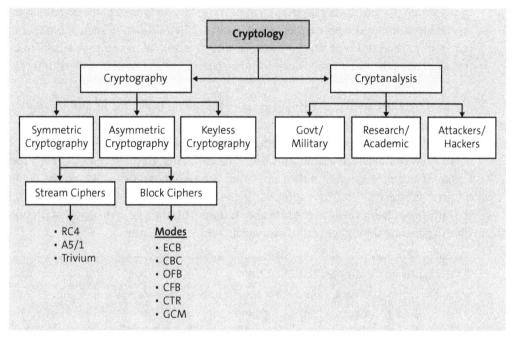

Figure 1.11 Types of Cryptology with Symmetric Encryption Expanded

Electronic Code Book Mode

Electronic code book (ECB) mode is the simplest form of the block cipher. It simply takes one block at a time, encrypts the entire block using a key, and generates the cipher text. The concept is very similar to a stream cipher but instead of 1 bit at a time, ECB encrypts one block at a time.

The size of the block depends on the underlying algorithm used for the encryption. For example, if the encryption algorithm is DES, then the block size is 64 bits, but if AES is used as an encryption algorithm, then the block size is 128 bits. (We'll discuss DES and AES in detail in Chapter 2. In real life use, the message or data is always larger than 64 bits or 128 bits, but for simplicity in this discussion, we'll assume that the block size for ECB is 64 bits or 8 bytes.

When the message is larger than the block size of 64 bits, then the message is divided into multiple blocks. If the message isn't exactly a multiple of 64 bits (8 bytes), then the last byte of the message is padded. *Padding* is the process of filling the empty bits or bytes with 1s and 0s. If the message has more than 1 byte of padding, then it's necessary to use the last byte as an indicator of the number of padded bytes. For example, say the message has only 4 bytes in the last block of 8 bytes. In this case, the following 3 bytes will be padded with 1s or 0s, and the last byte will indicate number 4. This will tell the receiver to delete the last 4 bytes after the message is decrypted.

The ECB mode works great and is very straightforward to implement. The ability of ECB to encrypt and transmit only one block at a time allows the sender to transmit blocks in any order. If a block is corrupted or lost in transmission, only one block of the message is lost and not the entire message. Moreover, the ECB can be set up to transmit blocks in parallel to speed up the transmission. This will allow multiple blocks to be transmitted at the same time. Blocks can be arranged at the receiving ends to reconstruct the message.

Although ECB has so many great advantages, it's hardly ever used in real-life applications because it has some serious vulnerabilities. One minor issue with the ECB isn't sending the entire message but sending chunks of the message, similar to seeing the email a few characters at a time. This would be annoying for the receiver. In daily life, we prefer an entire message at the same time. Another, more severe, issue is the deterministic nature of the ECB operation. ECB generates the same cipher text every time for the same plain text as long as the same key is used. This means, in theory, that it's possible to generate the cipher text of every plain text for a given key.

> **Why Call It an Electronic Code Book?**
> The electronic code book name comes from the code books Germans used during the First World War (refer to Section 1.1). The code book used by Germans had all the possible words and corresponding numeric codes. In ECB, theoretically, we can generate plain text and cipher text combinations for a given key. If the block size is b, it would be 2^b possible combinations or permutations. For a block size of 128, there could be 2^{128} possible permutations, and it isn't practically possible to create a book. However, conceptually, we can generate an electronic book of codes—very similar to what Germans did manually more than 100 years ago—hence the name.

The first issue, not receiving an entire message at the same time, is an inconvenience, but the second issue, generating the same cipher text for a given plain text, makes the ECB vulnerable. An attacker can possibly figure out or get some clue if he records the transactions over and over for a period. One way to resolve this problem is to change the key frequently. We won't go into the details of this vulnerability, but many introductory cryptography books have explained how this vulnerability can be exploited with an example of bank transactions (refer to [Schneier, 1994] and [Paar, 2010] for further explanation of how the attack could be carried out).

Cipher Block Chaining Mode

We learned in the ECB mode that it has two basic problems: (1) the deterministic approach of encrypting the blocks is vulnerable, and (2) the message is transmitted one block at a time. The cipher block chaining (CBC) mode is used to solve these issues. The deterministic behavior can be changed by adding randomness. This means if we feed different random numbers every time along with the plain text message to the encryption function, then the generated cipher text will be different—even for the same key. However, the important point is different random numbers. If the number is repeated, we end up with the same vulnerability as EBC.

1.3 Primer on Ciphers

In CBC mode, as shown in Figure 1.12, the first block P_0 is XORed with a random number or IV. The output of XORed is fed to the block encryption function f_e to generate the cipher text C_0. The subsequent blocks of plain text, P_1 to P_n, use the previous cipher text C_{n-1}. This basically kills two birds with one stone—it uses a random number as an input and adds the blocks of messages sequentially—hence the name *block chaining*. Both problems of EBC mode are solved!

Figure 1.12 CBC Mode

Mathematically, we can rewrite the previous explanation as follows:

$C_0 = f_e (P_0 \oplus IV)$
$C_n = f_e (P_n \oplus C_{n-1})$ where $n \geq 1$

The decryption process for CBC is very similar. First, feed the cipher text through the block decryption function and then XORed the first block, C_1 with IV, and the rest of the blocks of the message with C_{n-1}. The key here is to not repeat the IV.

Initialization Vector

The IV or a seed value is a number that is added to the keystream to generate randomness. The IV can be generated in several ways:

- Have a fixed value IV. However, this defeats the purpose because it will make the CBC also deterministic like EBC.
- Use an up-counter. The counter increments with every new block in the message starting with 0.

- Associate the IV with the clock timing, IP address, MAC address, or some truly random variable.
- Use a nonce (a number used only once). The nonce by nature must be unique. The IV isn't a secret and can be transmitted in clear text; however, the integrity of the IV must be protected.

Precautions must be taken in using the CBC mode. If the same IV and key combination is repeated with the same plain text, the cipher text will be duplicated. Chaining also poses some challenges in CBC mode. The chaining creates dependencies and decryption of blocks must happen in the same order. Second, in CBC mode, because of the chaining, the error propagates, meaning if the block C_n introduces an error in one bit, the part of the block will be corrupted or generate somewhat unreadable cipher text, and all subsequent blocks, C_{n+1}, will have an error for that bit. (This is opposed to EBC mode where an error occurs in a block, and only that block is impacted.)

Output Feedback Mode

As we learned in Section 1.3.3, stream ciphers are fast and efficient but lack robust security. One of the fundamental weaknesses of stream ciphers, especially with a synchronous stream cipher, is its vulnerability to the uniqueness of the keystream. If the seed values are repeated, the cipher text for a given plain text will be the same and will make the message vulnerable. This problem is solved using the block cipher to generate the random keystream.

As shown in Figure 1.13, the block cipher is used to generate keystream.

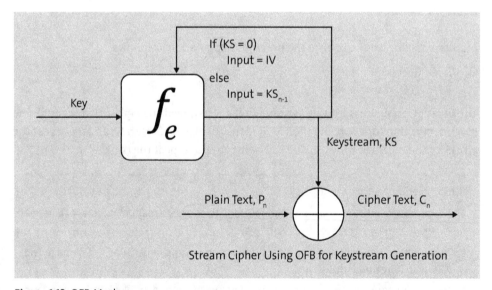

Figure 1.13 OFB Mode

The idea is simple. The *IV* is used for the first time to generate the first cipher block. The subsequent blocks use the previous cipher (keystream KS_{n-1}) to generate the cipher text. These cipher texts are used as keystreams in the stream cipher. This is a synchronous stream cipher because the keystream doesn't depend on the plain text or the cipher text.

It's important to note that the block cipher in OFB mode isn't really encrypting the message. The message is still encrypted using a stream cipher. That is why when decrypting the keystream generation on the receiving side, the block cipher in OFB mode still uses the same encryption function and not the decryption function. The decryption is performed by the stream cipher.

Cipher Feedback Mode

The cipher feedback (CFB) mode is very similar to the OFB mode discussed previously. The only difference is that the CFB mode generates the keystream for an asynchronous stream cipher instead of a synchronous stream cipher, as in the case of OFB mode. The asynchronous stream cipher, as discussed in Section 1.3.3, uses plain text or cipher text in keystream generation. This is exactly what the CFB mode is trying to achieve.

As shown in Figure 1.14, the first block of the block cipher is encrypted using the *IV*. The output cipher text of this block cipher, KS_0, is used as a keystream for the stream cipher. The stream cipher encrypts the plain text message *P* using the first cipher block keystream, KS_0, and generates cipher text *C*. This cipher text, C_{n-1}, is used with CFB to encrypt the subsequent blocks.

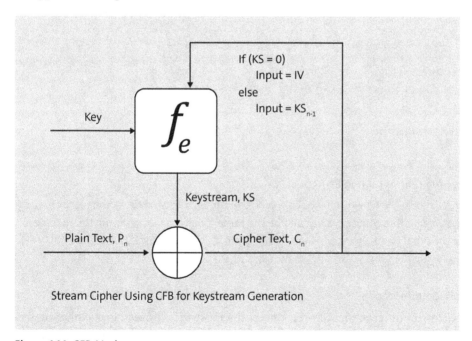

Figure 1.14 CFB Mode

1 Fundamentals of Cryptography

Counter Mode

The counter (CTR) mode of a block cipher is also used with a stream cipher in key generation. The difference in this case, compared to OFB and CFB modes, is that CTR mode doesn't rely on the cipher text. It generates a keystream simply using a counter. Unlike OFB and CFB, CTR doesn't use any feedback, and that makes CTR faster than CFB and OFB.

As shown in Figure 1.15, the CTR mode concatenates an IV with the counter to add the randomness. This makes CTR mode probabilistic or nondeterministic. The counter can increase with some constant value after encrypting every block. The counter value and IV aren't secret and can be shared in clear text with the receiver. However, the integrity of the IV and counter must be preserved.

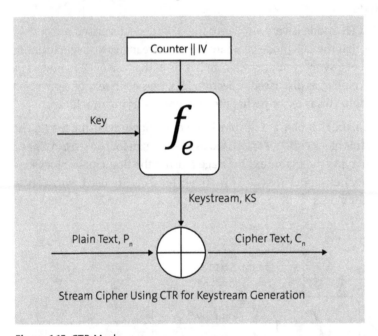

Figure 1.15 CTR Mode

Can the Counter Value Be Repeated?

In CTR mode, the IV, which is typically a nonce, can be the same for all the blocks in the message but the counter value can't be the same for all the blocks within the message. On the other hand, the nonce can't be the same for two messages, but the counter values can be the same between the messages.

Galois/Counter Mode

The Galois/Counter mode (GCM) is one of the most widely used block cipher modes. The modes of operation we learned so far are used to encrypt the message and provide confidentiality. However, an attacker can manipulate the cipher bit, and the receiver won't know. The receiver will see some gibberish in the message but won't be able to confirm if that is what the sender transmitted or if it was manipulated by an attacker. The GCM provides this integrity check for the message. GCM computes a MAC and appends a checksum or a tag to the encrypted message. The GCM uses CTR to provide confidentiality and Galois field theory for authenticity and integrity, hence the name Galois/Counter mode.

What Is a Galois Field?

Cryptography performs mathematical operations over a finite set of numbers. However, we want this finite set large enough that the numbers aren't repeated, not easily at least. A Galois field, named after the mathematician Evariste Galois, is one such finite field. A Galois finite field is represented as

$GF(p^n)$

where p is a prime number, and n is an integer.

In cryptography, a Galois field is used with symmetric and asymmetric cryptography algorithms, such as AES and elliptic curve cryptography (ECC). To learn more about how a Galois field is used with cryptography, refer to [Paar, 2010].

To recap, CTR mode is a block cipher mode that encrypts the data one block at a time. The counter is incremented after encrypting every block. The counter is concatenated with a nonce *IV* to add randomness. The counter is incremented and fed to the block cipher encryption function. The encryption key is also fed to the encryption function. The output of the encryption function is XORed with the plain text, *P*, to generate the cipher text, *C*. At this stage, the encryption or confidentiality part of the GCM is achieved.

GCM then computes the tag or MAC to provide authenticity and integrity. To achieve this, GCM performs the Galois multiplication. Breaking this down mathematically is beyond the scope of this book, but we'll explain the process at a high level. This multiplication is a product of additional authenticated data (AAD), the *H*, and a constant. The *H* is derived by encrypting a block of 128 bits with all values to 0. The product is XORed with cipher text as well as with the (IV||Cntr0). The last part, XORing of *IV* and counter value of *0*, is to provide authenticity to IV. This creates the tag that is then appended to the ciphered message.

At the receiving end, the cipher text is decrypted using the CTR mode. The receiver also recomputes the tag and compares it with the tag in the message to validate the authenticity and integrity of the message.

Because GCM encrypts the message and adds authenticity, it's known as an *authenticated encrypted mode*.

> **Cipher War: A Struggle for Survival?**
>
> In the early days of modern cryptography, stream ciphers were very promising and were a popular choice. However, it appears that the popularity of stream ciphers is fading away against the block ciphers. GSM technology in mobile phones is replacing the use of the A5/1 stream cipher with the A5/3 block cipher. Our wireless technology uses WAP2, the block cipher, instead of WEP, which uses the RC4 stream cipher. This shift is due to several factors:
>
> - Weaknesses were discovered in stream ciphers. Block ciphers provide better security.
> - The use of stream ciphers was driven by their simplicity and efficiency. However, modern technology has helped block ciphers become very efficient and easy to implement.
> - The choice of block cipher as a standard made more people use block ciphers, which ultimately fueled the growth.

1.4 Summary

The chapter gave a quick walk-through of the history of cryptography and discussed the fundamental concepts of security and cryptography. It explained why cryptography is required and how it can help us. We reviewed classical cryptography to get insight into how secrecy was achieved historically and why it doesn't stand a chance against modern technology. We discussed modern cryptography and how it uses technological advancements to protect our data and information.

A detailed discussion of classical and modern ciphers provides insight into stream and block ciphers. We learned briefly about several stream ciphers and the various modes of operation of block ciphers.

> **Key Takeaways from This Chapter**
>
> - Learned the fundamentals and basic concepts of cryptography
> - Refreshed our knowledge of modulo arithmetic and random number generators
> - Reviewed classical versus modern cryptography as well as the difference between encoding, hashing, and encryption
> - Discussed various ciphers, and took a deep dive into the modern stream ciphers and block ciphers

This chapter introduced the concepts of symmetric and asymmetric cryptography. The next two chapters will discuss symmetric and asymmetric cryptography in detail. You'll learn about symmetric cryptography, its advantages and weaknesses, and various algorithms for implementing symmetric encryption in the next chapter.

Chapter 2
Symmetric Cryptography

Symmetric cryptography is not only one of the oldest forms of cryptography but also the first widely used and implemented cryptographic mechanism. Classical cryptography was based purely on symmetric cryptography, and modern cryptography relies heavily on it for data encryption. In this chapter, we'll discuss symmetric cryptography and examine various symmetric cryptography algorithms in detail.

What is the commonality between Caesar sending a message to his war general on the battlefield, Zimmerman sending the telegram to German forces, and us talking on our cell phones? The war general, German forces, and our cell tower all must know the same secret as the sender of the message to understand or decipher the message. In Chapter 1, we learned the history and development of cryptography over the centuries. We saw that methods of secret communication varied widely over the years. However, there is one common thread throughout the development of cryptography—the sender and receiver both must know the secret used for encrypting the message. That is the core of symmetric cryptography.

In this chapter, we start by discussing why symmetric encryption is so lucrative and what makes the use of asymmetric encryption necessary in Section 2.1. The chapter will analyze and discuss in depth various symmetric encryption algorithms in Section 2.2, such as the Data Encryption Standard (DES), Triple DES, International Data Encryption Algorithm (IDEA), Advance Encryption Standard (AES), and AES finalists.

> **Chapter Highlights**
> - Fundamentals of symmetric cryptography
> - The architecture of various symmetric algorithms, such as DES, triple DES, and IDEA
> - Design and implementation of AES

2.1 Primer on Symmetric Cryptography

To recap the basic concept that we learned in Chapter 1, symmetric cryptography uses common knowledge or shared secrets to encrypt and decrypt the message. The sender of the message and the receiver of the message must have knowledge of the same secret key. All the historical ciphers that we studied in Chapter 1, Section 1.1.1, are of symmetric cryptography. Going back to our example of the Caesar cipher, the sender, Caesar, and the receiver, the war general, both must know the number of alphabet shifts used. Otherwise, the receiver won't be able to decipher the message. The same is true for the Vigenère cipher and other classical ciphers.

Fast-forward to modern cryptography. Symmetric ciphers were among the first modern ciphers to protect data and information in the digital age. As we learned earlier, stream ciphers and block ciphers are types of symmetric cryptography; they use the same keystream or key bits to encrypt and decrypt.

Symmetric cryptography is also referred to as *private key* and *shared key* cryptography, to distinguish it from public key or asymmetric cryptography. Refer to Chapter 1, Section 1.2.5, for more information on these terms. In this section, we'll dive deeper into the strengths, weaknesses, and use cases of symmetric encryption.

Let's start by establishing the strengths of symmetric encryption, as follows:

- **Speed**
  The first and very important advantage of symmetric encryption is the speed. The process of encryption is very fast. This is true not only for stream ciphers but also for block ciphers. Block ciphers are also significantly faster than asymmetric ciphers. Speed is the most attractive feature of symmetric encryption to users.

- **Security**
  Symmetric encryption is very secure. If we use the key as long as the message and a new key with every message, the cipher is almost unbreakable. As we learned in Chapter 1, Section 1.2.7, this unbreakable concept is achieved with the one-time pad. Although this isn't possible to implement in practice, the keystream generation of symmetric cipher comes close to this.

- **Mathematics**
  The mathematics used in implementing symmetric encryption is relatively simple and makes software implementation easier.

- **Hardware implementation**
  Symmetric encryption can be achieved by hardware implementation. In fact, the design of initial block ciphers such as Lucifer and DES was for hardware implementation only, which makes symmetric cryptography even faster.

- **Post-quantum resistance**
  Symmetric cryptography is post-quantum resistant, meaning that symmetric cryptography, based on what we know now, will be able to withstand attacks by quantum computers. This makes symmetric cryptography quantum resistant. The security of symmetric encryption will be weakened against the quantum algorithms, but using a larger key size will provide the necessary security even in the post-quantum era. You'll learn more about post-quantum cryptography in Chapter 10.

Now, let's move on to the weaknesses of symmetric encryption:

- **Key sharing**
  One of the major problems or weaknesses of symmetric cryptography is key sharing or secret sharing. Since both parties must have the same key, it's difficult to share the key securely. They must have a secure channel to share the key, or the sharing must happen offline.

- **Not scalable**
  Sharing the key becomes complex when more parties are involved in the communication. It becomes very difficult to share the same key with several people and maintain security. Plus, the number of keys required also increases exponentially.

- **Vulnerable (if designed poorly)**
  Symmetric encryption can be cracked if the keystream is repeatable or predictable. Proper randomness (initialization vector [IV]) must be added to generate the key.

- **Lacks integrity and nonrepudiation**
  Symmetric encryption offers confidentiality only. It doesn't detect a *man-in-the-middle* attack; that is, if an attacker somehow taps the message and adds some random information, then the receiver won't know that the message is tempered. However, the receiver may not be able to understand the complete message. This means symmetric cryptography doesn't offer any integrity to the message. Moreover, the receiver has no idea if the message came from the said sender. Anyone who has access to the key can send the message, which means symmetric encryption lacks nonrepudiation.

- **Key management**
  Symmetric encryption with multiple people requires that all parties have multiple keys to communicate. That means if one person leaves the communication, everyone else who is involved with that person must replace the key; otherwise, the person who left can still access the messages. This forces the administrator to generate a new key when someone leaves. If the turnover is high, then generating and distributing a new key every time someone leaves becomes an administrative nightmare.

> **How Many Keys Are Required?**
>
> How many keys are really required in symmetric encryption? The question doesn't look very difficult because we need only 2 keys for 2 parties and 3 keys for 3 parties involved in symmetric encryption. However, the problem grows exponentially as the number of participants increases. As shown in Figure 2.1, for 10 participants, we need 45 keys, but for 50 participants, we need 1,225 keys. Things are almost out of control with 100 or more participants. Because everyone involved in the communication must have the same key, the number of keys required grows exponentially.

2 Symmetric Cryptography

The following is a general formula to calculate the number of keys required by symmetric encryption, where K = number of keys and N = number of participants in communication:

$K = [N \times (N - 1)] \div 2$

Now let's use this formula to calculate the number of keys, K, required for a given N. The exponential effect of the number of keys needed is shown in Figure 2.1.

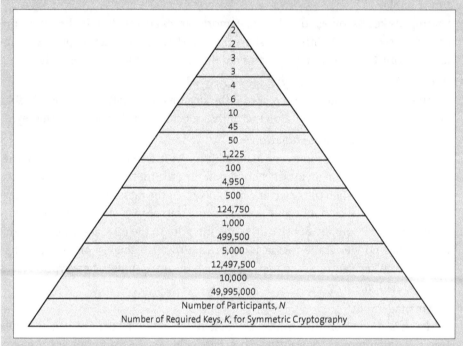

Figure 2.1 Exponential Growth of Required Symmetric Encryption Keys

Taking both the strengths and weaknesses of symmetric encryption into account, there are several common use cases where symmetric encryption is implemented:

- Better performance and efficiency make symmetric cryptography very useful in encrypting databases and other large data storage.
- Low overheads make symmetric cryptography a go-to encryption method for small devices and applications. This is one reason cell phones use symmetric encryption.
- The robust security of symmetric cryptography is very lucrative to the financial industry. Banks and other financial institutions use symmetric encryption to protect sensitive data such as personally identifiable information (PII).
- The use of hardware to implement symmetric cryptography provides an extra layer of security and efficiency. As it runs on dedicated hardware, the process is faster and more secure. This makes it an easy choice for the military and other sensitive information protection.

2.2 Symmetric Key Algorithms

In this section, we'll discuss the different symmetric key algorithms in detail. But first, let's review what came before them.

The world changed significantly after the Second World War. More and more people started flying, and the travel speed translated into a fast-paced lifestyle. People started using technology in day-to-day life, including TVs, VCRs, and computers. The use of computers in daily business transactions raised another question: How do we keep the data secure? And that gave birth to modern cryptography.

IBM is credited with introducing computerized cryptography or modern cryptography to the world. Horst Feistel and his colleagues at IBM started the work in the late 1960s. The block cipher, Lucifer, was developed by Feistel and others and released by IBM in 1973 [Feistel, 1973][Sorkin, 1984]. Lucifer was probably the first cipher to encrypt computer data. Lucifer used the confusion and diffusion properties to generate the cipher text.

> **Claude Shannon**
>
> Claude Shannon proposed confusion and diffusion properties in his paper "Communication Theory of Secrecy Systems" [Shannon, 1949]. Simply put, confusion is substitution while diffusion is a permutation, as we discussed in Chapter 1, Section 1.2.8. Confusion or diffusion alone isn't enough to make cipher text robust against attacks. Both must be used (perhaps, several times) to generate the strong cipher text. Lucifer and DES are designed based on this principle.

Lucifer performs 16 rounds to encrypt on the 128-bit block with key sizes up to 128 bits. The mechanism Lucifer uses to scramble the data in each round is known as the Feistel network, which is the basis of DES, as we'll cover in the next section.

Lucifer didn't achieve the commercial success its inventors hoped for, but it has provided the foundation for the symmetric key algorithms we'll discuss in the following sections—particularly DES.

2.2.1 Data Encryption Standard

The Data Encryption Standard (DES) holds a special place in cryptography as it ruled the world of cryptography for almost a quarter of the 20th century. DES did a remarkable job holding up against most attacks until it was replaced by the Advanced Encryption Standard (AES) in 2001.

Let's dive into DES in the following sections, starting with the background and moving into an in-depth exploration of the implementation and decryption processes.

Background

In the late 1960s and early 1970s, interest in cryptography was growing outside military and government applications. The banking and finance industry was looking for ways to encrypt data, but there was no reliable, efficient, and economical option. The US federal agency witnessed a growing use of computers and other digital resources and realized a need to protect digital information and communication channels.

In 1973, the National Institute of Standards and Technology (NIST)—then known as the National Bureau of Standards (NBS)—issued a call for proposals (CFP) and invited submissions for cryptographic algorithms. To be selected as a standard, the algorithm must meet a set of criteria:

- The cryptography algorithm must be secure, easy to understand and implement, sharable, and confidential.
- The key should determine confidentiality, not the algorithm.
- The implementation should be economical and efficient.
- The algorithm must be exportable.
- It must be compatible with electronic devices such as computers.

Although most of these criteria are common sense now, they were important to specify back then. The NIST had no idea what kind of cryptography skill level was out there in the private sector and what kind of response to expect.

The response to CFP indicated that there was significant work going on in the cryptography community, and NIST received several good candidates, but there was no winner. A year later, NIST issued another CFP, and this time, NIST had a winner. You guessed it right—Lucifer by IBM. Lucifer met all the design criteria required by NIST. NIST on their part did its due diligence and asked the National Security Agency (NSA) for evaluation as well as invited the general public for comments.

After a lengthy process of almost three years, NIST adopted a modified Lucifer as the DES. The official description of the standard was published in the Federal Information Processing Standards (FIPS) Pub 46 in early 1977.

DES was a US federal government standard. Those who were doing business with any US federal agency were required to use it. Private sector businesses and other governments around the world still could use any cryptographic algorithm they wanted. However, the first formal release of any cryptographic algorithm to the private sector piqued everyone's interest. More and more people started studying and using it. Widespread usage made DES more economical and affordable compared to other algorithms. Ultimately, DES became the most studied algorithm, and in the 1970s and 1980s, DES was probably the most widely used cryptography algorithm.

> **DES, IBM, and Patent**
>
> NIST took a good step to make DES more easily available by working out a deal with IBM to grant nonexclusive, royalty-free use by anyone. Per the FIPS:
>
> *Cryptographic devices implementing this standard may be covered by U.S. and foreign patents issued to the International Business Machines Corporation. However, IBM has granted nonexclusive, royalty-free licenses under the patents to make, use and sell apparatus which complies with the standard. [FIPS 46-2, 1993]*
>
> The "royalty free" phrase truly attracted most businesses.

NIST evaluated DES every five years, conducting due diligence and opening the standard for public comments. In 1983, 1988, and 1993, there was no better alternative to DES, and it was reaffirmed.

Implementation

DES is a symmetric block cipher. Unlike stream ciphers, DES takes a block of 64-bit plain text and generates a 64-bit cipher text. DES uses a 56-bit key. The same key, with a reverse schedule, is used to decrypt the message, which makes DES a symmetric cipher. The key schedule is discussed later in this section.

DES uses the basic properties of confusion and diffusion to encrypt the message. DES performs this confusion and diffusion several times. The simple combination of substitution and permutation is considered a round (or iteration). DES performs 16 rounds of substitution followed by a permutation process to generate the cipher text. In a nutshell, DES uses the same concept as classical ciphers—substitution and permutation!

> **Product Cipher versus Iterative Cipher**
>
> When an encryption algorithm uses more than one property to generate cipher text, it's known as a product cipher. The product cipher becomes an iterative cipher when the product cipher performs the same operations multiple times. For example, DES performs the substitution and permutation in one round, so it's a product cipher—a product of confusion and diffusion. However, DES also repeats this process 16 times—goes through 16 rounds—so it's also considered an iterative cipher.

Figure 2.2 shows the high-level implementation of the DES algorithm. The algorithm can be summarized as follows [FIPS 46-2, 1993]:

1. DES receives a 64-bit block of plain text, P.
2. It performs the initial permutation on the plain text, P. This is a bitwise permutation and basically interchanges (or cross wires) the position of the bits.
3. The round begins after the initial permutation.

4. The 64-bit plain text block is divided into two 32-bit blocks—left block, L_0, and right block, R_0.
5. Function, f, takes the right block, R_0, and subkey, K_0, as two inputs to the function and generates an output, f_0.
6. The output of the function, f_0, and left block, L_0, are XORed.
7. The output of XOR and the right block, R_0, are swapped.
8. That means, at the end of the round, the right block, R_0, becomes the left block, L_1, for the next round, and the output of XOR, which was the left block, L_0, becomes the right block, R_1, for the next round.
9. This concludes one round. At the end of the first round, we have a product cipher.

DES goes through 16 such rounds. After 16 rounds, DES performs a final permutation, which reverses the initial permutation.

Mathematically, we can explain the operation as follows:

$L_i = R_{i-1}$
$R_i = L_{i-1} \oplus f(R_{i-1}, K_i)$

This is a complicated process, but bear with me. We'll clear up any confusion by taking a closer look into these operations.

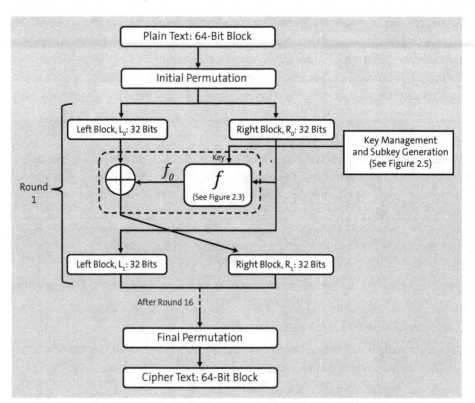

Figure 2.2 Architecture of the Data Encryption Standard

2.2 Symmetric Key Algorithms

From the round 1 steps, the following key points are worth highlighting:

- The left block, L_O, is encrypted, and not the right side, R_O.
- R_O passes through the function f. The function f performs confusion and diffusion, not encryption.
- Encryption happens as a result of XORing L_O and the output of function f_O. The process of encryption takes place inside the box with dotted lines in Figure 2.2. This encryption process is very similar to stream ciphers (refer to Chapter 1, Section 1.3.3, for more information on stream ciphers). The output of the function, f_O, works as a keystream for the stream cipher, which encrypts the plain text L_O.
- Only the subkey, K_O, is 48 bits. Everything else in the product cipher is 32 bits.

Architecture of Function f

The next question is, what happens inside the function f? Let's zoom inside the function to understand its workings.

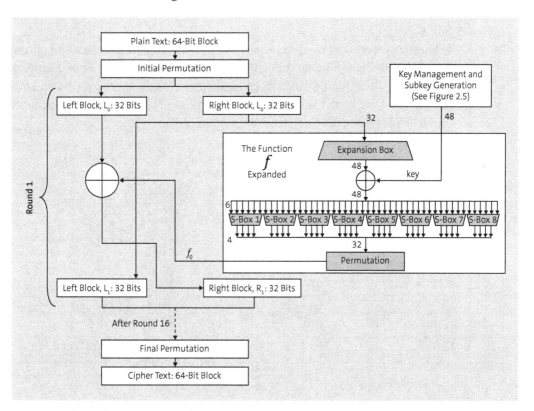

Figure 2.3 The Architecture of Function F

The architecture of the function is shown in Figure 2.3. The right half or the right block, R_O, of the plain text serves as input to the function f. The 32-bit block first passes through the expansion box, E. The output of expansion box E is XORed with subkey K_i. The XORed output is fed to the substitution box (S-box), S.

Lastly, a final permutation is performed on the output of the S-box to generate the keystream or the function output, f_O.

This is a lot to take in and understand in only a few lines. Let's dissect this further:

- **Expansion box E**
  As shown in Figure 2.3, expansion box E takes the 32-bit plain text input and expands these bits into 48-bits—hence the name expansion box. This happens by a simple permutation. Half of the input bits, 16 bits out of 32 bits, are connected to or passed through two outputs. The 32 bits are divided into 8 groups of 4. Each 4 bits group is expanded into a group of 6 bits. At this stage, it's ensured that input bits aren't duplicated in the same 6 bits group. The output of the expansion box, E, is a 48-bit block.

- **XOR operation**
  The next step in calculating function f is XORing the 48-bit output of expansion box E with 48-bit subkey K. This is a simple XOR operation and generates a 48-bit output.

- **S-box S**
  The selection process of the S-box output is described in Figure 2.4. The 48-bit output of XOR is divided into 8 groups of 6 bits each and fed to 8 S-boxes, S-box 1 through S-box 8. Each substitution, S-box, takes 6 bits as input (8 × 6 = 48) and generates 4 bits of output (8 × 4 = 32 bits). The 4-bit output from the 6-bit input is selected using a set of predefined tables. Each S-box table has 16 columns and 4 rows.

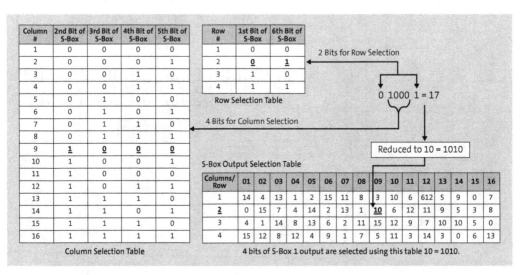

Figure 2.4 The S-Box Process to Select 4-Bit Output

For example, as shown in Figure 2.4, S-box 1 has the 6-digit number 010001. We convert from binary to decimal and get 17. The most significant bit (MSB) and the least significant bit (LSB) are used for row selection. The MSB and LSB of 17, 010001, are 0 and 1. The 01 converts to row number 2 in the row selection table. The middle 4 bits of 17, 0 **1000** 1, are used for column selection. In this example, the middle four digits are 1000, which converts to row number 9.

Now select the ninth column and second row from the S-box output selection table. This will give the decimal number 10, which in binary is 1010. Only output selection table for S-box 1 is shown in this example. There are eight such tables with pre-defined values, one for each S-box. For the remaining tables, refer to the NIST official publication [FIPS 46-2, 1993].

> **Old Wine in a New Bottle!**
> What happens in S-boxes is conceptually similar to how the Caesar cipher worked. In the Caesar cipher, one letter was replaced with another—like A with D and B with E, and so on—while in S-boxes, one binary number is replaced with another. The difference is, thanks to digital electronics, DES offers a much higher degree of complexity and randomness of substitution, which makes DES a lot more difficult to crack compared to classical ciphers.

- **Permutation box**
  The 32-bit output of S-boxes (8 × 4) is fed to the permutation box, P. A bitwise permutation happens in the permutation box. The output of permutation box P is the output of function f_O and is used in encrypting the left block using a XOR operation. Output f_O can also be seen as the keystream for the XOR operation.

Subkey Scheduling

The next piece of the puzzle in encrypting with DES is generating a subkey for each round. The key scheduling and generation process is divided into three parts: first permutation (permutation choice 1), rotate left operation, and second permutation (permutation choice 2).

Although DES takes 64 bits or 8 bytes as a key input, it effectively uses 56 bits. In the first permutation, DES drops the 8th bit from every byte. That will take out 8 bits and convert the key from 64 bits to 56 bits. The 8th bit of each byte can be used for the parity check, which verifies whether the key (64 bits) is error free. A different 48-bit subkey is generated for each round from these 56 bits.

After extracting the 56 bits from 64 bits, the 56 bits are divided into two blocks of 28 bits each and rotated left. Although officially it's called shift left, the shift is cyclic, meaning the shifted bit isn't dropped but fed back again at the end. During 16 rounds in the DES encryption process, a total of 28 shifts or rotations happen. The bits are rotated by 1 bit in the 1st, 2nd, 9th, and 16th rounds while rotated 2 bits per round in the remaining 12 rounds. That means 24-bit shifts from the 12 rounds and 4-bit shifts from the other four rounds make it a total of 28-bit shifts. At the end of the 16th round, the first bits of both halves are back to their first position. This means the key bits are at the reset or the start position for the next round of encryption as well as for decryption.

In the second (or final) permutation, 48 bits from 56 bits are selected to generate the subkey for that round.

Figure 2.5 shows the entire DES process, with the subkey scheduling on the right side.

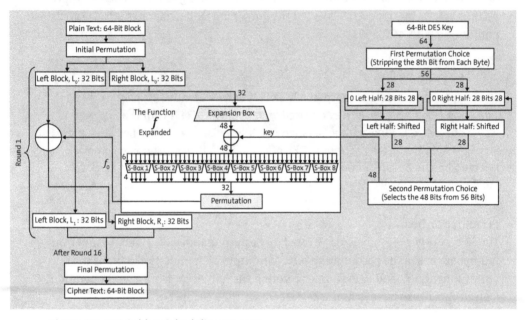

Figure 2.5 DES Subkey Scheduling Process

Decryption

We implemented the DES algorithm for encryption. However, it's equally important to understand and implement the DES decryption process. If we can't decipher and read the cipher text, the cipher texts are useless. The DES decryption process is very simple—it literally reverses the steps of the encryption process.

First, initial permutation is performed on the 64-bit cipher text block. This, basically, is reversing the final permutation at the end of the 16th round. Next, DES starts the first round of decryption. The first round of decryption is exactly the same as the first round of encryption. It divides the cipher text into two blocks of 32 bits: L_0 and R_0. However, L_0 is L_{15}, and R_0 is R_{15} (the data from the last round of encryption). The subkey for the first round of decryption is the same as the K_{15} from the last round of encryption. The process continues in the reverse order. The last round of decryption basically uses the data and key from the first round of encryption. At the end of the 16th round, DES performs the bitwise permutation to generate the plain text.

DES in Practice

DES is effectively implemented with hardware because the permutations used in DES are easily achieved using cross-wiring. This isn't the case for the software implementation, which has been slower and not as efficient. This is why DES is mostly implemented using hardware.

The security of DES, and for that matter any cipher, depends on the key. DES has a 56-bit key, which means it can have 2^{56} possible key combinations! Obviously, this was a large number for any computer in the 1970s and 1980s. Eventually, cryptographic researchers and cryptanalysts published analytical and brute-force attacks on DES. The differential analytical attack was published in 1990 [Biham, 1991], and the linear analytical attack was published in 1993 [Matsui, 1993]. However, both of these attacks weren't very effective since a differential analytical attack needs 2^{47} and a linear analytical attack needs 2^{43} combinations to potentially break the DES cipher. (Since these attacks didn't pose a major threat to DES, NIST reconfirmed DES as a federal standard again in 1993.) The real blow came to DES from the brute-force attack. The increasing processing speed and reduced cost of hardware made it feasible to compute the 56-bit DES key and crack the DES algorithm.

> **Deep Crack**
>
> Electronic Frontier Foundation (EFF) designed hardware that launched a brute-force attack on DES to calculate the 2^{56} key combinations and find the key in 56 hours! In 1998, EFF built the Deep Crack for $250,000 USD [EFF, 1998]. The same hardware would have cost $20 million USD in 1976 when DES was accepted as a standard. EFF's DES cracker was nicknamed Deep Cracker after IBM's Deep Blue—IBM's chess expert.

2.2.2 Triple DES and DES Variants

DES is a robust encryption algorithm, but it can be cracked. How can we make DES resistant to attacks? There are several ways to make DES more secure, which we'll discuss in the following sections.

Key Whitening

One way to protect DES is by using the process known as *key whitening*. In simple terms, key whitening is the process of adding (XORing) a 64-bit key to the plain text before the DES algorithm and adding another 64-bit key to the cipher text generated by the DES algorithm. The concept of key whitening is shown in Figure 2.6. Mathematically, this is as follows:

2 Symmetric Cryptography

$C = K_1 \oplus E_{k\text{-}DES}(P \oplus K_2)$

where

C = 64-bit block of cipher text

P = 64-bit block of plain text

K_1 and K_2 = 64-bit whitening keys

$K\text{-}DES$ = 56-bit DES encryption key

The *DES-X* encryption algorithm uses this key whitening technique, which increases the effective key size for the algorithm and adds security against brute-force attacks. However, the technique doesn't protect the cipher against analytical attacks.

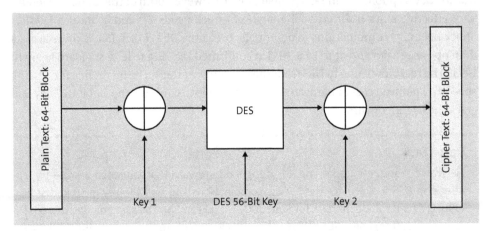

Figure 2.6 DES-X and Key Whitening

The Invention of Key Whitening

This process of adding extra layers of keys before and after the DES encryption was first proposed in 1984 and used by Ron Rivest who named the technique key whitening. The DES-X algorithm is added to BSAFE cryptographic libraries in the late 1980s [van Tilborg, 2005].

It's important to note that Rivest used the whitening technique with DES, but this technique can be (and is) used with any block cipher to protect against exhaustive key search attacks. The key whitening doesn't help with differential analysis types of attacks.

Double Encryption

As the name indicates, double encryption is simply doing the encryption twice. In this algorithm, encryption is performed by the block cipher twice with different keys, K_1

and K_2. The first block cipher performs the encryption using key K_1, and the output of the first block cipher is again encrypted with key K_2.

Theoretically, this should be twice as difficult to brute force since the exhaustive key search is doubled. The attacker must first crack the first cipher and then do the whole process again for the second cipher. However, that isn't the case in practice. The attacker can divide and conquer. The attacker can first do the key search for the first block cipher and create a lookup table for the keys and corresponding cipher text. Next, the attacker can use the same keys to decrypt the second block cipher. The attacker continues with the process until they find a match. This technique is known as a *meet-in-the-middle* attack. If single encryption takes 2^k searches, then the double encryption takes 2^{k+1}. So, there is no advantage to this method.

Triple Encryption

Triple encryption simply encrypts the same message three times with three keys, significantly increasing security because the meet-in-the-middle attack doesn't work. One good example of triple encryption is *Triple DES*.

Triple DES, also referred to as *3DES*, is simply using the DES algorithm three times. Mathematically speaking:

$C = (E_{k1-DES}(E_{k2-DES}(E_{k1-DES}(P))))$

Here, C is the cipher text, and P is the plain text. E_{k1-DES} are the three keys for each DES cipher.

Triple DES is implemented as encryption only (3DES: all three 3 DES ciphers perform encryption) or encryption-decryption-encryption (3DES-EDE), depending on how each DES cipher within 3DES is implemented. Moreover, there is a choice of using three different keys for each DES cipher with 3DES or using two keys (using one key for all three would defeat the purpose). The effective key length for 3DES could be 168 or 112, depending on whether three different keys are used or two.

Triple DES is very effective against analytical attacks and brute-force attacks. It's also resistant to meet-in-the-middle attacks. Triple DES is more efficient with hardware implementation than software.

2.2.3 International Data Encryption Algorithm

International Data Encryption Algorithm (IDEA) is a symmetric block cipher. IDEA was first proposed by Xuejai Lai and James Massey [Lai, 1991] to overcome the short key length of DES and with an expectation of succeeding DES for the encryption standard. The initial version of IDEA was named Proposed Encryption Standard (PES). However, after the demonstration of differential analytic attack, the PES was enhanced against such attacks and relabeled as IPES—Improved Proposed Encryption Standard. Eventually, IPES was rebranded as IDEA.

2 Symmetric Cryptography

IDEA is a 64-bit block cipher with a 128-bit key size. IDEA takes eight rounds and an output transformation to encrypt the 64-bit plain text block and the same algorithm for decryption. Recall our discussion on DES from Section 2.2.1: DES divides 64-bit plain text into two blocks of 32-bits each and encrypts one block in each round. IDEA, meanwhile, divides the 64-bit plain text block into four 16-bit blocks, P_1 to P_4. The implementation of IDEA is shown in Figure 2.7.

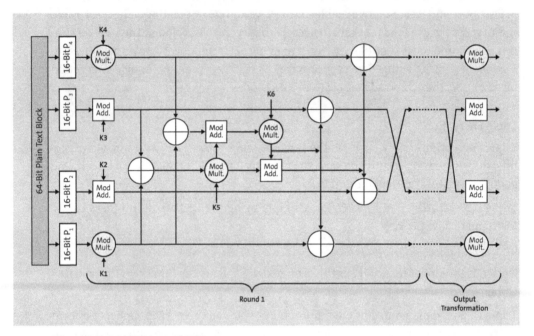

Figure 2.7 Architecture of IDEA

In each round, the four 16-bit plain text blocks go through a series of mathematical operations that include XORing, modulo 2^{16} additions, and modulo $2^{16} + 1$ multiplications. Additionally, each round uses six subkeys. Four of those six subkeys, K_1 to K_4, are mathematically included in each 16-bit block of the plain text at the beginning of each round. Keys K_1 and K_4 are multiplied with the first block P_1 and the fourth block P_4 of the plain text, respectively, while the second key, K_2, and the third key, K_3, are added to the second block, P_2, and third block, P_3, of the plain text, respectively. The outputs of these four mathematical operations are XORed. The remaining two subkeys, K_5 and K_6, are multiplied with the outputs of XOR operations. Each multiplication product is again XORed with two blocks. Lastly, blocks P_3 and P_2 are swapped. The four blocks are passed onto the next round.

IDEA performs eight such rounds. The confusion and diffusion properties are provided by XOR operation, addition, and multiplication mathematics in each round. At the end of the eighth round, IDEA performs the output transformation. Four subkeys are used in the output transformation instead of six subkeys in the previous eight rounds.

These four subkeys are added and multiplied with the four blocks, and, finally, the middle two blocks are swapped again. This completes the encryption process.

IDEA uses the same algorithm for decryption with a reverse key schedule. The key schedule is relatively simple in IDEA. IDEA needs 52 subkeys; that is, six subkeys for eight rounds ($8 \times 6 = 48$) and four subkeys for the last round. Each subkey is 16-bit. The 128-bit key is divided into eight subkeys. Six of these eight subkeys are used in the first round, and the remaining two keys are used in the next round. The 128-bit key is rotated 25 bits to the left, and again eight subkeys are generated. These keys are used in the next two rounds. The process continues until the end of the encryption process.

The simple key schedule is easy to implement and use but creates a threat, although the risk isn't very high. The IDEA algorithm was cracked, and attacks were published [Biham, 2015]. The DES algorithm was hardware based and used physical (using wires) methods to create confusion and diffusion, while IDEA uses math to provide the confusion and diffusion properties to the algorithm. IDEA was designed with the aim of being the next encryption standard and was very promising in the early years, but IDEA was never chosen as the standard or even made it to the list of finalists for AES.

2.2.4 Advanced Encryption Standard

By the mid-1990s, it was evident that DES couldn't withstand the exhaustive key search attacks, and with the fast development of CPU power and speed, it was almost certain that DES was approaching the end of its usefulness. DES was replaced by Triple DES; however, as we discussed in Section 2.2.2, 3DES has its own share of problems. The primary concern with Triple DES was the difficulty with software implementation. The industry was moving rapidly toward more software implementation than hardware, and the encryption standard needed to support that trend. Other disadvantages of Triple DES were slower speed, small block size, and the key size. All these reasons led NIST to look for a better alternative.

We'll explore the selection of AES next, and then dive into the implementation and decryption processes.

Background

In 1997, NIST began the process of searching for a DES successor [Smid, 2021]. However, the process was much different this time than selecting DES back in the 1970s. This time, the process was more transparent. NIST announced a formal call to invite various algorithms. NIST outlined many criteria that the algorithm must meet to qualify for consideration. Apart from the obvious, security and efficiency, some of the main criteria were as follows:

- The cipher type must be symmetric and a block cipher.
- The block size must be 128 bits.

2 Symmetric Cryptography

- Variable key sizes must be used, starting with a 128-bit key.
- Software implementation specifications are required—particularly in Java and ANSI C.

NIST received more than 20 submissions and selected 15 for review by 1998. All 15 submitters were given an opportunity to present their algorithms and answer questions. Following the presentations, NIST opened the submission for review and comments. At the end of the review process, NIST selected five candidates as AES finalists. The five finalist algorithms were: Twofish, Mars, RC6, Serpent, and Rijndael. After another round of review, at the end of the process, NIST selected Rijndael as the AES, proposed by Joan Daemen and Vincent Rijmen.

Implementation

Rijndael is a 128-bit symmetric block cipher with variable key sizes of 128 bits, 192 bits, and 256 bits [FIPS 197, 2001]. In this book, we'll use the name AES, the NIST-approved implementation of Rijndael. The very high-level structure of AES is shown in Figure 2.8.

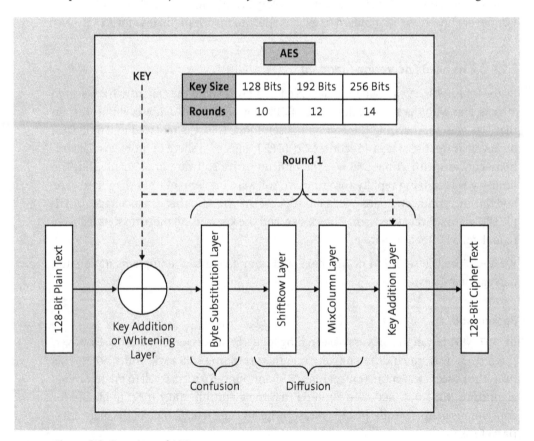

Figure 2.8 Overview of AES

The number of encryption rounds depends on the key size: the longer the key, the more rounds. For the 128-bit key, AES encrypts plain text in 10 rounds; for the 192-bit key, AES takes 12 rounds; and for the highest key size of 256 bits, AES takes a maximum of 14 rounds. It's important to note that DES used 16 rounds of encryption while AES can securely encrypt in fewer rounds. This is because DES encrypts only half the plain text block every round. It encrypts only 32 bits each round while AES encrypts all 128 bits in each round.

The first step in AES is XORing the key to the plain text block. This is the first half of the key whitening process that we discussed earlier. This is known as the *key addition layer*. This happens only once before the first round of encryption begins. The first round of encryption begins right after the key whitening process. The *byte substitution* is the first layer of each round. The input bytes are substituted with values from a lookup table—the same as in the case of the Caesar cipher and DES. The lookup table is precomputed with mathematically generated values. This process "causes" confusion because this layer is replacing one value with another to obscure the relationship between the cipher text and the key. The next two layers in each round provide diffusion by shifting rows and mixing columns. In these layers, bytes are changing positions and providing the permutation. The *ShiftRow layer* shifts the bytes in a row while the *MixColumn layer* uses a matrix operation. The last round is slightly different since in the last round there is no MixColumn layer. This is to provide symmetry in encryption and decryption.

> **Long Live the SP-Network**
>
> In cryptography, the substitution-permutation network (*SP-network*) is a combination of substitution and permutation techniques. Most of the ciphers in modern cryptography implement confusion and diffusion properties by using the SP-network approach. The AES cipher is no exception.

The last operation of each round is to add the subkey. The key schedule generates a subkey for the key whitening layer as well as for each round. The last operation of the last round again adds the key. This is the second half of the key whitening process.

Let's dive deeper into the architecture of the AES cipher, as shown in Figure 2.9. The AES cipher operates at a byte level (as a reminder, a byte is made up of 8 binary bits—1s or 0s). This is different from DES, which operates at a bit level. The 128-bit plain text block is divided into 16 bytes and arranged in a 4×4 array. This is known as the *state*. Each state has 16 bytes. Four bytes make a word—in other words, each state has four words. Data in AES passes in the state form from one layer to another and from one round to another. Consider the example shown in Table 2.1.

2 Symmetric Cryptography

Message	Binary	Hexidecimal
CryptographyBook*	01000011 01110010 01111001 01110000 01110100 01101111 01100111 01110010 01100001 01110000 01101000 01111001 00100000 01000010 01101111 01101111 01101011	43 72 79 70 74 6F 67 72 61 70 68 79 42 6F IF 6B
128-bitsKeyOfAES	00110001 00110010 00111000 00101101 01100010 01101001 01110100 01110011 01001011 01100101 01111001 01001111 01100110 01000001 01000101 01010011	31 32 38 2D 62 69 74 73 4B 65 79 4F 66 41 45 53

* Although there should be a space in the English message Cryptography Book, for this example, the space is eliminated. The binary equivalent of the message doesn't show the space. This is to keep the phrase 16 bytes long. Otherwise, the space would make the message 17 bytes long.

Table 2.1 AES Cipher Example

Let's dive into the layers of the AES with our example in mind:

- **Key addition layer**

  As we discussed previously, this is a key whitening layer. In this layer, 16 bytes of input plain text are XORed with the 16 bytes of the first subkey. The subkey used here is known as the "round key," probably because it's used at the beginning of the round. The round key is simply the original key copied into the first subkey.

  Using the example, Figure 2.9 shows how the key addition layer works. The message and the key are arranged in a 4×4 state table, and the XOR output also produces the 4×4 state table.

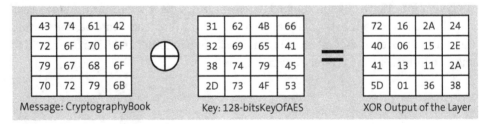

Figure 2.9 The Key Addition Layer

2.2 Symmetric Key Algorithms

- **Byte substitution layer**

  The byte substitution layer introduces the confusion property in AES. Byte substitution happens using a lookup table. The binary values of 8 bits are converted into two-digit hexadecimal (hex) values. The first digit of the hex value represents the row, and the second digit represents the column in the lookup table. Figure 2.10 shows the continuation of our example. Use Table 2.2 [FIPS 197, 2001] to look up the value of the first byte 72. For 72, look up the seventh row and second column. That gives the value 40. The bye substitution layer will replace 72 with 40.

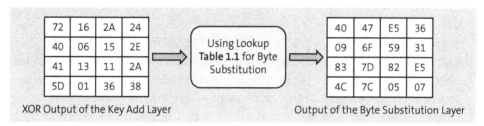

XOR Output of the Key Add Layer   Output of the Byte Substitution Layer

Figure 2.10 The Byte Substitution Layer

	0	1	2	3	4	5	6	7	8	9	A	B	C	D	E	F
0	63	7C	77	7B	F2	6B	6F	C5	30	01	67	2B	FE	D7	AB	76
1	CA	82	C9	7D	FA	59	47	F0	AD	D4	A2	AF	9C	A4	72	C0
2	B7	FD	93	26	36	3F	F7	CC	C4	A5	E5	F1	71	D8	31	15
3	04	C7	23	C3	18	96	05	9A	07	12	80	E2	EB	27	B2	75
4	09	83	2C	1A	1B	6E	5A	A0	52	3B	D6	B3	29	E3	2F	84
5	53	D1	00	ED	20	FC	B1	5B	6A	3B	BE	39	4A	4C	58	CF
6	D0	EF	AA	FB	43	4D	33	85	45	F9	02	7F	50	3C	9F	AD
7	51	A3	40	84	92	9D	38	F5	BC	B6	DA	21	10	FF	F3	B2
8	CD	0C	13	EC	5F	97	44	17	C4	A7	7E	3D	64	5D	19	73
9	60	81	4F	DC	22	2A	90	88	46	EE	B8	14	DE	5E	0B	DB
A	E0	32	3A	0A	49	06	24	5C	C2	D3	AC	62	91	95	E4	79
B	E7	C8	37	6D	8D	D5	4E	A9	6C	56	F4	EA	65	7A	AE	08
C	BA	78	25	2E	1C	A6	B4	C6	E8	DD	74	1F	4B	BD	8B	8A
D	70	3E	B5	66	48	03	F6	0E	61	35	57	B9	86	C1	1D	9A
E	E1	F8	98	11	69	D9	8E	94	9B	1E	87	E9	CE	55	28	DF
F	8C	A1	89	0D	BF	E6	42	68	41	99	2D	0F	B0	54	BB	16

Table 2.2 AES Byte Substitution Lookup Table

The byte substitution layer is very similar to the S-box in DES (Section 2.2.1). They both introduce confusion. However, there is a big difference in the way the byte substitution layer is implemented. First, AES has only one lookup table for all 16 bytes, while DES had eight different S-box lookup tables for eight S-boxes. Second, S-box lookup tables were given with fixed values. We didn't know the math behind these values in each lookup table. On the other hand, the AES lookup table is populated using the Galois field property. Refer to Chapter 1, Section 1.3.4, for more information on Galois fields.

Each byte of the message has 8 bits. That means each byte has 256 (2^8) possible values. This makes it easy to populate a 16×16 table with 256 unique values. These values are populated using Galois field (2^8) inverse and affine mapping [Paar, 2010]. These values are fixed. This makes the software implementation of AES very efficient because the table can be prepopulated and used over and over again. During the process of encryption, the software function simply looks up the value from the table. On the flip side, the values of the lookup table need to be calculated every time in the hardware implementation.

- **ShiftRow layer**

    The ShiftRow layer just does that—it shifts the row. In doing so, it forms the first part of the diffusion property. At the end of the byte substitution layer, the message values are replaced or substituted with different values. The ShiftRow layer uses this substituted value as input. It shifts the byte to the left in the state array (technically it's a rotate left as the byte enters back from the right) and generates the new state array. The first row of the state array stays unshifted. The second row is shifted left by 1 byte. So, the first byte becomes the last byte, and the second byte shifts left one position and moves to the first position. The third and fourth bytes move to the second and third position, respectively.

    This operation is shown in Figure 2.11 for the ShiftRow layer with our example. Just to recap, our 128-bit plain text message is "CryptographyBook". The shifted bytes are in bold.

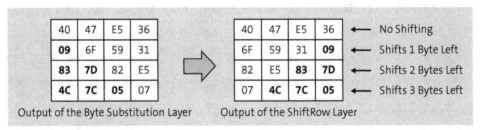

Figure 2.11 The ShiftRow Layer

- **MixColumn layer**

    The next layer in the AES round is the MixColumn layer. This layer introduces a greater degree of diffusion by mixing columns. This forms the second part of the diffusion property. As shown in Figure 2.12, the output matrix or state array from the ShiftRow is multiplied with a matrix of constant values.

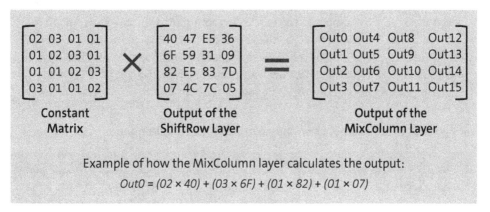

Example of how the MixColumn layer calculates the output:
$Out0 = (02 \times 40) + (03 \times 6F) + (01 \times 82) + (01 \times 07)$

Figure 2.12 MixColumn Layer

This matrix arithmetic uses Galois field (2^8). The finite number range in the Galois field (2^8) is as follows:

$\{00000000.... X^7 + X^6 + X^5 + X^4 + X^3 + X^2 + X + 1\}$

In addition, the multiplication process isn't as complicated as it looks. In the constant matrix, 01 is *1*, 02 is *X*, and 03 is *X + 1*. The software implementation of AES can use a lookup table to find the multiplication with 02 and 03 values. The irreducible polynomial is used if the multiplication product is larger than $X^7 + X^6 + X^5 + X^4 + X^3 + X^2 + X + 1$, then it's reduced by a polynomial $P(x) = X^8 + X^4 + X^3 + X + 1$.

As a result of MixColumn, the diffusion effect is significant. One bit change in 1 byte is spread across several bytes. This makes it very difficult for cryptanalysts. Just to recap, the MixColumn layer is present in all rounds except the last round.

- **Key addition layer**

  This is the last step in any round. In this step, the output of the MixColumn layer (or the ShiftRow layer for the last round) and the subkey are XORed. This is very similar to the round key addition at the beginning of the encryption process, except here a subkey is used instead of the copy of the actual key.

  The subkey is calculated using the key expansion schedule for each round. The details of the key expansion schedule are explained next.

The Key Expansion Schedule

The AES algorithm supports three key lengths—128 bits, 192 bits, and 256 bits. Depending on the key size, AES needs 10, 12, or 14 rounds, respectively, to encrypt the 128-bit block of plain text. If there are *n* number of rounds, AES generates n+1 subkeys. For a 128-bit key size, AES generates 11 (10+1) keys; for a 192-bit key, it will generate 13 subkeys; and for a 256-bit key, it will generate 15 subkeys.

2 Symmetric Cryptography

> **Why More Rounds as the Key Size Increases?**
>
> AES encrypts the entire plain text block of 128 bits in every round, and, as a result, it needs fewer rounds compared to the DES algorithm. At this stage, you might ask, shouldn't there be fewer rounds as the key size increases? Why does AES need more rounds to encrypt the data as the key size increases? The answer is, AES doesn't need more rounds to encrypt the data; however, it needs more rounds to randomize the key. That's why the longer the key, the more rounds it needs.

Let's go back to our example key, shown in Table 2.3 as a reminder.

Key	Binary	Hexadecimal
128-bitsKeyOfAES	00110001 00110010 00111000 00101101 01100010 01101001 01110100 01110011 01001011 01100101 01111001 01001111 01100110 01000001 01000101 01010011	31 32 38 2D 62 69 74 73 4B 65 79 4F 66 41 45 53

Table 2.3 Example Key

If we put these 16 bytes of keys in a state array (a 4×4 matrix), then each column in the matrix will form a word. In other words, the key has 4 words—W_0, W_1, W_2, W_3. These 4 words are copied as is to create the first subkey, K_0. Subkey K_0 is known as the *round key*. Next, for the first round, subkey K_1 is generated and that needs another 16 bytes or 4 words, and the process continues. For 128-bit key size, the AES encryption algorithm needs 11 subkeys or 44 words—W_0 to W_{43}. The process of expanding 4 words, W_0 to W_3, to 44 words, W_0 to W_{43}, is called *key expansion*. In this section, we'll learn the key expansion process, as shown in Figure 2.13.

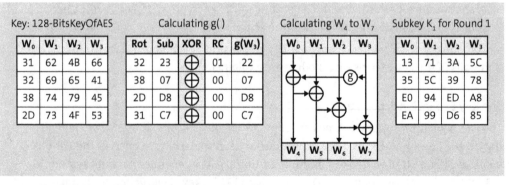

Figure 2.13 Key Expansion

Subkey K_0 has four words, W_0 to W_3, as shown on the left side of Figure 2.13. The next subkey is generated using these four words. The last word, W_3, of subkey K_0 is used to generate the function $g()$. The output of function $g()$ and the word W_0 are XORed to generate W_4. W_5 is simply the XORing of W_4 and W_1. W_6 is the XORing of W_5 and W_2, and W_7 is the XORing of W_6 and W_3. The entire process is shown in Figure 2.13 under "Calculating W_4 to W_7". Mathematically, this can be explained as follows:

$W_4 = W_0 \oplus g(W_3)$
$W_5 = W_1 \oplus W_4$
$W_6 = W_2 \oplus W_5$
$W_7 = W_3 \oplus W_6$

The million dollar question is this: How do you calculate $g()$? Function $g()$ can be calculated as follows:

1. Take the first word of the previous subkey—in this example, it's W_0.
2. Rotate left by one position. This will move each byte of the word to the left.
3. Next, substitute the rotated word with values from the lookup table (see Table 2.2). This is the same table used for the byte substitution layer.
4. XOR the substituted word with the round coefficient (RC). RCs are predefined constants. See Table 2.4 for RC values. The RC is an 8-bit value and an element of the Galois field. The RC is added to the first byte only. Other bytes have a 00 value for RC.
5. The output of XOR from the previous step is $g(W_3)$. This value is used to calculate W_4.

Once all four words for subkey K_1 are calculated, $g(W_7)$ is calculated for subkey K_2. As explained previously, the XOR operation is used to calculate the remaining words for the subkey. The process continues until all 44 words are calculated.

	R01	R02	R03	R04	R05	R06	R07	R08	R09	R10
Byte0	01	02	04	08	10	20	40	80	18	36
Byte1	00	00	00	00	00	00	00	00	00	00
Byte2	00	00	00	00	00	00	00	00	00	00
Byte3	00	00	00	00	00	00	00	00	00	00

Table 2.4 Round Coefficients

For the key size of 192 bits, a total of 13 keys (12 subkeys + 1 round key) are needed. The process of computing the subkeys is similar. It computes 52 words in eight iterations. In this case, the key iterations don't correspond to the number of rounds (which are 12 for 192-bit key size). Each iteration generates 6 words, but only 4 words are used for each subkey. For the key size of 256 bits, a total of 15 (14 subkeys + 1 round key) keys are needed. The process of computing the subkeys is similar to a 128-bit key size. It computes 60 words in seven iterations. In this case, the key iterations don't correspond to the number of rounds (which are 12 for 192-bit key size). Each iteration generates 8 words, but only 4 words are used for each subkey. Remember, in both cases, 192-bit key size and 256-bit key size, the first 4 words are simply copied from the key since the round key is the copy of the actual key.

There are two ways to generate the key. One is to precompute the subkeys. This speeds up the encryption process and is very efficient, but it takes up memory space. This method is generally used with systems with bigger devices such as PCs and servers. On the other hand, the subkeys can also be generated during the encryption (and decryption) process. This slows down the process but there is less reliance on memory/hardware. This method is used with smaller devices.

Decryption

Symmetric ciphers use the same key to encrypt and decrypt the message. Because of this, the algorithm itself stays very similar on both ends. The AES algorithm is no exception. The process of decryption is very similar to what we learned so far in encryption. The only difference is these processes are inversed. In other words, the byte substitution layer becomes the inverse byte substitution layer, and the ShiftRow layer becomes the inverse ShiftRow layer, and so on. A high-level decryption of AES is shown in Figure 2.14.

The AES decryption algorithm takes in a 128-bit block of cipher text. As we've established, decryption is basically the inverse of encryption; in round 10 of encryption, there is no MixColumn layer, so in the first round of decryption, there is no MixColumn layer. The first round starts with the key addition layer, followed by the inverse ShiftRow and inverse byte substitution layer. However, from the second round to the last round, all layers—key addition, inverse byte substitution layer, inverse ShiftRow layer, and inverse MixColumn layer—are present for decryption. There is another key addition layer after the last round. All the key addition layers in the decryption algorithm are the same as encryption—they all use the XOR operation. That is because the inverse of the XOR operation is the same as a regular XOR operation. Needless to say, the number of rounds is exactly the same as in the case of encryption. That is because a symmetric algorithm uses the same key to encrypt and decrypt. If the 128-bit key was used for encryption, the decryption will have the same key, and, as a result, the decryption process will have the same number of rounds.

The decryption process starts with a key addition layer in round 1. Since the entire process is reversing, the 4 words of the key expansion schedule are used in the first round. In the case of a 128-bit key, there will be 10 rounds and 11 subkeys. (For simplicity, throughout the explanation of the decryption process, it's assumed that there are 10 rounds.) That means the original key will be expanded into 44 words. The first decryption round will use the last 4 words as K_0—from W_{40} to W_{43}.

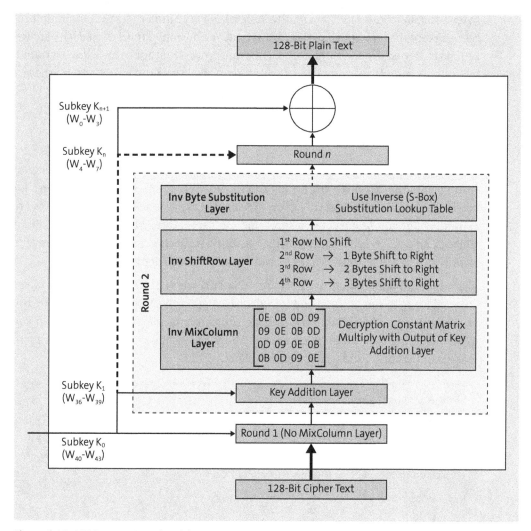

Figure 2.14 AES Decryption Algorithm

In the second round, subkey K_1 will use the next four words, W_{36} to W_{39}. The output of the key addition layer is multiplied with a constant matrix in the inverse MixColumn layer. The matrix multiplication process is the same as explained in the encryption process. The only difference is the values of the constant matrix, as the matrix is used in reverse order. Next, rotate right or cyclic shift right is performed on the output of the inverse MixColumn layer. In the inverse ShiftRow layer, the matrix rows are shifted to the right instead of the left. As in the case of encryption, there is no shift in the first row. The second row is shifted right by 1 byte. The third row is shifted right by 2 bytes. The last row is shifted right by 3 bytes. Lastly, in the inverted byte substitution layer, the bytes of the output of the inverted ShiftRow layer are substituted with another set of bytes using a lookup table. Again, this process is very similar to the encryption process—except the lookup table used is different. The output of the inverted byte substitution layer is used in the third round. The same process continues until the last round. After all the rounds are completed, the output of the last round is XORed with the last subkey K_{n+1} or $K11$ for 10 rounds. The output of the XOR operation is the 128-bit block of plain text. This concludes the decryption process.

AES in Practice

AES has been the encryption standard for more than 20 years, and as of the time of writing (summer 2024), it's still strong and secure. Over the years, there has been a lot of research and claims about attacks on AES. Most of these attacks aren't practical or are extremely difficult to execute. In short, AES isn't something we need to worry about while designing security controls—that is, until someone can implement Grover's algorithm using quantum computers (we'll discuss that in depth in Chapter 10).

AES is best implemented using software, especially when precomputed lookup tables are used for byte substitution and for calculating $g()$ in the MixColumn layer. The use of precomputed tables makes the software implementation very efficient. Even key expansion schedules can be precomputed, and subkeys can be used in the software to make the encryption and decryption even faster.

> **Python Makes It Easy!**
>
> Many programming languages offer AES implementation as a module or function that makes the implementation of AES very easy. Python has a cryptography tool kit, PyCrypto, which offers AES as part of the crypto library and can be used without much effort. You simply need to run this command:
>
> `Pip install pycrypto`
>
> Once PyCrypto is installed, you can create a Python file and import AES, along with other necessary packages, using the following command:
>
> `From Crypto.Cipher import AES`
>
> Of course, you'll have to write the complete program, but this is just to show that there's no need to implement the AES algorithm itself. The difficult part is done for you. All you, as a developer, need to know is how to use it correctly. For information, check out *https://pypi.org/project/pycrypto/*.

Hardware implementation of AES is also possible, and with modern technology such as application-specific integration circuits (ASICs), it's also efficient. But DES was designed primarily for hardware implementation while AES isn't designed with that intention.

Theoretically, AES can be implemented with all block cipher modes, but it's more commonly implemented with CTR and GCM modes. See Chapter 1, Section 1.3.4, for more information on block cipher modes.

2.2.5 AES Finalists

As discussed in the previous section, about 15 proposals were accepted as part of the call for encryption algorithms for the new encryption standard. Out of these 15, the top 5 were selected to be the finalists. The Rijndael, Serpent, Twofish, RC6, and Mars ciphers emerged as finalists. The Rijndeal cipher was the ultimate winner with a maximum (86 positive and 10 negative) votes and was named the AES, as we discussed in the previous section.

> **Finalists from around the Globe**
>
> The Rijndael cipher was proposed by two Belgium natives—Joan Daemen and Vincent Rijmen. It's interesting to note that in spite of other finalists who had some big names, like Bruce Schneier, Eli Biham, and Ron Rivest, as well as the IBM team who developed DES, Rijndael was named the winner. The development of Mars, RC6, and Twofish was done by US citizens, while Serpent was developed by people from the United States, Israel, and Norway. The point is that a US government agency gave a fair shot to everyone who had an innovative idea.

In this section, we'll briefly discuss the other four finalists.

Serpent

The Serpent [Biham, 1998] is a 128-bit symmetric block cipher. Serpent offers three key sizes—128-bit, 192-bit, and 256-bit. Serpent, like DES, is based on the Feistel network. It encrypts 32 bits every round. Serpent uses the classic substitution-permutation (SP-network) strategy.

At a very high level, Serpent's algorithm is composed of the following steps:

1. Initial permutation (IP)
2. 32 rounds of substitution and permutation consisting of subkey mixing, substitution with S-boxes, and a linear transformation
3. Final permutation

The last round, just like AES, is slightly different. The last round doesn't have linear transformation but instead has an extra step of key mixing. The initial permutation and final permutation, just like DES, don't add any cryptographic value.

In each round, the key is mixed using a 128-bit XOR operation. Next, 128 bits are divided into four blocks of 32 bits each. The output of the XOR operation is passed through the series of 32 S-boxes. These 32 S-boxes are basically nonlinear substitution tables. In each round, there are 32 identical S-boxes. However, there are eight different tables (each 16 bits long) used in each round. Next, a linear transformation is performed on the S-box output. The linear transformation is used to optimize the avalanche effect—a small change creates a larger impact on the result.

Serpent was designed for security. The developers of Serpent believed that they could have achieved encryption in 16 rounds. However, their goal was to make sure that the cipher withstands cryptanalyst attacks even with the rapid technological developments and to achieve that security Serpent was designed with 32 rounds. (DES was broken because the CPU and memory power increased so much between the 1970s and 1990s. Developers didn't want to see that happening to Serpent in 10 or 15 years.) Security was definitely a plus point for Serpent compared to almost all AES finalists—and that is why, probably, Serpent received the least number (only seven) of negative votes. On the flip side, speed was a disadvantage for Serpent. That is where, probably, Serpent lost the battle for the AES crown against Rijndael.

Serpent can be implemented in software and hardware. Serpent isn't patented and can be used without any licensing fees.

Twofish

Twofish was developed by a team of cryptographers: Bruce Schneier, John Kelsey, Doug Whiting, David Wagner, Chris Hall, and Niels Ferguson. Twofish [Schneier, 1998] is a symmetric block cipher with a block size of 128 bits and a variable key size of 128, 192, and 256 bits. Although it was placed third in voting, according to developers [Ferguson, 2010], it falls between Rijndael and Serpent in terms of performance and security. It's a lot faster than Serpent, but it's less secure, like all other finalists, compared to Serpent.

Figure 2.15 shows the high-level functionality of the Twofish algorithm. At the core of the algorithm, Twofish uses the same Feistel network as DES. The 128-bit plain text block is divided into four 32-bit blocks. Each 32-bit block is known as a word (a word has 32 bits or 4 bytes). The key is added to each word before the first encryption round and after the last encryption rounds to achieve the key whitening effect. The outputs of words W_0 and W_1 are used as the input to the eight S-boxes in function f. Four bytes from each word are fed to each S-box as input. The key-dependent S-box of Twofish makes it a lot more secure against differential and linear cryptanalyst attacks compared to other Feistel network–based encryption algorithms.

Another nice feature of Twofish is the use of the maximum distance separable (MDS) matrix, which is a fixed-value matrix that provides diffusion.

The last step in function f is to arithmetically add two subkeys. The output of the function is XORed with the remaining two words—W_2 and W_3.

Like other Feistel network ciphers, Twofish encrypts only half the plain text in each round. The output of round 1 is swapped before being provided to round 2 as an input. This is to make sure that the other half is encrypted in the next round. The process continues until the last round. There is one more swap after the 16th round. After the last swap, all four words are XORed with subkeys for the whitening purpose.

2.2 Symmetric Key Algorithms

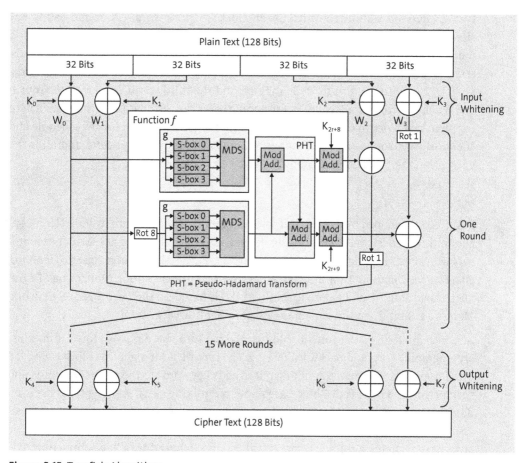

Figure 2.15 Twofish Algorithm

Twofish can be implemented very efficiently using software or hardware. This is a very elegant cipher and can be used in a variety of applications, including applications where the keys are changing frequently. The cipher also is a good fit for applications with very little or no memory.

RC6

The RC6 team includes Ron Rivest, Matt Robshaw, Ray Sidney, and Yigun Lisa Yin. RC6 is viewed as a modification or extension of RC5. RC6 [Rivest, 1998] is a parameterized cipher with variable block size, key size, and rounds. The cipher is represented as follows:

RC6(w,r,b)

Here, w is the size of the block, r is the number of rounds, and b is the key size in bytes.

For AES, RC6 was submitted with a 128-bit block size (four words), 20 rounds, and a key size of 128, 192, and 256 bits. RC6 is based on a modified Feistel network. The second and fourth words pass through function f and are XORed with the first and the third words, respectively. At the end of the round, the words swap—the second and fourth words swap with the first and third while the first and the third words swap with the fourth and second words, respectively. The swapping takes care of encrypting the other half of the plain text block in the next round. Subkeys are generated using a key schedule and added in each round. The RC6 cipher also adds the round key before and after the encryption to have the key whitening effect.

MARS

MARS was developed by a team of cryptographers at IBM. The team included Don Coppersmith who designed DES with Horst Feistel in the early 1970s at IBM. Other members were, in alphabetic order, Carolynn Burwick, Edward D'Avignon, Rosario Gennaro, Shai Halevi, Charanjit Jutla, Stephen M. Matyas Jr., Luke O'Connor, Mohammad Peyravian, David Safford, and Nevenko Zunic. MARS is based on the Type-3 Feistel network. Mars is a 128-bit, 32-round variable key size cipher [Burwick, 1999].

At a very high level, the functionality is divided into three phases: forward mixing, cryptographic core, and backward mixing. The input 128-bit plain text block is divided into four words. The forward mixing phase adds keywords to the plain text words and then performs 8 rounds of S-box–based unkeyed mixing (i.e., the encryption isn't used during the mixing). The cryptographic core phase has 16 rounds of keyed transformation—8 forward keyed transformations and 8 backward keyed transformations. The last phase includes 8 rounds of backward mixing followed by a key subtraction. The unkeyed forward and backward mixing protects against chosen plain text and chosen cipher text attacks by creating enough confusion that the attacker wouldn't know the relationship between the plain text and the cipher text.

According to the team of developers, MARS offers better security than Triple DES, but it's faster in performance than single DES. MARS can be implemented with software or hardware and is very efficient and compact in both implementations.

2.3 Summary

In this chapter, we learned about symmetric cryptography. Symmetric ciphers are used today to protect most of the data. The cryptography tree from Chapter 1 can be expanded further to include the symmetric ciphers. The revised tree is shown in Figure 2.16.

The basic concept of symmetric cryptography hasn't changed since the Caesar cipher. At that time, substitution and permutation were used, and today, modern ciphers work on the same principles. However, modern technology made it possible to take this

2.3 Summary

concept to the next level. This chapter reviewed how modern cryptography made its way into commercial space and how symmetric cryptography was at the core of the transformation.

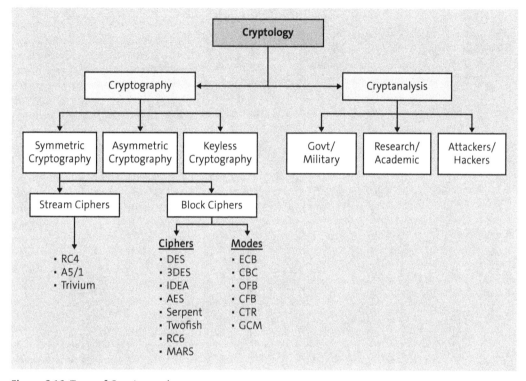

Figure 2.16 Type of Cryptography

The chapter introduced the first ever commercial cipher, DES, and its variants. Although DES isn't used much, it builds a solid foundation on how symmetric cryptography works. The variant, Triple DES is still in use! The second half of the chapter focuses on AES — the popular symmetric cipher in use today. We learned the architecture of AES and how it can be implemented and used. The chapter ended with a brief comparison of other AES finalists.

> **Key Takeaways from This Chapter**
> - Reviewed fundamentals of symmetric cryptography and its strengths and weaknesses
> - Studied early block ciphers and their variants
> - Learned how AES works and how it can be implemented today

The symmetric cipher is only part of the story. Asymmetric cryptography is the other part of the story, which we'll discuss in the next chapter.

Chapter 3
Asymmetric Cryptography

Unlike the long history of symmetric cryptography, which goes back a couple of thousand years, the history of asymmetric cryptography is only 50 years old. Asymmetric cryptography is relatively new to the field of cryptography but has gained a lot of attention in a short time. Asymmetric cryptography wasn't invented to compete against symmetric cryptography but rather to complement it. In this chapter, we'll introduce asymmetric cryptography and take a detailed look at various asymmetric cryptography algorithms.

Protecting data in transit has become more important than ever since the internet has become an integral part of society. Asymmetric cryptography, the protagonist of this chapter, plays a critical role in protecting data in transit. Asymmetric cryptography is also known as public key cryptography. Like most inventions, asymmetric cryptography was born out of necessity—the necessity to overcome the key sharing weakness of symmetric cryptography.

> **Chapter Highlights**
> - Fundamentals of asymmetric cryptography
> - Mathematics necessary to understand and implement asymmetric cryptography
> - Architecture and implementation of various asymmetric algorithms

In this chapter, you'll learn all about asymmetric cryptography. Section 3.1 starts with an explanation of what asymmetric cryptography is, why we need it, and what we can do with it. The section also covers the mathematics necessary to understand asymmetric cryptography. In Section 3.2, we'll learn about various asymmetric encryption algorithms used today, such as RSA, Diffie–Hellman key exchange algorithm, Elgamal, and elliptic curve cryptography (ECC).

Unlike symmetric cryptography, asymmetric cryptography is heavily based on mathematics. In this chapter, we'll also study mathematics—but only the part that is absolutely necessary to understand, implement, and use asymmetric cryptography. The chapter will also discuss the best approach and ways to implement asymmetric algorithms.

> **A Word on Mathematics**
>
> The mathematics throughout this book and in this chapter is used for explanation purposes. We may have taken the liberty to use nonstandard mathematical notations or to simplify for demonstration purposes. This is to make the explanation or theory behind the mathematics easy to understand.

3.1 Primer on Asymmetric Cryptography

Cryptography is categorized based on the symmetry of the encryption key and encryption-decryption algorithms. Symmetric cryptography algorithms use the same key for encryption and decryption, while asymmetric cryptography algorithms use different keys for encryption and decryption. Additionally, symmetric cryptography algorithms have the same encryption and decryption processes, except the decryption executes in the reverse order, while asymmetric cryptography uses the same steps for encryption and decryption. Another important difference is how the key is generated. In symmetric cryptography, the key is generated, usually by pseudorandom number generator (PRNG) or some similar mechanism, while in asymmetric cryptography key is computed.

The use of cryptography for commercial purposes became more necessary by the 1970s. Many banks and financial institutions started using Data Encryption Standard (DES) to encrypt the data. However, the million-dollar question was how to share the key with their other branches and offices. Banks actually sent keys in a safe; they actually had people flying around the country every day to just distribute the key [Singh, 2000]. But this wasn't a manageable solution. Symmetric cryptography was secure, efficient, and useful, but it needed a proper key distribution mechanism. That's when asymmetric cryptography was born.

Asymmetric cryptography has two keys: public key and private key. The public key, as the name implies, is available for public consumption and known to everyone. The private key, on the other hand, is private to the owner of the key. The sender uses the receiver's public key to encrypt the message. Then the encrypted message can be shared over the unsecured network. Upon receiving the message, the receiver uses his own private key to decrypt the message. This process was shown in Chapter 1, Section 1.2.5 (refer to Figure 1.5).

The concept of public key cryptography was first proposed by Whitfield Diffie, Martin Hellman, and Ralph Merkle in 1976 [Diffie, 1976]. Since then, asymmetric, or public key encryption, played a prominent role in the way we communicate today. You may be wondering, if asymmetric cryptography is good, why not continue using asymmetric cryptography only? Why do we need centuries-old symmetric cryptography? The answer lies in the strengths and weaknesses of asymmetric cryptography, which we'll discuss next.

> **Cryptography Is All About Keeping Secrets!**
>
> Although the credit for discovering public key cryptography goes to Diffie, Hellman, and Merkle, British government documents released in the late 1990s revealed that the concept of public key cryptography was first discovered by James Ellis, Clifford Cocks, and Graham Williamson in the early 1970s at the UK's Government Communications Headquarters (GCHQ). However, it was classified information at the time, and no patent was filed or research paper could be published. After all, cryptography is all about keeping secrets [Singh, 2000]!

3.1.1 Properties of Asymmetric Cryptography

Symmetric cryptography provides confidentiality to the data or message, but asymmetric cryptography has many other properties that are very useful. In this section, we'll discuss strengths, weaknesses, and use cases of asymmetric cryptography.

Let's start with the strengths of asymmetric cryptography, as follows:

- **Key distribution**
  Key distribution is one of the major advantages of asymmetric cryptography. As we saw in Chapter 2, key distribution and management can quickly become a nightmare in symmetric cryptography. This problem is solved with asymmetric cryptography by using two different keys for encryption and decryption.

- **Confidentiality**
  Another advantage of asymmetric cryptography is confidentiality. Asymmetric cryptography provides a high level of security. Not to say that symmetric ciphers aren't secure, but asymmetric encryption algorithms, such as RSA, are very difficult to crack (i.e., until we implement Shor's quantum algorithm—you'll learn more about Shor's algorithm and quantum computing in Chapter 10). In fact, it hasn't been cracked yet—you'll learn a lot more about it in Section 3.2.1. Apart from difficult math, in asymmetric cryptography, the user's private key is never shared. (Compare that to symmetric cryptography in which the user has to share the key for decryption, creating the possibility of compromising the key.)

- **Authenticity**
  Asymmetric cryptography can be used for digital signatures. The sender of the message can sign the message and then encrypt the signature with his private key. The recipient of the message can decrypt the signature. This provides authenticity to the message—the receiver can confirm that the message has really come from the said person.

- **Nonrepudiation**
  Asymmetric cryptography is used to sign the message with a digital signature, so the sender can't deny the fact. This provides a nonrepudiation property or advantage.

Asymmetric cryptography offers many advantages over symmetric cryptography, but it isn't a magic wand to solve all our problems. Asymmetric cryptography has the following disadvantages:

- **Slow speed**
  One of the major disadvantages of asymmetric cryptography is the slow speed of encryption and decryption. Asymmetric cryptography takes a significantly longer time to encrypt and decrypt the data. That alone is a good enough reason to continue using symmetric cryptography.

- **Resource intensive**
  Asymmetric cryptography is heavily based on mathematics. The computations are large and involved. This takes a lot of computing resources. This is another big negative. Plus, because of the heavy resource consumption, it can't be used with smaller devices or at least is difficult to use with smaller devices.

- **Risk of being "locked out"**
  If the owner of the key loses the private key, then there is no way to recover from that. The data is lost. (Not that a symmetric key can't be lost, but because they exist in a pair, in a worst case scenario, it can be "borrowed" from the receiver.)

- **Trust in public key**
  As we learned previously, asymmetric cryptography has the advantage of two keys and the public key can be shared openly. However, how do we trust the public key? Is there a way to validate that the public key really belongs to a particular person, or that it hasn't been tampered with? One solution is to use certificates (you'll learn all about certificates in Chapter 6). However, the certificate authority (CA) must be trustworthy and, more importantly, secure.

- **Key length**
  The major reason for the strong security of asymmetric cryptography is its large key lengths. That makes it very difficult to crack. However, this is also a drawback because the longer keys can be difficult to manage and slow down the execution.

Thanks to the wide-ranging properties of asymmetric cryptography and considering its strengths and weaknesses, it has a variety of applications and uses. Some of the key use cases are as follows:

- **Symmetric key sharing**
  Asymmetric cryptography is used to share (or send) the symmetric key securely—without flying people around the country. In fact, it turns out that the primary use

of asymmetric cryptography is to send the symmetric encryption key over the unsecured network. The process is shown in Figure 3.1 and can be summarized in the following steps:

1. Person A shares his public key with everyone—Person B gets this public key.
2. Person B encrypts the message, the symmetric key, using person A's public key, and sends the encrypted message, the symmetric key, back to person A.
3. Person B decrypts the message using his private key and retrieves the symmetric key (the message). Now both parties have the shared key for symmetric encryption and decryption.
4. From this point on, the data is encrypted and decrypted using symmetric encryption. The message or data could be encrypted and decrypted using Advanced Encryption Standard (AES) or the symmetric cipher of your choice.

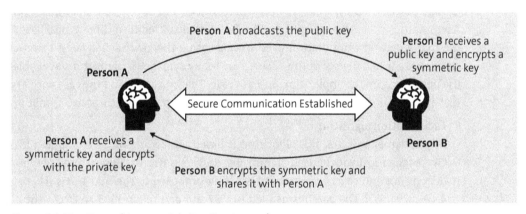

Figure 3.1 Use Case of Asymmetric Key Cryptography

- **Digital signature**
  Asymmetric cryptography is used with digital signatures in signing messages. As discussed earlier, this provides authenticity and nonrepudiation to the messages. However, the role of the private key and the public key is (sort of) reversed. The digital signature is encrypted using the private key and decrypted or verified using the public key. We'll discuss digital signatures in Chapter 4, Section 4.3.

- **Cryptocurrency**
  Cryptocurrencies, like Bitcoin, use asymmetric cryptography to authenticate and authorize transactions. In fact, asymmetric cryptography makes cryptocurrency dealing very secure. The cryptocurrency transaction isn't complete until it's verified by both keys. Moreover, it also provides nonrepudiation so there's no denying the transaction later. You'll learn about how cryptography is used with cryptocurrency in Chapter 8.

- **PKI**

  Asymmetric cryptography is central to the public key infrastructure (PKI). PKI allows us, through the certificates, to trust public keys and communicate, transact, and browse securely over the internet. You'll also learn all about PKI in Chapter 6, Section 6.4.

- **Security**

  This is an obvious use case—the main goal of any cryptosystem is to provide confidentiality and security. Asymmetric encryption provides a very high degree of security to the data.

The later chapters explain most of these use cases in depth except security, which is explained in Section 3.2 as we discuss each algorithm.

Symmetric cryptography, as we learned in Chapter 2, is divided into two types of ciphers based on how the encryption is performed—stream ciphers and block ciphers. Asymmetric cryptography, on the other hand, is classified into three types, based on the mathematical computation used in calculating the key. Each type of symmetric cipher has a wide range of options. There are many symmetric algorithms available for use depending on the application and use case. That certainly isn't the case with asymmetric algorithms. The three major types of cryptography algorithms are as follows:

- **Factorization algorithm**

  As the name indicates, this algorithm is based on factorizing a very large number. One of the most widely used algorithms, RSA, uses this mechanism to encrypt and decrypt the message. This method was first proposed in the mid-1970s [US patent 4,405,829, 1977]. The patent was filed in 1977 and granted in 1983. In 2000, the algorithm was released in the public domain.

- **Discrete logarithms algorithm**

  This class of algorithms is based on discrete logarithm math problems. The discrete logarithm algorithm, which was also proposed in the mid-1970s, is also used in the Diffie-Hellman and Elgamal encryption algorithms [Diffie, 1976] [Elgamal, 1985].

- **Elliptic curve algorithm**

  The elliptic curve forms the third type of asymmetric algorithm. In this type, the elliptic curve is used to derive the public and private keys. This class was proposed in the mid-1980s [Koblitz, 1987] [Miller, 1985].

We'll discuss all of these algorithms throughout this chapter. Other asymmetric algorithms, such as lattice-based, McEliece, and so on are also very robust but were ahead of their time. These algorithms aren't used currently; however, they are making their way back with the looming threat of quantum computers. We'll discuss these algorithms in Chapter 10.

3.1.2 Introductory Mathematics

As mentioned earlier, asymmetric cryptography is heavily based on mathematics. It's important that we have at least a high-level understanding of some of the mathematical concepts used in asymmetric cryptography. Asymmetric cryptography primarily uses modular arithmetic, Euclidean algorithm, Extended Euclidean algorithm, Euler's (Phi) function, and primality theorems. All of these are part of number theory. Modular arithmetic isn't discussed here as we've already studied it in Chapter 1, Section 1.2.9. Some concepts, such as elliptic curves, are explained with the algorithm and not discussed here.

> **What Is Number Theory?**
>
> Number theory is the mathematics or study of positive numbers—also known as integers, whole numbers, or real numbers—and the relationship between these numbers. It starts with basic operations like addition and subtraction and moves on to higher arithmetic, modular arithmetic, the Pythagorean theorem, and the list goes on and on. Cryptography relies on number theory because it works on a finite set of positive integers.

Euclidian Algorithm

The Euclidean algorithm is a factorizing algorithm that calculates the greatest common divisor (GCD) of two prime numbers. In other words, the Euclidean algorithm calculates the largest number that divides two integers without any remainder.

The Euclidean algorithm uses the simple technique of breaking down a large difficult problem into multiple small problems. It takes two very large numbers and solves recursively for the GCD. To do so, it subtracts the smaller number from the larger number and repeats the process of subtraction until the two numbers are equal, which is the GCD. For example:

$gcd(21, 35) = (35 - 21, 21) = (14, 21)$
$gcd(21, 14) = (21 - 14, 14) = (7, 14)$
$gcd(14, 7) = (14 - 7, 7) = (7, 7)$

The process of subtracting can take many steps when the numbers are large, or the difference is too big between them. One way to expedite the process is by dividing the numbers or using modular arithmetic. The process reaches the end or reveals the GCD when the remainder is zero. Mathematically, the process can be written as follows:

$gcd(x,y) = (y, x \bmod y)$ when $y \neq 0$
If $y = 0$, then $gcd(x,y) = x$

3 Asymmetric Cryptography

Let's understand with an example of finding a GCD of 21 and 35:

$gcd(21, 35) = (35, 21 \bmod 35) = (35, 21)$
$gcd(35, 21) = (21, 35 \bmod 21) = (21, 14)$
$gcd(21, 14) = (14, 21 \bmod 14) = (14, 7)$
$gcd(14, 7) = (7, 14 \bmod 7) = (7, 0)$

Another approach to find the GCD is to use the division algorithm. The division algorithm is just a different way to express the factoring problem:

$x = y \times q + r$

Here, x and y are two integers, q is a quotient, and r is a remainder. This is known as a division algorithm because it's derived from the basic division problem:

$x/y = q + r$ (writing it in the form of an equation: $x = y \times q + r$)

Let's understand with an example of finding a GCD of 21 and 35, assigning the larger number to y (35) and the smaller number to x (21):

$y = x \times q + r$
$35 = 21 \times 1 + 14$
$21 = 14 \times 1 + 7$
$14 = 7 \times 2 + 0$

All of these approaches are considered part of the Euclidean algorithm.

Extended Euclidean Algorithm

The Extended Euclidean algorithm is used to express the GCD as a linear combination of two numbers. The formula is as follows:

$gcd(x, y) = ax + by$

Let's consider an example:

$gcd(284, 774) = a \times 284 + b \times 774$

First, let's calculate the GCD using the division formula:

$gcd(284, 774)$

$\quad y = x \times q + r$
$\quad 774 = 284 \times 2 + 206 \quad$ (i)
$\quad 284 = 206 \times 1 + 78 \quad$ (ii)
$\quad 206 = 78 \times 2 + 50 \quad$ (iii)
$\quad 78 = 50 \times 1 + 28 \quad$ (iv)
$\quad 50 = 28 \times 1 + 22 \quad$ (v)
$\quad 28 = 22 \times 1 + 6 \quad$ (vi)
$\quad 22 = 6 \times 3 + 4 \quad$ (vii)
$\quad 6 = 4 \times 1 + 2 \quad$ (viii) $\quad gcd(284, 774)$
$\quad 4 = 2 \times 2 + 0$

Here, we've solved the left side of the equation using the division algorithm, where $y = 774$ and $x = 284$.

Now, solve the right side of the Extended Euclidean algorithm for (284, 774):

$gcd(284, 774) = a \times 284 + b \times 774$
$\quad\quad 2 = a \times 284 + b \times 774$
$\quad\quad 2 = 6 - 4 \times 1$ *replacing 2 with (viii) above*
$\quad\quad 2 = 6 - 1 \times (22 - 6 \times 3)$ *replacing 4 with (vii) above*
$\quad\quad 2 = 4 \times 6 - 1 \times 22$
$\quad\quad 2 = 4 \times (28 - 1 \times 22) - 1 \times 22$ *replacing 6 with (vi) above*
$\quad\quad 2 = 4 \times 28 - 5 \times 22$
$\quad\quad 2 = 4 \times 28 - 5 \times (50 - 1 \times 28)$ *replacing 22 with (v) above*
$\quad\quad 2 = -5 \times 50 + 9 \times 28$
$\quad\quad 2 = -5 \times 50 + 9 \times (78 - 1 \times 50)$ *replacing 28 with (iv) above*
$\quad\quad 2 = -14 \times 50 + 9 \times 78$
$\quad\quad 2 = -14 \times (206 - 78 \times 2) + 9 \times 78$ *replacing 50 with (iii) above*
$\quad\quad 2 = -14 \times 206 - 14 \times (-78 \times 2) + 9 \times 78$
$\quad\quad 2 = -14 \times 206 + 37 \times 78$ *replacing 78 with (ii) above*
$\quad\quad 2 = -14 \times 206 + 37 \times (284 - 1 \times 206)$
$\quad\quad 2 = -51 \times 206 + 37 \times 284$ *replacing 284 with (i) above*
$\quad\quad 2 = -51 \times (774 - 2 \times 284) + 37 \times 284$
$\quad\quad 2 = (-51) \times 774 + (39) \times 284$ *solved for gcd*

$gcd(x, y) = ax + by$
$gcd(284, 558) = a \times 284 + b \times 774$ can be written as:
$gcd(284, 558) = (39) \times 284 + (-51) \times 774$

Here, we've solved the right side of the equation by substituting the left side with the GCD value calculated previously and keep going in the reverse order to solve for a and b.

We'll use the Extended Euclidean algorithm in calculating the modular multiplicative inverse.

Modular Multiplicative Inverse

Before learning about the modular multiplicative inverse, we should recap the multiplicative inverse. The *multiplicative inverse* is the number that gives the product of 1 when multiplied by the original number. We can use modular arithmetic and the Extended Euclidean algorithm to calculate the multiplicative inverse of any number. The formula is as follows, where x is the number and i is the multiplicative inverse:

$x \times i = 1 \bmod m$
$xi \bmod m = 1$

For example:

$35i \bmod 8 = 1$

3   Asymmetric Cryptography

Let's find the multiplicative inverse of 35 with respect to mod 8:

$gcd = (x, y) = gcd(35, 8) = gcd(8, 35)$
$$y = xq + r$$
$$35 = 8 \times 4 + 3$$
$$8 = 3 \times 2 + 2$$
$$3 = 2 \times 1 + 1 \quad \rightarrow \quad \text{The gcd}$$

The GCD of these two numbers must be 1; otherwise, x doesn't have a modular multiplicative inverse.

Now, let's apply the Extended Euclidean algorithm to find the multiplicative inverse:

$gcd(8, 35) = ax + by = a \times 8 + b \times 35$
$$1 = 3 - 2 \times 1$$
$$1 = 3 - 1 \times (8 - 3 \times 2) = -8 + 3 \times 3$$
$$1 = -8 + 3 \times (35 - 8 \times 4) = -8 \times 13 + 3 \times 35 = -104 + 105$$

Finally, substitute the values of x, m, and i in the modular multiplicative equation:

$35i \bmod 8 = 1$
$35 \times 3 \bmod 8 = 1$
$105 \bmod 8 = 1$

The modular multiplicative inverse of 35 with respect to mod 8 is 3.

Chinese Remainder Theorem

As we'll see in RSA cryptography in Section 3.2.1, the decryption process in RSA cryptography is made faster by using the Chinese remainder theorem. In a nutshell, the Chinese remainder theorem can be explained as if $N = n_1, n_2, n_3$, where all n are coprimes, then we can calculate $x \bmod N$ from $x \bmod n_1$, $x \bmod n_2$, and $x \bmod n_3$. For example, if n_1, n_2, and n_3 are 3, 5, and 7, then $N = 3 \times 5 \times 7 = 105$.

The value of $x \bmod N$ (i.e., $x \bmod 105$) can be calculated using $x \bmod 3$, $x \bmod 5$, and $x \bmod 7$.

It's Prime Time!

First, a quick refresher: a number is considered a prime number if it's divided by 1 and itself—without any remainder. The rest of the numbers are known as composite numbers. The largest prime number has 24,682.048 digits [GIMPS, 2018]!

The GCD of two prime numbers is always 1. If p and q are any prime numbers, then the GCD of p and q is always 1, as follows:

$gcd(p,q) = 1$ if p and q are prime numbers

However, the reverse isn't true—meaning two numbers whose GCD is 1 don't have to be prime numbers. And that leads us to the definition of coprime.

> The numbers are said to be *coprime* when numbers don't have any common factor between them other than 1. Needless to say, all prime numbers are coprime since 1 is the only common factor, however, the definition of coprime is applicable to composite numbers also. For example, 24 and 35 are coprime since the only common factor between them is 1; that is, GCD is 1. And, yes, you guessed it right—1 is coprime to every number!
>
> There are mathematical ways to check if the number is a prime number; these methods or tests are called primality tests.

Euler's Totient

Euler's totient or Euler's phi (ϕ) function is another very useful mathematical concept in asymmetric cryptography. Euler's totient, $\phi(n)$, gives the number between 1 and n, which are coprime with n. For example:

$\phi(n) = \phi(9) = \{1,2,3,4,5,6,7,8,9\} = \{1,2,4,5,7,8\} = 6$, i.e., $\phi(9) = 6$

In this example, n is 9. The function ϕ gives all the coprime numbers with 9. The example shows, there are six numbers—*1,2,4,5,7,8*—that are coprimes with 9.

However, Euler goes a step further and makes our lives easy by stating that, if p and q are prime numbers, then we can find n as follows: $\phi(n) = (p-1)(q-1)$.

Let's consider an example where $p = 3$ and $q = 3$:

$\phi(n) = (p-1)(q-1)$
$\quad\quad = (3-1)(3-1) = 2 \times 2 = 4$

In this example, we're substituting the values of p and q to find the value of $\phi(n)$. This property of Euler's function is also used in RSA key generation.

This can formerly be defined as, if p is a prime number, then *gcd* of (p,q) is 1 (see the box on prime numbers) and $\phi(p) = p - 1$ as long as $0 < q < p$.

Another useful property of Euler's function is $\phi(p \times q) = \phi(p) \times \phi(q)$, where p and q are coprime.

The ϕ of a product is equal to the ϕ of two prime multipliers used in the product. For example:

$\phi(p) = \phi(3) = 3 - 1 = 2$. $\phi(q) = \phi(5) = 5 - 1 = 4$. $\phi(5) \times \phi(3) = 4 \times 2 = 8$
$\phi(p \times q) = \phi(15) = \{1,2,4,7,8,11,13,14\} = 8$

Substitute the values of p and q. The example proves that the product of $\phi(p)$ and $\phi(q)$ gives the same result as $\phi(pq)$.

The last property of Euler states that if p is a prime number and k is greater than zero, then:

$p^k = p^k - p^{(k-1)}$

For example, if $p = 2$ and $k = 3$, then

$2^3 = 2^3 - 2^{(3-1)} = 8 - 4 = 4$

Here, we've substituted the values of p and k in the preceding equation.

> **Is Modern Cryptography Still in 300 BC?**
>
> Most of the mathematics used in asymmetric cryptography was first discovered/proposed a very long time ago. For example, modular arithmetic, Euclidean algorithm, prime numbers, and Euler's function all were initially conceptualized around 300 BC. Just because the concept is around for so long, it doesn't make it easy to solve! That's why modern cryptography uses these concepts.

3.2 Asymmetric Cryptography Algorithms

In this section, we'll learn about various public key or asymmetric algorithms, including RSA, Diffie-Hellman, Elgamal, and ECC. Asymmetric cryptography is heavily based on the concepts or difficulty of math problems so it's almost impossible to avoid math, but as promised, we'll try to minimize the detailed explanation and use of mathematics.

> **Speed versus Strength**
>
> Public key cryptography relied on the difficulty in solving mathematical problems. Right now, even with hundreds of the most efficient computers, it takes years to solve these problems. In other words, we're capitalizing on the slow speed of computers. The strength of public key cryptography is the slow speed of available hardware. However, as we'll learn in Chapter 10, quantum computers may solve hard to crack math problems. And that is why, eventually, we'll have to find other options. After all, how long can we bank on someone else's weakness?

3.2.1 RSA Algorithm

RSA is named after its inventors, Ron Rivest, Adi Shamir, and Leonard Adleman, who proposed the RSA public key scheme in 1977 [Rivest, 1978]. The RSA encryption scheme is, probably, the most widely used encryption scheme now—although we don't have any official number or survey, most internet communication uses the RSA encryption/decryption mechanism. You'll be surprised to know that such a successful (and mathematically difficult) cryptosystem was conceptualized in just one night as Ron Rivest was unable to sleep [Singh, 2000]! Although the initial draft of the paper was drafted by

Rivest, he shared the credit with Shamir and Adleman, probably because they were part of the discussion that eventually helped Rivest write the paper. In the initial draft, Rivest alphabetically listed the names of Adleman, Rivest, and Shamir and named the cipher ARS. However, later Shamir and Adleman insisted and made him change the order of the names. We now know the cipher as RSA [Singh, 2000].

The RSA cipher is based on the *one-way function* theory: it's easy to solve a problem in one direction, but reversing the solution to get the original value is very difficult or practically impossible. This philosophy is at the core of RSA cryptography: it uses a multiplication/factorization approach. It's possible to calculate a product of two numbers, even very large numbers; however, factoring the result to find the original numbers is computationally intensive. If the resulting product is very large, then it's almost impossible to find the original numbers.

Trapdoor Function

In cryptography, one-way functions are known as *trapdoor functions*. Basically, these functions are easy to calculate in one direction but difficult or almost impossible to calculate in the other direction. The RSA public key system is the best and easiest to understand example—since most of us can relate to middle school math. In RSA, two prime numbers are multiplied to generate a number; however, it's very difficult to factor this number to calculate the original prime numbers.

It's important to note that RSA uses only prime numbers in the process in order to use the properties of number theory to make the computation easy for legitimate users and difficult for hackers to crack.

In this section, we'll describe the step-by-step process to implement RSA. The process is divided into two parts: the key generation process and the encryption-decryption process.

Key Generation

RSA's key generation process is usually a one-time process. The process is only required for the key owner or the one who keeps the private key. Figure 3.2 shows the key generation process, along with the mathematical concepts used at each stage. We can break this down into the following five steps:

1. Choose two prime numbers, p and q. The chosen prime numbers must be very large—at least 512 bits each. If the numbers are smaller, the resulting product won't be large enough and can be factored to crack the private key. We could use any letters for these two prime numbers, but p and q are commonly used with RSA.

2. Calculate n, the product of p and q, as follows:

   $n = p \times q$

3. Next, calculate φ(n). If you recall from our earlier discussion, this is Euler's totient. The function gives us total coprime numbers for n:

   φ(n) = (p − 1) × (q − 1)

4. Choose the public key number or public key exponent, e. Public key exponent e must be chosen to meet the two requirements: (1) it must be between 1 and φ(n), and (2) it must be coprime with n and φ(n). Mathematically, it can be written as follows:

   Choose e such that:

   1 < e < φ(n) and

   gcd(e, φ(n)) = 1

   Exponent e is also known as the public exponent or encryption exponent. Public exponent e and the product of two prime numbers, n, form the public key. So, (e, n) is the public key.

5. Calculate private key d such that it satisfies the following condition:

   d × e = mod φ(n)

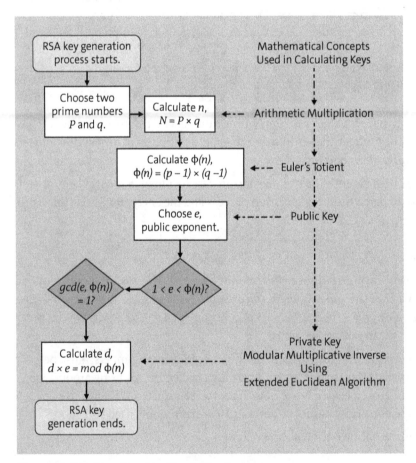

Figure 3.2 RSA Algorithm Key Generation Process

3.2 Asymmetric Cryptography Algorithms

This condition uses the modular multiplicative inverse property. Exponent *d* is also known as the private exponent or decryption exponent. Only the owner or the generator of the key should have this value.

These five simple steps are jam-packed with lots of math and important information. Let's review some of the most important considerations:

- Exponents *e* and *d* should be kept relatively small to help speed up the encryption and decryption process.
- You should be able to choose multiple pairs of *e* and *d* for a given *n*. That means there should be several public and private key pairs for a given *n*. This will keep the algorithm resistant to brute force.
- The *gcd(e, ϕ(n))* = *1* condition in choosing *e* confirms that there is always a modulo inverse for *e* and that the *d* exists. This property mathematically ties public and private keys.
- These five steps are computed only once in the process. Once the keys are generated, only encryption and decryption functions are computed.
- Although not necessary, it's possible to create two different public exponents—one for the encryption and the other for the digital signature. We'll discuss digital signatures in Chapter 4, Section 4.3.

Encryption-Decryption

Next, we'll learn the RSA encryption and decryption processes, which are very simple. The encryption process can be defined as generating cipher text *C* from plain text *P* using the following public key:

- $K_{pub} = (e, n)$
- $C = P^e \bmod n$

This is your RSA encryption function. To encrypt the message, you only need two values, *e* and *n*. Using these values with plain text *P* and performing modular arithmetic generates cipher text *C*.

The decryption process can be defined as generating plain text *P* from cipher text *C* using the following private key:

- $K_{prv} = (d)$
- $P = C^d \bmod n$

This is your RSA decryption function. To decrypt the message, you only need two values, *d* and *n*. Using these values with cipher text *C* and performing modular arithmetic generates plain text *P*.

Remember: the encryption and decryption processes use *mod n*, while the key generation process uses *mod ϕ(n)*.

3 Asymmetric Cryptography

Now let's understand this with an example. We'll use very small numbers so that most of the math can be performed mentally while reading the book:

1. Choose two prime numbers, p and q:

    $p = 3$ and $q = 7$

2. Calculate n by multiplying p and q:

    $n = p \times q = 3 \times 7$

    $n = 21$

3. Calculate $\phi(n)$:

    $\phi(n) = (p - 1) \times (q - 1) = (3 - 1) \times (7 - 1)$

    $\phi(n) = 2 \times 6$

    $\phi(n) = 12$

4. Choose public key exponent e such that $1 < e < \phi(n)$ and $gcd(e, \phi(n)) = 1$.

    There are only two numbers that are coprime with $n = 21$ and $\phi(n) = 12$: 5 and 11 (other than 1, of course). Let's choose 5 to keep the calculation simple:

    $e = 5$

    $gcd(5,12) = 1$ and $1 < 5 < 12$

5. The public key is $(e, n) = (5, 21)$. Numbers 5 and 21 can be shared with everyone.

6. Calculate the private key exponent d such that $d \times e = mod\ \phi(n)$, or we can say that $d \times e\ mod\ \phi(n) = 1$:

    $e = 5$ and $\phi(n) = 12$

    $(d \times 5)\ mod\ 12 = 1$

    $d = 17$

    $5 \times 17 = 85\ mod\ 12 = 1$

Another example of d is 5 ($5 \times 5 = 25\ mod\ 12 = 1$). However, that makes e and d the same numbers. RSA works even if e and d have the same values. We chose the next available number to avoid any confusion.

Since we already calculated that $gcd(e, \phi(n)) = 1$, we've ensured that that there is an inverse multiplicative of e. For a very large number, the modular multiplicative inverse can be calculated using the Extended Euclidean algorithm.

The initial setup to perform RSA encryption and decryption is now complete. Here are the results for our example:

- The public key is (5, 21).
- The private key is (17).

The next step is to perform encryption and decryption. For encryption, use the following formula:

$C = P^e\ mod\ n$

Here, C is the cipher text, n is 21, and e is 5. Let's assume that plain text P is 2:

$C = 2^5 \bmod 21$

$C = 32 \bmod 21$

$C = 11$

The final encrypted message or cipher text is 11. This can now be transmitted. Even if a hacker sees this number, there's no way to find out the real number, as long as the *d* is kept secret from the hacker. For decryption, use the following formula:

$P = C^d \bmod n$

Here, $n = 21$, $d = 17$, and $C = 11$:

$P = 11^{17} \bmod 21$

$P = 2$

Implementation Considerations

Real-life implementation of the RSA algorithm is very complicated. The following important points must be considered before attempting to implement the RSA algorithm:

- The encryption and decryption processes are very simple but calculating the exponent of a very large number takes a long time. The performance can be improved by keeping a low public exponent, but the private key exponent must be kept very large—roughly in the range of 1,024 bits or more. One way to calculate the exponent faster is by using the square-and-multiply algorithm, which is a basic process that adds and multiplies the values but is very useful in calculating the exponent. Refer to [Paar, 2010] for more explanation.

- The RSA algorithm isn't foolproof and has its weaknesses. For example, it's deterministic. This means that for a given key, the relationship between the plain text and the cipher text is always the same. In addition, if the private exponent, *d*, and plain text message itself both have small values, then the cipher can be cracked. Another weakness of the RSA algorithm is its malleability. Malleability simply means that the attacker can tweak the cipher text itself without even finding out the plain text or the key. This can be done if the format of the data is predictable.

- One way to strengthen the RSA cipher is to use padding. *Padding* is the process of adding some unpredictable or randomness to the plain text before encryption. This is conceptually very similar to the block cipher modes where we add IV to create the randomness. One method is Optimized Asymmetric Encryption Padding (OAEP), which was suggested by Bellare and Rogaway [Bellare, 1995] and was added to the RSA Laboratories in the Public Key Cryptography Standards (PKCS).

- RSA can be implemented using both hardware and software. However, most internet usage is based on software implementations. Even with modern high-powered

CPUs and huge memory capabilities, RSA is still very slow compared to symmetric encryption. In fact, when RSA was proposed in 1977, computers weren't powerful enough to perform these computations. By slow, it means the encryption times are in the range of a few milliseconds or less. However, this takes a significantly longer time compared to symmetric encryption when used to encrypt large databases.

> **RSA Challenge**
>
> RSA Laboratories invited cryptanalysts, cryptography researchers, mathematicians, and other cybersecurity enthusiasts to participate in the RSA Factoring Challenge in 1991 and rewarded prize money to the person or the team upon successfully factoring.
>
> The challenge was simple: for each RSA number, n, there are two prime numbers/factors, p and q, and $n = p \times q$. The challenge is to find the prime numbers p and q from given, n.
>
> The challenge formerly ended in 2007. However, researchers are still actively trying to factor larger and larger numbers.
>
> As of February 2020, a 250-decimal digit (829 bits) number was the largest number to be factored! If you're interested, there is still time to prove your mettle and grab some fame!
>
> $n =$
> *2140324650240744961264423072839333563008614715144755017797754920881418023447140136643345519095804679610992851872470914587687396261921557363047454770520805119056493106687691590019759405693457452230589325976697471681738069364894699871578494975937497937*
>
> $p \times q =$
> *64135289477071580278790190170577389084825014742943447208116859632024532344630238623598752668347708737661925585694639798853367*
> ×
> *33372027594978156556226010605355114227940760344767554666784520987023841729210037080257448673296881877565718986258036932062711*
>
> The complete details about the challenge are available online [The Register, 2001].

3.2.2 Diffie–Hellman–Merkle Key Exchange Algorithm

Although the RSA cryptosystem is more commonly used, the concept of asymmetric key cryptography was first publicly proposed by Whitfield Diffie, Martin Hellman, (and Ralph Markle) in 1976 [Diffie, 1976]. Diffie knew way ahead of most people that to secure the data and to effectively use cryptography, we must solve the key distribution problem. When Diffie was invited to speak at IBM's Watson Laboratories, he met with Alan Konheim, IBM's cryptographic expert at the time. He recommended Diffie to speak with Martin Hellman. Dr. Hellman was a professor at the University of California,

Berkeley. Diffie enrolled at the university as a student, and they began the quest for the key. Ralph Merkle later joined them.

> **What's in a Name?**
>
> It turns out that the key exchange concept was conceived by Merkle before Diffie and Hellman [Merkle, 1978]. In fact, Martin Hellman acknowledged that:
>
> *The system I called the ax1x2 system in this paper has since become known as Diffie–Hellman key exchange. While that system was first described in a paper by Diffie and me, it is a public key distribution system, a concept developed by Merkle, and hence should be called "Diffie–Hellman–Merkle key exchange" if names are to be associated with it. I hope this small pulpit might help in that endeavor to recognize Merkle's equal contribution to the invention of public key cryptography. [Hellman, 2002]*

The Diffie–Hellman–Merkle key exchange algorithm, like the RSA algorithm, also heavily depends on the concept of mathematics and the one-way or trapdoor function. The RSA cryptosystem uses the difficulty of factorization, while the Diffie–Hellman–Merkle key exchange algorithm uses the difficulty of finding the exponent. The concept of the algorithm is commonly explained by the color analogy, which we'll keep brief here to set the stage because many YouTube channels and websites already have used this analogy to explain the algorithm. The concept is using primary colors to create secondary colors, as follows:

- Assume that Person A and Person B agree on a public color that is known to everyone.
  - Public color: YELLOW
- Person A and Person B also select their secret color. This color is only known to them.
  - Person A selects RED as a secret color
  - Person B chooses BLUE as a secret color.
- Next, Person A mixes the secret RED color with the public color YELLOW to produce ORANGE.
  - Person A: RED + YELLOW = ORANGE
- Person B does the same thing on his end, mixing the secret color BLUE with the public color YELLOW to create GREEN.
  - Person B: BLUE + YELLOW = GREEN
- Person A shares ORANGE with Person B and Person B shares GREEN with Person A using an unsecured medium such as the public internet. (Or in terms of our analogy, exchanges ORANGE and GREEN color paint cans without lids.)
- At the end of the transfer:
  - Person A ends up with GREEN.
  - Person B ends up with ORANGE.

- Now Person A and Person B add their secret color to the color they received from the other person.
  - Person A: GREEN + RED = Light Brown (Marigold)
  - Person B: ORANGE + BLUE = Light Brown (Marigold)
- Person A and Person B end up with the same Marigold color!

There are a few important points to derive from this analogy:

- Everyone knows the public color, YELLOW, and the resulting mixers—GREEN and ORANGE.
- Only Person A and Person B know their respective secret colors.
- After the transfer, both Person A and Person B arrive at the same Marigold color.
- This acts like a one-way function—it's hard to separate out the original colors from the mixers. If someone eavesdrops and get the public known values, Yellow, Orange, Green, and Marigold, the eavesdropper still can't find the original secret colors—Red and Blue.

The math used in the Diffie–Hellman–Merkle system is conceptually very similar—it's easy to find an answer if you know the exponent, but it's very difficult the other way around. The Diffie–Hellman–Merkle system of key exchange uses the discrete logarithm.

Let's review the definition of the *discrete logarithm*: if x is an integer and is also relatively prime to y and m, there is only one exponent, e, that satisfies the condition:

$x = g^e \bmod p$

This is represented as follows:

$e = dlog_{g,p}(x)$

Calculating x is easy if we know e, but calculating e from x is very difficult and time-consuming. Discrete logarithms help us solve for e, but it's almost impossible if the number is very large.

Let's go back to our example of Person A and Person B. We can once again break down this algorithm into five steps, as shown in Figure 3.3:

1. **Decide on public parameters.**

    Person A and Person B agree on two public parameters: p and g.
    - p is a large prime number.
    - g is a generator, and it's a primitive root to $\bmod p$.

    The primitive root can be defined as if g is a primitive root of $\bmod p$, then $g^n \bmod p$ will result in distinct values, where $n = \{1,2,3...\}$. To simplify, the primitive root ensures that two different n values won't result in the same answer for $g^n \bmod p$; every n will have a different value.

    Public parameters p and g are YELLOW in our color analogy.

2. **Choose the private keys.**
   Person A and Person B choose their private keys, k_a and k_b respectively, such that:
   - $k_a = \{2, 3, 4, \ldots, p-2\}$
   - $k_b = \{2, 3, 4, \ldots, p-2\}$

   Private keys, k_a and k_b are private colors RED and BLUE, respectively.

3. **Calculate the public keys.**
   Person A and Person B calculate their public keys α and β, respectively:
   - Public key of Person A: $\alpha = g^{ka} \bmod p$
   - Public key of Person B: $\beta = g^{kb} \bmod p$

   Here, k_a and k_b are the private keys chosen in step 2 while p and g are the public parameters chosen in step 1.

   Person A and Person B mix their private colors with public colors to generate the shared mixer: GREEN and ORANGE.

4. **Exchanging the public keys.**
   Once the public keys are calculated, Person A and B share their public keys with each other. At the end of the exchange, Person A has β, and Person A has α.
   - Person A receives $\beta = g^{kb} \bmod p$.
   - Person B receives $\alpha = g^{ka} \bmod p$.

   This is similar to exchanging GREEN and ORANGE colors with each other.

5. **Derive shared encryption key K_e.**
   Person A and Person B can now use their private keys and the other person's public key to derive the shared encryption key:
   - Person A derives a shared encryption key: $K_{eA} = \beta^{ka} \bmod p$
   - Person B derives a shared encryption key: $K_{eB} = \alpha^{kb} \bmod p$

Person A and Person B are now ready to use the key, K_e, with the symmetric cipher of their choice and start communicating securely.

To continue with our analogy, Person A and Person B add their private colors to GREEN and ORANGE mixers that they have received to generate the third common color—MARIGOLD, which is a shared key.

> **Mission Mocha**
>
> If you have difficulty understanding math or trouble following the color analogy, we have another analogy. This one is special for the coffee lovers out there.
>
> Let's assume that you and your friend have milk, and this is public information. You have the coffee grounds, and your friend has the cocoa powder. Only you and your friend know about your grounds and powder, respectively. Here, milk is a public parameter while coffee grounds is the secret key kept with you, and cocoa powder is the secret key kept with your friend.

3 Asymmetric Cryptography

> You mix milk and coffee while your friend mixes milk and cocoa. Once done with mixing, you and your friend exchange the cups. Now your friend adds cocoa powder to your cup, and you add coffee to your friend's cup.
>
> Now you and your friend have the same drink—mocha. And the process is one way! If an attacker were to steal your cup or your friend's cup, the attacker won't be able to take apart the coffee grounds or cocoa powder from the mocha. No reverse operation is possible.
>
> And this mocha won't taste any good, but you get the point!

Let's prove what we preach. One way to prove that $K_{eA} = K_{eB}$ is to substitute the values, as follows:

$$K_{eA} = \beta^{ka} \bmod p \qquad K_{eB} = \alpha^{ka} \bmod p$$
$$\quad = (g^{kb})^{ka} \bmod p \qquad \quad = (g^{ka})^{kb} \bmod p$$
$$\quad = g^{kbka} \bmod p \qquad \quad = g^{kakb} \bmod p$$
$$K_e = K_{eA} = K_{eB}$$

Here, K_e is the shared secret or the encryption key, while K_{ea} and K_{eb} are the derived encryption keys by Person A and Person B. This proves that Person A and Person B both have the same encryption key K_e.

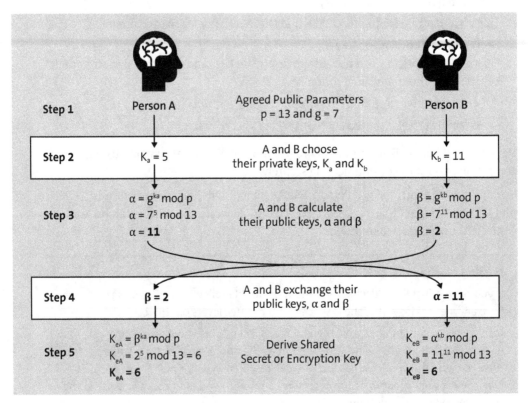

Figure 3.3 Diffie-Hillmen-Merkle Key Exchange

Next, we can plug in some small values to see how it works:

1. **Decide on the public parameters.**
   Person A and Person B agree on two public parameters: $p = 13$ and $g = 7$. The generator g is a primitive root of mod 13.

2. **Choose private keys.**
   Person A and Person B choose their private keys, $k_a = 5$ and $k_b = 11$, respectively.

3. **Calculate public keys.**
   Person A and Person B calculate their public keys, α and β, respectively, by substituting the values of g, p, k_a, and k_b in the following equations:
   - Public key of Person A:

     $α = g^{ka} \bmod p$

     $α = 7^5 \bmod 13 = 11$

     $α = 11$
   - Public key of Person B:

     $β = g^{kb} \bmod p$

     $β = 7^{11} \bmod 13 = 2$

     $β = 2$

4. **Exchange public keys.**
   - Person A receives $β = 2$.
   - Person B receives $α = 11$.

5. **Derive shared encryption key K_e.**
   Person A and Person B can now use their private keys and the other person's public key to derive the shared encryption key.
   - Person A derives a shared encryption key:

     $K_{eA} = β^{ka} \bmod p$

     $K_{eA} = 2^5 \bmod 13 = 6$

     $K_{eA} = 6$
   - Person B derives a shared encryption key:

     $K_{eB} = α^{kb} \bmod p$

     $K_{eB} = 11^{11} \bmod 13 = 6$

     $K_{eB} = 6$

We've now proven that $K_e = K_{eA} = K_{eB} = 6$.

The Diffie–Hellman–Merkle key exchange algorithm can be extended to more people or parties. For example, if three people, A, B, and C, want to create a shared encryption key, then the protocol can be extended to three (or even more) people. The concept remains the same.

Persons A, B, and C agree on public parameters p and g. They also decide on their private keys, k_a, k_b, and k_c. Person A shares public key α with Person B. Person B calculates public key β and shares it with Person C. Person C shares public key γ with Person A (or the next person, if there are more than three). At this stage, everyone has one public key. However, for three people to communicate, they need two other public keys. To achieve this, the whole process is repeated. In the second round, Person A passes original key α plus the new public key received from Person C, γ. So, person B has public keys from Person A and Person C. The cycle continues until everyone has public keys from all other participants. Once this is done, each participant has to derive the shared key by using their own private key as explained in step 5.

The Diffie–Hellman–Merkle key exchange is very simple and secure if the proper parameters are chosen. The prime number, p, should be very large—between 2,096 and 4,192. Any larger value isn't recommended because it doesn't add to the protocol's security but can increase computation time and degrade performance. The value for the generator, g, can be kept small. One way to optimize the performance is by precomputing the public keys.

Once the shared encryption key is calculated, any symmetric cipher can be used to improve the encryption efficiency.

3.2.3 Elgamal Cryptosystem

The Elgamal cryptosystem (also known as the Elgamal [or ElGamal] encryption system) is an extension of the Diffie–Hellman–Merkle key exchange. The algorithm was proposed by Taher Elgamal in 1985 [Elgamal, 1985].

The primary difference between the Elgamal algorithm and the Diffie-Hillman-Merkle key exchange is that the former performs key exchange as well as encryption, while the latter is used to simply exchange the keys. Apart from this, the Elgamal algorithm executes steps slightly differently.

Figure 3.4 shows the concept of the Elgamal encryption system. Let's go back to our example of Person A and Person B to understand the Elgamal algorithm using the following steps:

1. Person B generates the keys and shares the public parameters:
   - Person B decides on two public parameters, p and g:
     - p is a large prime number.
     - g is a generator, and it's a primitive root to $mod\ p$.
   - Person B chooses the private key, k_b.
   - Person B calculates the public key, $β = g^{kb}\ mod\ p$.
   - Person B shares/publishes p, g, and β.

3.2 Asymmetric Cryptography Algorithms

In this step, Person B acts independently—this is different from the Diffie–Hellman–Merkle algorithm where Person B and Person A decide mutually on public parameters and exchange the public keys.

2. Person A chooses the private key and calculates the public key:
   - Person A chooses the private key, k_a.
   - Person A calculates the public key, $\alpha = g^{ka} \bmod p$.

   In this step, Person A acts independently. Person A grabs the public parameters and Person B's public key from the directory and calculates the public key, α.

3. Person A generates a shared encryption key and encrypts the message:
   - Person A derives the shared encryption key: $K_e = \beta^{ka} \bmod p$.
   - Person A encrypts the message: $C = P_t \times K_e \bmod p$.

   At this stage, Person A has all the information to calculate the encryption key and encrypt the message.

4. Person A sends the ciphered message and the public key (C, α) to Person B.

   Person A sends the encrypted message along with the public key to Person B. This is another major difference between the Elgamal algorithm and the Diffie–Hellman–Merkle algorithm. In the Diffie–Hellman–Merkle algorithm, Person A's public key is shared with Person B as soon as it is calculated, while in the Elgamal algorithm, it's shared only when the message is sent.

5. Person B calculates the encryption key and decrypts the message:
   - Person B derives a shared encryption key: $K_e = \alpha^{kb} \bmod p$.
   - Person B decrypts the message: $P_t = C \times K_e^{-1} \bmod p$.

   Person B has all the information to calculate the encryption key, K_e, and decrypt the message. The decryption process uses a combination of two techniques: the square-and-multiply method to calculate the exponent, and the Extended Euclidean algorithm to calculate the modular multiplicative inverse.

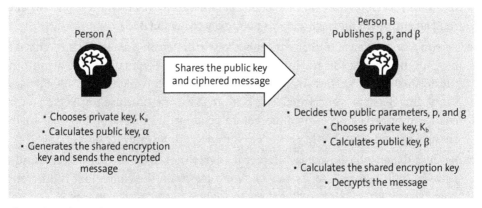

Figure 3.4 Elgamal Encryption

Next, let's plug in some small values in the equations to see how it all works. We'll stick to the same numbers we used for the Diffie–Hellman–Merkle algorithm:

1. Person B generates the keys and shares the public parameters:
   - Person B decides on two public parameters, $p = 13$ and $g = 7$.
   - Person B chooses the private key, $k_b = 11$.
   - Person B calculates the public key, $\beta = g^{kb} \bmod p = 7^{11} \bmod 13 = 2$.
   - Person B shares/publishes $(p, g, \beta) = (13, 7, 2)$.
2. Person A chooses the private key and calculates the public key:
   - Person A chooses the private key, $k_a = 5$.
   - Person A calculates the public key, $\alpha = g^{ka} \bmod p = 7^5 \bmod 13 = 11$.
3. Person A generates a shared encryption key and encrypts the message:
   - Person A derives the shared encryption key, $K_e = \beta^{ka} \bmod p = 2^5 \bmod 13 = 6$.
   - Person A encrypts the message, $C = P_t \times K_e \bmod p = 2^5 \bmod 13 = 20 \times 6 \bmod 13 = 3$.

   C is the cipher text, and P_t is the plain text. (P_t is used to distinguish the plain text from the prime p.)
4. Person A sends the ciphered message and the public key (C, α) to Person B.
5. Person B calculates the encryption key and decrypts the message:
   - Person B derives a shared encryption key, $K_e = \alpha^{kb} \bmod p = 11^{11} \bmod 13 = 6$.
   - Person B decrypts the message, $P_t = C \times K_{e-1} \bmod p = 3 \times 6^{-1} \bmod 13 = 20$.

You may be wondering what benefit this algorithm offers over the Diffie–Hellman–Merkle system. That's a fair question. There are some minor but important differences between the two algorithms:

- In the Elgamal algorithm, Person B isn't sharing the public key only with Person A and Person C. Person B is publishing the public parameters and the public key on the directory. That means anyone, not only Person A or Person C, who wants to communicate with Person B can grab the public parameters and Person B's public key and send an encrypted message with his public key to Person B.

- Person A doesn't need to share the public key with Person B ahead of time. This is a big plus. Because of this modification over the Diffie–Hellman–Merkle algorithm, the Elgamal algorithm has an ephemeral encryption key, which means it's changed every time the data is encrypted. This is more like a one-time use key. Person A can change private key K_a every time before sending the message. This essentially changes Person A's public key as well as the encryption key every time.

- Because the encryption key is different for every message, the attacker can't find a pattern. So, if Person A sends an exact message twice, the cipher text is different. This makes it difficult for cryptanalysts to analyze the message patterns.

However, one downside to consider is that the Elgamal algorithm isn't a very popular choice because it's slower.

A variant of the Elgamal algorithm is used as the DSS. We'll discuss more about digital signatures in Chapter 4, Section 4.3.

3.2.4 Elliptic Curve Cryptography

Public key cryptography gained a lot of momentum because it helped solve the problem of key sharing. However, it has one major issue: performance. It's very slow. AES, a symmetric cipher, needs a 128-bit key to securely encrypt the data; RSA or Diffie-Hillman-Merkle/Elgamal, on the other hand, require a 2,048-bit or longer key. No matter how much advancement we see in the technology, users will feel a difference in performance between the 128-bit key and the 2,048-bit key. Reducing the key length of the public key algorithms isn't an option as it would reduce the security. A new cryptosystem was necessary to solve this problem.

In 1985, Neal Koblitz [Koblitz, 1987] and Victor S. Miller [Miller, 1985] independently proposed the idea of using an elliptic curve for public cryptography. While RSA depends on the difficulty of integer factorization and Diffie–Hellman–Merkle relies on discrete logarithm problems (DLPs), *elliptic curve cryptography (ECC)* is based on the coordinates on a mathematical curve. ECC, like other public key cryptosystems, also depends on one-way or trapdoor functionality.

ECC provides very efficient cryptography compared to RSA and Diffie–Hellman–Merkle algorithms, but the downside is the complexity of mathematics. We're very familiar with the concept of multiplication and integer factorization, but the concept of the elliptic curve isn't as simple as middle school math, and most of us haven't learned it. For this reason, ECC received a very slow response initially, but it has steadily gained momentum since the early 2000s. Modern technology has no excuse for not using ECC. It's fast, it's efficient, and, most importantly, it can be implemented in small devices like mobile phones and other gadgets. It also plays an important role in Bitcoin transactions (you'll learn more about it in Chapter 8).

It's important to note that neither Neal Koblitz nor Victor S. Miller proposed a brand-new concept of cryptography. Rather, they proposed alternate mathematics for difficult and efficient one-way function to implement with the existing public key cryptography (the Diffie–Hellman–Merkle key exchange).

In this section, we'll briefly introduce the elliptic curve and its usage in cryptography. The complete mathematics of the elliptic curve is very complex and beyond the scope of this book. Some recommendations are listed at the end of this section if you're interested in learning more about ECC.

Understanding the Elliptic Curve

A curve is simply a set of points on a plane. Usually, it's an equation with x and y values representing the coordinates. The simplest example of how points are plotted on a plane is a straight line. For example:

$y = mx + b$

This is an equation for a straight line, where m is a slope, and b is the intercept point on the y-axis.

One more example follows:

$y = x$

This is an equation for a straight line passing through the origin.

This is to show you how points—or x and y coordinates—can be used. But the line isn't a curve. Let's take another example—a circle. The circle is probably the simplest form of a curve. If we plot various values of x and y in the following equation, it will solve for the corresponding values of r, which is a radius of a circle:

$x^2 + y^2 = r^2$

If we introduce coefficients in the equation of a circle, then it will result in an ellipse.

Things get a bit difficult from here on. Cosines, sines, exponentials, parabolas, hyperbolas, and so on are different types of curves. Each of these curves has different underlying mathematical equations that translate into various shapes. Informally, we can say that geometry is a visual representation of algebra. (Another, albeit crude, analogy that might upset mathematicians is that algebra is Microsoft Word and geometry is Microsoft PowerPoint.) You get the point. So how about elliptic curves?

The elliptic curves are based on

$y^2 = x^3 + ax + b$

where $4a^2 + 27b^2 \neq 0$.

The elliptic curves can be defined over real numbers, rational numbers, complex numbers, and integer modulo p. Cryptography works over the mathematics of a finite set of numbers, and it uses the elliptic curve over the integer *mod p*.

Let's consider some important characteristics of elliptic curves. They offer symmetry over the x-axis or horizontal symmetry. This means that any point on one side of the x-axis can be reflected over the other side of the axis. Another characteristic is that any nonvertical line can intersect the curve at three points. These characteristics sound simple but are very important.

ECC works by calculating the number of points on the curve. These points are kept secret. The one-way property of the function is that it's easy to calculate these points but, given the final number, it's difficult to find or calculate the number of selected points on the curve. This may sound complex, but we'll walk through the high-level concepts that will help us understand.

Suppose there are points $P(x_p, y_p)$ and $Q(x_q, y_q)$ on the curve. How can we add these points? First, we draw a line connecting points P and Q. Next, we extend the line such that it intersects the curve again. From this intersected point, take the mirror point (i.e., on the other side of the x-axis); that point, R, is the addition of points P and Q. This is called *point addition* and is shown in Figure 3.5 on the left-hand side.

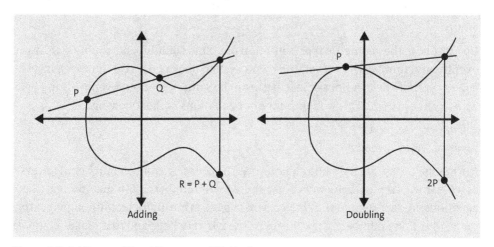

Figure 3.5 Adding and Doubling on an Elliptic Curve

Mathematically, we can represent the addition on the curve as $P + Q = R$ and $P \neq Q$.

The coordinates of point $R(x_r, y_r)$ can be computed as

$x_r = m^2 - x_p - x_q$
$y_r = m(x_p - x_r) - y_p$

where m is a slope of the line PQ, and $m = (y_q - y_p) / (x_q - x_p)$.

The computation changes if $R = P + Q$ when $P = Q$. In this case, it becomes $P + P = 2P$. We draw a tangent line through point P and extend it until it intersects the curve again. Take a mirror point from this intersection point on the x-axis. This resulting mirror point is $2P$. This is known as *doubling*, with the following formula:

if $P = Q$, $P + Q$ becomes $P + P = 2P$

The coordinates of point $R(x_r, y_r)$ can be computed as

$x_r = m^2 - x_p - x_q$
$y_r = m(x_p - x_r) - y_p$

where m is a slope of the line (tangent) PQ, and $m = (3x_p^2 + a) \div 2y_p$.

If we keep adding and doubling point P, it will keep increasing, 3P, 4P, 5P, and so on until eventually it will be nP. Now even if someone knows the value of the product n × P, it will be very difficult to derive the value of n. This one-way (trapdoor) functionality is what makes the elliptic curve very attractive for cryptography.

Stay with me for a few more lines of mathematics (it's almost over!). The points on the curve are added together, and this set of points meets all the conditions (closer, associativity, existence of an identity element, and existence of an inverse) to form a *group*. As these points are in the group, the resulting addition point, R, will also belong to the group.

Up until now, the geometric representation of points addition and doubling is shown over the real numbers. However, the cryptosystem doesn't use real numbers; instead, it uses the finite field. It's possible to perform the same operations over the finite field, with a prime p: $GF(p)$. This would change our equations to the following:

$x_r = m^2 - x_p - x_q \bmod p$
$y_r = m(x_p - x_r) - y_p \bmod p$

In a nutshell, these are the same equations as the equations for real numbers. However, as we learned, cryptography works on a finite set of numbers. With that in mind, these equations are represented with *mod p*, where p is a prime number. (Although cryptography uses finite numbers, the value of prime p is very large and that makes it almost unlikely to repeat the number.)

The next step would be to define a DLP over the elliptic curves. However, further mathematics is beyond our scope. Instead, we'll go straight to the key exchange problem.

Elliptic Curve Key Exchange

The elliptic curve key exchange is the same as the Diffie–Hellman–Merkle key exchange protocol, except for the underlying math. Let's go back to our example of Person A and Person B and go through the following steps:

1. **Decide on the public or domain parameters.**

   Person A and Person B agree on two public parameters, p and G:
   - p is a prime number.
   - G is a generator.

   Prime p, curve coefficients a and b, and G are domain (public) parameters.

2. **Choose the private keys.**

   Person A and Person B choose their private keys, k_a and k_b respectively.

3. **Calculate the public keys.**

   Person A and Person B calculate their public keys, α and β, respectively:

- Public key of Person A: α = k_a × G
- Public key of Person B: β = k_b × G

k_a and k_b are the private keys chosen in step 2, and G is the publicly known generator.

4. **Exchange the public keys.**
   Once the public keys are calculated, Person A and B share their public keys with each other, as follows. At the end of the exchange, Person A has β, and Person B has α.
   - Person A receives β = k_b × G.
   - Person B receives α = k_a × G.

5. **Derive the shared encryption key K_e.**
   Person A and Person B can now use their private keys and the other person's public key to derive the shared encryption key:
   - Person A derives a shared encryption key: K_{eA} = k_a (k_b × G).
   - Person B derives a shared encryption key: K_{eB} = k_b (k_a × G).

Person A and Person B are now ready to use the key, K_e, with the symmetric cipher of their choice and start communicating securely.

Implementation Considerations

Apart from the key exchange, ECC is also used for encryption and digital signature. Table 3.1 compares ECC with symmetric ciphers and asymmetric ciphers such as RSA and Diffie–Hellman–Merkle. RSA needs a 3,072-bit key size to achieve the same level of security that ECC achieves with only a 256-bit key size. No matter how fast technology and computation power are, it's hard to beat a 256-bit key size. In other words, to achieve the security level of symmetric ciphers, such as AES, with a 128-bit key, ECC needs a 256-bit key, and RSA needs a 3,072-bit encryption key.

Symmetric Cipher	RSA/Diffie–Hellman–Merkle	Elliptic Curve
80-bit	1,024-bit	160-bit
128-bit	3,072-bit	256-bit
192-bit	7,680-bit	384-bit
256-bit	15,360-bit	512-bit

Table 3.1 Key Size Comparison for Various Encryption Standards

Because of efficiency, the elliptic curve has become very popular over the past 10 years. It took a very long time to catch up, since historically, RSA was a much more popular and commonly used cipher. Apart from efficiency, security is another reason for ECC to gain popularity. It uses mathematics that is much more difficult to crack compared to factorization. This gives a very high level of security—as long as the right elliptic curve is used. The elliptic curve must be chosen very carefully. The curve is plotted by plugging in varies values in the corresponding equation. The coordinates or values that satisfy the curve provide the set of finite numbers. Two different mathematical approaches are used—a finite field to define the curve and a cyclic group formed by the points on the defined curve. This mathematical complexity increases the security of the cryptosystem.

ECC can be implemented using software or hardware. Because of the complexity, software implementations aren't as efficient as hardware implementations.

It's important to note that this is a very high-level overview of ECC. We've only scratched the surface of the mathematics used in ECC. If you aren't familiar or would like to dive deeper into the subject, we highly recommend exploring other sources. Some useful resources are [Paar, 2010], [Hankerson, 2004], and [Hoffstein, 2008].

3.3 Summary

Symmetric cryptography is very fast and very secure. Even quantum computers aren't expected to completely crack the symmetric ciphers. However, symmetric cryptography isn't useful if there is no easy way to share the key. That's where asymmetric cryptography comes into play. The cryptography tree from Chapter 1 and Chapter 2 is further expanded to include asymmetric cryptography. The cryptography tree shown in Figure 3.6 summarizes various types of asymmetric cryptography we covered in this chapter.

The first part of the chapter introduced asymmetric cryptography, including some mathematics. We then focused on the three primary techniques of asymmetric cryptography—integer factorization, discrete logarithm, and elliptic curve. A major part of the chapter dealt with the mathematics and implementation of the most popular asymmetric ciphers—RSA, Diffie–Hellman–Merkle, Elgamal encryption, and ECC.

> **Key Takeaways from This Chapter**
> - Reviewed fundamentals of asymmetric cryptography and its properties
> - Studied three primary techniques of asymmetric cryptography
> - Learned the implementation of popular asymmetric cryptosystems

3.3 Summary

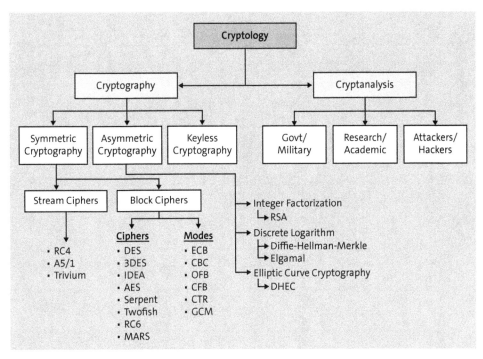

Figure 3.6 Types of Cryptography

In the next chapter, we'll explore other uses of cryptography—hash functions, digital signatures, and message authentication codes (MACs). As you're about to turn the page, here is an incentive for you: we'll have less mathematics going forward!

Chapter 4
Cryptography Services

The primary goal of cryptography is to provide confidentiality using encryption. Modern cryptography extends the use of cryptography to achieve much more than just confidentiality. Techniques such as hash functions, message authentication codes, digital signatures, and Merkle trees take the use of modern cryptography one step further. This chapter focuses on these techniques.

Cryptography was invented and used with one simple goal in mind: to allow only the intended recipient to read the message and hide it from everyone else. This was true 2,000 years ago, and it's true even today. The primary purpose of cryptography is confidentiality, that is, to conceal information from unauthorized users. As we've seen in earlier chapters, we have various techniques, such as symmetric encryption and asymmetric encryption, that provide confidentiality to the message or data. However, modern cryptography not only offers confidentiality but also offers integrity, authenticity, and nonrepudiation. These byproducts of cryptography are also known as *cryptography services* or *security services of cryptography*.

> **Chapter Highlights**
> - Recapping cryptography services: integrity, authenticity, and nonrepudiation
> - Understanding and implementing hash functions, message authentication codes, digital signatures, and Merkle trees
> - Architecture and implementation of commonly used algorithms such as message digest (MD), secure hash algorithm (SHA), and digital signatures

Confidentiality is keeping data secret from unauthorized users. We already learned several algorithms to implement confidentiality in Chapter 2 and Chapter 3. In this chapter, we'll cover the other three cryptography services: integrity, authenticity, and nonrepudiation.

The chapter is divided into four parts based on the topics it covers. Section 4.1 introduces hash functions, which provide integrity. You'll learn what a hash function is, explore its properties, and discuss key usages. We'll cover two major families of hash algorithms: MD and SHA. We'll dive deeper into MD5 and SHA family algorithms.

4 Cryptography Services

In Section 4.2, we'll present the concept of a message authentication code (MAC). We'll discuss how it's different from hash functions and digital signatures. You'll learn about implementing MAC algorithms.

In Section 4.3, we'll focus on the digital signature. Digital signature is a very important cryptography service. It's heavily used for authenticity and for nonrepudiation. You'll learn how a digital signature works as well as the architecture and implementation of the digital signature. The digital signature algorithms will be discussed in detail.

The concept of a Merkle tree and how it's used to digitally sign a large message is explained in Section 4.4.

4.1 Hash Functions and Algorithms

Hash functions are mathematical functions that convert an arbitrary length of data (bit string) into a short, nonreversible, and fixed-length bit string. The fixed-length output is also known as a hex value, a digest, or simply a *hex*. Hash functions are also known as auxiliary functions in cryptography because they offer a lot more when used with digital signatures, MAC, and other techniques than what they can offer on their own. Hash functions have no key, so they are also known as keyless cryptography.

From our security triad—confidentiality, integrity, and availability—we can say that confidentiality and integrity form a major part of data protection. Confidentiality is provided using encryption, and integrity is provided using the hash functions. As we've seen so far, traditional cryptography provides confidentiality but won't help if the message is changed or deleted in transit. The recipient has no idea if the message decrypted is the original message. That's where message integrity comes into play. Hash functions are used with other mechanisms to ensure message integrity using authenticity and nonrepudiation.

4.1.1 Primer on Cryptographic Hash Functions

In this section, we'll discuss the core requirements or properties of hash functions, some of the most popular applications of hash functions, and a brief overview of types of hash functions.

Properties of Hash Functions

What makes a hash function a cryptographic hash function? When a hash function ensures that it meets the following core cryptographic requirements or properties, as shown in Figure 4.1, it becomes a cryptographic hash function:

- **Arbitrary length input**
  The hash function must be able to take arbitrary length input. This is necessary when transmitting a long message or document. If the hash function restricts the

input size, then long messages must be divided into several chunks, and hash values must be calculated for each chunk. These values are then concatenated into a series of chunks and respective hash values. This slows down the process and introduces weakness. An attacker can replace one block with another, change the order, or even remove the block, and the recipient may not be able to find out. To avoid these weaknesses, cryptographic hash functions must accept arbitrary length input.

- **Fixed, short length**
  Based on the definition of the hash function; another obvious requirement is the ability of the hash function to generate a short but fixed length digest or a hash. The length of the hash function is a tricky decision. A very short length makes the function vulnerable to brute-force attacks, while longer lengths make the functions slow and inefficient.

- **Same hash value**
  The hash functions produce the same value every time for the same input message or data. The output hash value doesn't change as long as the input to the hash function doesn't change.

- **Efficiency**
  The hash functions are auxiliary functions, so it's important that they are very efficient. If hash functions need lots of computational power or time, then they aren't very useful.

- **One-way**
  All encrypted messages can be decrypted as long as you have the authorized keys. However, hash functions must not be "de-hashed." Once you generate the hash, there shouldn't be any way to get the original message back from the hash value. This one-way or irreversible property is also known as *preimage resistance*. This property sets the hash function apart from the other cryptographic mechanisms. The fact that the receiver can't get the original message from the hash function makes it useless for security purposes, but a good candidate for integrity checks.

- **Weak collision resistance**
  If two input messages generate the same hash, value then the hash function has a collision. The collision is considered weak if, when given a hash value, an attacker finds another input message that generates the same hash value. For example, if the attacker knows the message m_1 and the hash value of m_1, the attacker shouldn't be able to computationally find another message m_2 such that the hash value of m_1 is equal to the hash value of m_2. If an attacker does find the same hash values for two messages, then he can replace the original message, and the recipient may not know about the replacement. This property is also known as *second preimage resistance*.

- **Collision resistance**
  This is very similar to the weak collision resistance with a small deviation. In this case, the attacker is free to choose both messages m_1 and m_2. The attacker must trick the sender into signing the first message. This attack is a bit difficult to launch since

4 Cryptography Services

the attacker has some extra work to do. The hash function must be strong enough to withstand the collision resistance attack.

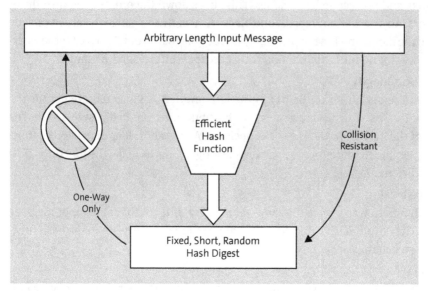

Figure 4.1 Properties of Cryptographic Hash Functions

Birthday Paradox

What are the chances that two people in a room have the same birthday? If the first person has a birthday of January 1st, then, theoretically, the 365th person could have the same birthday again. In the real world, however, the next person could also have the birth date of January 1st, or we may have to wait for the 1,317th person to get the same date of January 1st. But theoretically, the probability is 1 in 365. Now let's assume that we don't need to meet the exact date. We just want to find two people with the same date of birth. If the first person has the date of January 1st, and the next person who comes across has February 1st as the date of birth, now the third person has double the chances—could match with January 1st or February 1st—and the third person has double the probability. If we keep going like this, then there is a 50% probability of having 2 people in every 23 who have the same birthdays. This is called the birthday paradox.

How does the birthday paradox relate to the hash function? It explains the probability of collision. If the hash value is n-bit, then the collision can be found in 2^n brute-force attempts.

Checksum versus Cryptographic Hash Function

In my first job as a software developer, we developed embedded software for various microprocessor-based applications and then flashed the software into the controller.

> The final software package that we flashed into the controller had a cyclic redundancy check (CRC) checksum. After flashing the controller, our first task was to check the checksum in the controller and make sure it's the same as it was before we flashed the software into the controller. This is to confirm that the software wasn't accidentally changed or corrupted while being flashed.
>
> At the core, the concept of the hash function and checksum is the same. Both are mathematically computed, based on the input message, and used to check the integrity of the data. However, the checksum or noncryptographic hash functions are very short (typically 32 bits), so there is a high probability of collision and can be cracked with brute-force attack. The purpose is different, though. The checksums are used to see the accidental change in the software or the file. They aren't intended to verify the data integrity against malicious attackers.

Common Applications of Hash Functions

Hash functions are common, and most of us interact with a hash function almost every day without realizing it. Following are some examples of where and how hash functions can be used:

- **Password verification**

  Hash functions are used to store the password securely on servers. When users create an account, they choose a password. This password must be stored on the server and compared every time the user logs in. Obviously, it's unsafe to store the password in plain text because if the server is hacked, all passwords are up for grabs or even a malicious insider can see everyone's password. Another option is to encrypt the passwords. Encryption adds some computation time and makes the login process slow. Moreover, using this method adds difficulty in managing and/or securely storing the encryption key. If the encryption key is compromised, then all passwords are compromised. Encryption becomes a single point of failure.

  Hashing the password solves all of these problems. When a user creates a password for the first time, the password is hashed and stored on the server. Subsequently, every time the user logs in, the password is hashed and compared with the stored hashed value. Even if the server is hacked or some malicious insider logs in to the server, passwords aren't compromised because they are hashed. There is no way for the hacker to retrieve the original value from the hash value thanks to the one-way property of the hash function. Additionally, thanks to the collision resistance property of hash functions, no two passwords will have the exact same hash values. There is no chance of mixing the passwords.

  The only potential problem of using hash values to store passwords is when two users create the same passwords. Because the passwords are the same for two users, the hash values must be the same. This problem is easily handled through salting and peppering techniques.

4   Cryptography Services

> **Adding Spices to Passwords**
>
> *Salting* is a process of adding random bits to the password. The password and salt are hashed and stored in the database. *Peppering*, on the other hand, is a secret word added to the password at the time of hashing, but it's not stored with the password. It's stored, usually, in the configuration file or other system files.
>
> Because the secret value isn't stored in the same database as the password, even if the password database is compromised, it's hard to launch an attack as the secret value is missing.

- **Virus detection**
  Another popular use of the hash function is in virus detection. When the file or data is stored on the server, the file could be infected by viruses. This could change or damage the file. However, when the user tries to access or download the file next time, he has no idea about this. This situation can be solved using hash functions. The file or data can be passed through the hash functions, and the hash value is computed while storing the file. The user can download or access the file later and again run through the hash functions. The user can then compare the hash values. If the hash values aren't the same, then the file is infected or changed. Otherwise, the file is safe to use.

- **Message authentication code (MAC)**
  The MAC authenticates the message by using the hash function. The hash functions used in MAC are based on symmetric encryption algorithms, and they use an encryption key to verify or authenticate the message. MACs and their implementation are discussed in detail in Section 4.2.

- **Digital signature**
  Digital signatures are one of the most important uses of the hash functions, providing nonrepudiation. The implementation and usage of digital signature is discussed in Section 4.3.

Hash Function Algorithms

The concept of hash functions isn't new. The idea of using hash functions for data science was first floated in the 1950s by Hans Luhn [Stevens, 2018]. The initial goal was compressing large data files to save memory space. Noncryptographic hash functions have been in use since the early 1960s. The use of cryptographic hash functions was proposed by Michael Rabin in 1978 [Rabin, 1978].

Hash algorithms are basically mathematical compression functions that take in an arbitrary length message and generate a fixed-length, random output. In practice, the arbitrary input string is divided into a predetermined block size. Each block is then passed through the compression function. Depending on the algorithm, the process goes through several iterations. After each iteration, the values are passed on to the next iteration using chaining. At the end, a fixed, compressed hash digest or a hash is generated. This iterative process to generate the hash value was first proposed independently by Ralph Merkle [Merkle, 1990] and Ivan Damgård [Damgård, 1989] in 1989. However, the process is credited to both and known as the *Merkle-Damgård construction*.

There are two popular types of hash function algorithms or construction in practice [Pattanayak, 2012]:

- **Block cipher-based hashed functions**
  It turns out that the process of dividing input messages into smaller length blocks, processing iteratively, and using chaining to generate the final value is very similar to how block ciphers work. Block ciphers such as Data Encryption Standard (DES) and Advanced Encryption Standard (AES) can be used to construct hash functions. The advantage of this approach is a high level of security, and, for the most part, existing work can be reused. However, on the flip side, the construction isn't very efficient compared to the dedicated ciphers. Another disadvantage is that it creates a single point of failure. Anything that goes wrong with the underlying block cipher can bring down both confidentiality and integrity.

- **Dedicated hash functions**
  Dedicated hash functions also use the block cipher concept under the hood; however, these block ciphers are designed with hash functionality in mind. As a result, they are a lot more efficient compared to block cipher-based hash functions. Over the past 30 years, many dedicated hash function algorithms have been proposed. All the popular hash functions, historical as well as one in use today, are dedicated hash functions. Some examples of dedicated hash functions are MD4, MD5, SHA-1, SHA-2, and SHA-3. We'll discuss MD5 and SHA-3 algorithms in the next sections.

4.1.2 Message Digest Algorithms

Ron Rivest developed the message digest (MD) family of hashing algorithms. The first in the series, MD2, was introduced in 1989 [Linn, 1989] [Kaliski, 1992]. MD2 takes an arbitrary length input and generates a 128-bit hash. The MD2 algorithm wasn't a huge commercial success.

Ron Rivest was inspired by the Merkle-Damgård proposal in 1989. Rivest used Merkle-Damgård's principle and proposed the MD4 hash algorithm in 1990 [Rivest, 1990]. It produces a 128-bit digest. MD4 uses 512 block size and computes the hash or a digest in three rounds. Each round has 16 iterations. The algorithm was robust, but soon after the release, attacks on MD4 were published. Although at the time, none of the attacks were able to completely crack the MD4 hash function algorithm, Rivest enhanced the algorithm.

Rivest proposed the strengthened algorithm, MD5, in 1991 [Den Boer, 1991] [Dobbertin, 1998]. The MD5 hash algorithm isn't used much today, but it has a lot of historical importance, like DES, and it forms a good base in understanding SHA. We'll discuss the SHA series of algorithms in Section 4.1.3.

MD5, like MD4, also takes 512-bit block size and produces a 128-bit hash value, but it increases security slightly with an additional round: MD5 has four rounds compared to MD4, which has three rounds. MD5 is basically the MD4 algorithm with an extra turn of the screw. The functionality of MD5 can be summarized in the following steps:

1. **Padding bits**
   MD5 can accept messages of any length, but it processes 512 bits at a time. This 512-bit chunk is called a *block*. The first step in the hashing process using MD5 is to make sure that the length is 64 bits short of 512 bits. For example, if the message is 400 bits long, then we must pad 48 bits (512 – 64 = 448).

   The padding starts with the first bit set to 1 and the rest of the bits filled with 0s. In our example:

   [message 400 bits] + [padding of 48 bits = 1 followed by 47 0s] = Total of 448 bits

2. **Padding the length**
   In this step, the goal is to make sure that the message length is a multiple of 512. Simply add 64 bits to the message padded in step 1.

3. **Performing the main loop**
   In this step, registers A, B, C, and D are initialized, and the actual process begins. These registers function as buffers to hold the data during the processing. Each 512-bit block goes through four rounds, and each round has 16 iterations. The final values are stored in registers A, B, C, and D. These stored values are used in chaining.

4. **Processing multiple blocks**
   The chaining registers store the final value of the previous rounds and are used as the initial value for the next block to be processed. After processing all the blocks, the final 128-bit hash value is stored in registers A, B, C, and D.

Figure 4.2 shows the high-level processing of the message. The arbitrary length message is processed 512 bits at a time. Each 512-bit block is passed through the MD5 hash function. The output of the first block is provided to the next block as an initial value.

Next, let's zoom into the H_{MD5} (512-bit) function. In the MD5 hash function, a 512-bit message block is processed through four rounds. The 512-bit message is divided into 16 words of 32 bits each (16 × 32 = 512). Each round processes the four words (32 × 4 = 128) of the message along with a constant and a dedicated nonlinear function for that round. The first round receives registers A, B, C, and D values with the initialized values. Subsequent rounds receive the values from the previous rounds. Figure 4.3 shows how the 512-bit block is processed by the four rounds. At the end of the fourth round, the output of registers A, B, C, and D are added using modulo addition with the original values (initialized values for the first block and output of the previous block for other blocks). The output of these modulo additions is fed to the next block. If this is the last block, these values are stored as the 128-bit hash value.

4.1 Hash Functions and Algorithms

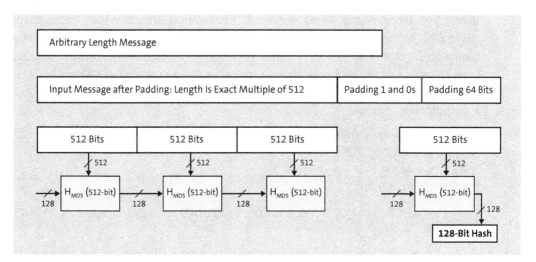

Figure 4.2 Overview of the MD5 Hash Function

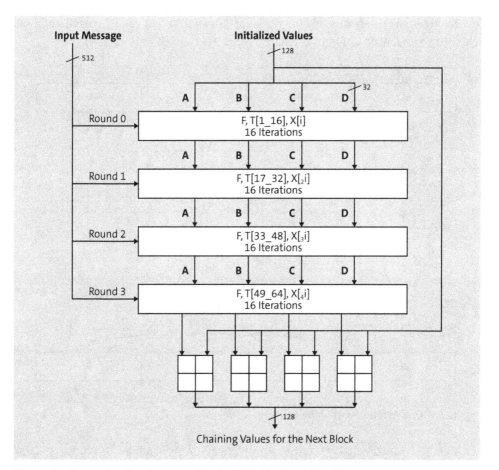

Figure 4.3 Processing One Message Block within the MD5 Algorithm

Now, let's further zoom in. What happens in each round? In each round, one fourth of the block (4 of the 16 words or 128 bits of the 512 bits) performs the computation with a nonlinear function and a given constant T. The function is computed using the values from three out of the four registers and stored in the fourth register. There are four functions, F, G, H, and I, and a different function is performed each round. The four functions are as follows:

$F(B,C,D) = (B \text{ \&\& } C) \mathbin{||} (\text{-}B \text{ \&\& } D)$
$G(B,C,D) = (B \text{ \&\& } D) \mathbin{||} (C \text{ \&\& } \text{-}D)$
$H(B,C,D) = B \oplus C \oplus D$
$I(B,C,D) = C \oplus (B \mathbin{||} \text{-}D)$

Here, $\oplus$ = XOR, && = AND, || = OR, and - represents NOT or negation. B, C, and D are 32-bit words.

Figure 4.4 shows the operation of 1 iteration in one round. In each round, there are 16 such iterations. For each iteration, the function, the part of the message (four words), and the values for A, B, C, and D received from the initialization or from the previous round all remain the same. Only the constant T has a different value for each iteration. There are a total of 64 different constant values (1 for each iteration) used.

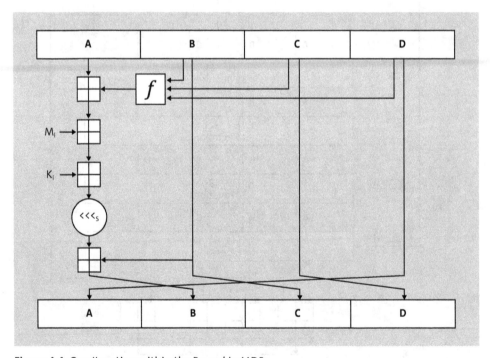

Figure 4.4 One Iteration within the Round in MD5

An MD5 hash function has 2^{64} collision-resistance. The probability of finding a collision of a hash function is 50% or half of the total number of output bits. For a 128-bit hash value a birthday attack needs to try 2^{64} input values before it finds a collision.

MD5 was one of the most popular hash functions in the 1990s and was very widely used in the industry. MD5 started showing some weakness from the early days [Dougherty, 2008], however, and the US National Institute and Standard Technology (NIST) started working on standardizing the hashing algorithm.

Although the design of MD2, MD4, and MD5 was very similar, MD2 was optimized for 8-bit processors, while MD4 and MD5 were optimized for 32-bit processors to keep pace with the changing technology.

In 2008, Rivest and others proposed another version of MD algorithms—the MD6 hash algorithm. MD6 can produce a variable length hash value or a digest up to 512 bits. The algorithm is based on the Merkle tree structure [Merkle, 1988], explained in Section 4.4. The MD6 algorithm was submitted to NIST as part of the SHA-3 selection competition but wasn't selected.

4.1.3 SHA Family of Algorithms

In the 1990s, the MD5 algorithm was gaining lots of momentum and was starting to be used in many applications. However, the MD5 algorithm was under scrutiny by researchers and industry leaders. Given the potential weaknesses of MD5, NIST developed another standard under the name of *Secure Hash Algorithm (SHA)* and published the initial version in 1993 with a 160-bit hash digest. NIST found some vulnerabilities in SHA and soon revised and released the modified version, SHA-1, in 1995. After this point, the original SHA was also referred to as SHA-0.

SHA-1 also has a 160-bit digest or a hash value and was quickly adopted widely in the industry. With published attacks on MD5 and with SHA-1 basically an extension or an improved and enhanced version of MD5, in 2001, NIST released several variants of SHA-1 with larger hash values or digest sizes: SHA-256, SHA-382, SHA-512. In 2004, NIST added another variant, SHA-224. Later in 2012, SHA-512/224 and SAH-512/256 were added to the list of SHA-1 variants. These SHA-1 variants—SHA-224, SHA-256, SHA-384, SHA-512, SHA-512/224, and SHA-512/384—are collectively known as SHA-2. SHA-1 has been deprecated, but SHA-2 is still the official version and is widely in use by the US government and in the industry.

All of these hash algorithms are still considered "MD4 family" algorithms because they all are built on the same principle: Merkle-Damgård construction. Additionally, all versions of SHA are designed by NIST.

In 2012, NIST released another version, SHA-3, which is neither built on Merkle-Damgård construction nor built by NIST. We'll cover SHA-3 in Section 4.1.4.

In this section, we'll explain the implementation of SHA-512 because learning this algorithm will help you understand all other SHA-2 variants. SHA-512 is an enhanced version of SHA. It takes arbitrary length input and generates a 512-bit hash or a digest. It's based on the Merkle-Damgård principle, and it's very similar to MD5 and SHA-1 algorithms in construction except for some minor differences to accommodate the larger hash size. At a high level, MD5 described in the previous section and SHA-512 have the following differences:

- SHA-512 operates on eight 64-bit words instead of four 32-bit words as in the case of MD5.
- To process one 512-bit message block, the MD5 algorithm goes through 4 rounds with 16 iterations in each round, while SHA-512 goes through 80 rounds to process the one input message block of size 1,024.
- SHA-512 uses the same equation in all rounds. On the other hand, MD5 has four different equations, and only one of these four equations is used in every round.
- SHA-512 expands the message block to create 80 64-bit words.

Let's dive deeper into the SHA-512 algorithm. An overview is shown in Figure 4.5, which you can refer to throughout this explanation.

SHA-512 takes any arbitrary length message but then divides the input message into blocks of 1,024 bits. The total length must be an exact multiple of 1,024. But messages aren't always an exact multiple of 1,024 bits in length, so this requires padding. The last 128 bits of the last 1,024 block of the message are reserved to indicate the length of the message. The remaining 896 bits are padded with first 1 and the rest 0s. Let's break down an example where the message has 1,783 bits:

- The closest multiple to 1,783 of 1,024 is 2,048.
- You then subtract the last 128 reserved bits: 2,048 – 128 is 1,920. The message is 1,783, so it's 137 bits short of 1,920.
- The message must be padded with 1 and 136 0s to bring the total number of bits to 1,920.
- The last 128 bits indicate the length of the message, which, in this example is 1,783. In the message, this 1,783 decimal will be in binary 0000 0110 1111 0111 (06F7 in hexadecimal). All the remaining bits will be filled with 0s.
- This 1,024-bit block of messages along with the eight 64-bit words are fed to the compression function. For the first block, the eight 64-bit words are fed the initialized values. These initialization values are also called initial hash values. The initial hash values for SHA-512 algorithms are as follows [FIPS 180-4, 2015]:

    $A = 6a09e667f3bcc908$
    $B = bb67ae8584caa73b$
    $C = 3c6ef372fe94f82b$
    $D = a54ff53a5f1d36f1$
    $E = 510e527fade682d1$
    $F = 9b05688c2b3e6c1f$
    $G = 1f83d9abfb41bd6b$
    $H = 5be0cd19137e2179$

These words are used as a buffer throughout the process and are designated by A, B, C, D, E, F, G, and H. These words, or buffers, sometimes also referred to as variables, work as chaining variables between the two input message blocks when the message is longer than 1,024 bits.

- The output of the function is added with the same eight 64-bit words, A through H. The output of the first function is added to the initialized values, while the output of the rest of the functions is added to the output of the previous function. This addition is performed using modular addition, 2^{64}.

- The function for each input message block of 1,024-bit goes through 80 rounds. Each round takes three inputs: a 64-bit word from the input message, eight 64-bit buffers (A through H), and a constant K. Each round gets a different constant value for K, from K_0 to K_{79}. The output of each round is stored in the buffer variables and passed to the next round.

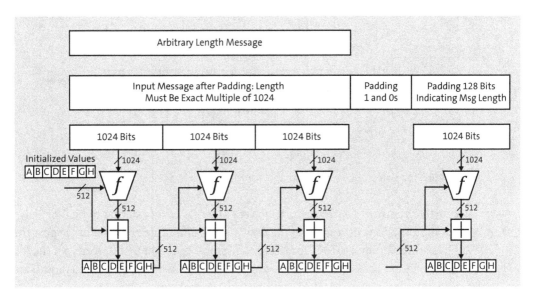

Figure 4.5 Overview of the SHA-512 Algorithm

- The input message block is only 1,024 bits and can feed only the first 16 words, W_0 to W_{15}, for the first 16 rounds out of 80 rounds. The function in the SHA-512 algorithm expands the 1,024-bit message block into 80 64-bit words. This process is very similar to the key expansion/scheduling logic in the symmetric block cipher algorithms (see Chapter 2, Section 2.2.4). The equation used to expand the input message, M, to 80 64-bit words, W, is given as follows [FIPS 180-4, 2015]:

for $0 \leq t \leq 15$, $\{W_t\} = M_t$
for $16 \leq t \leq 79$, $\{W_t\} = \sigma_1^{(512)} (W_{t-2}) + (W_{t-7}) + \sigma_0^{(512)} (W_{t-15}) + (W_{t-16})$

Figure 4.6 shows how rounds are processed in each function. The output of the last round, round 80, is added to the initial buffer values of the round (either the initialized values or the values from the output of the previous round) using modular addition, 2^{64}. The output of each function is stored in the buffers A through H.

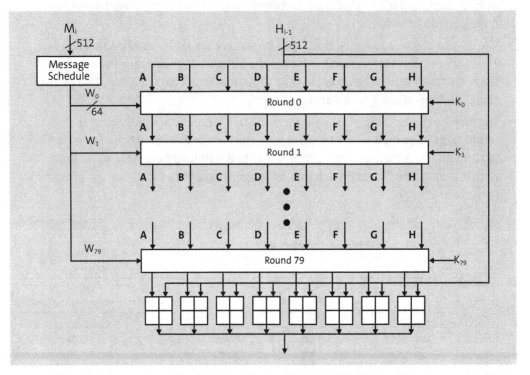

Figure 4.6 Processing of One Message Block within the SHA-512 Algorithm

If we further zoom into each round, the overall process is very similar to MD5 or SHA-1 with some variations necessary to make this algorithm more robust and support the 512-bit hash. Each round, 0 to 79, processes and stores the eight 64-bit words per the following equations. Look at Figure 4.7 as you go through the set of equations. T_1 and T_2 are the two dotted line circles, and A through H are the registers.

$A = T_1 + T_2$
$B = A$
$C = D$
$D = C$
$E = D + T_1$
$F = E$
$G = F$
$H = G$

where:

$T_1 = H + \Sigma_1^{512}(E) + Ch(E, F, G) + K_t^{512} + W_t$

where:

$Ch(E,F,G) = (E \;\&\&\; F) \oplus (\neg E \;\&\&\; G)$

$\Sigma_1^{512}(E) = ROTR^{14}(E) + ROTR^{18}(E) + ROTR^{41}(E)$

$T_2 = \Sigma_1^{512}(E) + Maj(A, B, C)$

where:

$Maj(A, B, C) = (A \;\&\&\; B) \oplus (A \;\&\&\; C) \oplus (B \;\&\&\; C)$

$\Sigma_1^{512}(E) = ROTR^{28}(E) + ROTR^{34}(E) + ROTR^{39}(E)$

ROTR is a rotate right or circular shift right on the 64-bit word. The superscript number on the ROTR indicates the number of shifts.

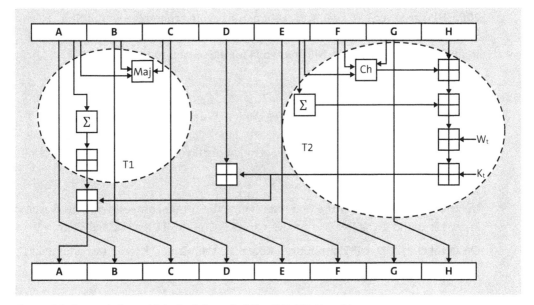

Figure 4.7 Computation within Each Round of the SHA-512 Algorithm

The functionality of each round is shown in Figure 4.7. The dotted lines show the computation of T_1 and T_2. The results of these computations or each round are stored again in the A through H buffers. The buffers pass the values on to the next round for the calculations, and the whole process repeats with a new 64-bit word and new constant K. The process continues for 80 rounds for each 1,024-bit message block. After the last block of the message is processed, these buffers will have the 512-bit hash value or a digest.

4.1.4 SHA-3 Algorithm

SHA-3 is the latest hash algorithm released by NIST. There were a couple of compelling reasons for NIST to start thinking about the next SHA version. First, SHA-0 was already vulnerable, and SHA-1 started showing signs of weakness; a theoretical paper [Wang, 2005] was presented in 2004 showing that SHA-0 could be vulnerable, and, in 2005, it was theorized that the same attack could be extended on SHA-1. Second, MD5, SHA-0, SHA-1, and SHA-2 were all based on the Merkle-Damgård construction. If MD5, SHA-0, and SHA-1 can be vulnerable, there is a possibility that SHA-2 will also eventually be vulnerable because it's also based on the same principle. Lastly, SHA-0, SHA-1, and SHA-2 were all designed by NIST, which leads to concerns about privacy and the potential overreach of the federal government.

NIST decided to organize a competition like an open call for design ideas. In late 2007, NIST invited researchers and the crypto community to submit their design ideas to be the next SHA standard—SHA-3. By the close of the submission deadline in October 2008, NIST received a whopping 64 submissions from around the world! NIST selected 51 of these 64 submissions for review in the first round. After one year of review and input from the community, NIST picked 14 for the second round in 2009.

> **Then and Now**
>
> In 1998, NIST organized a similar competition for AES, as we discussed in Chapter 2, Section 2.2.4. NIST received only around 20 submissions compared to 64 in 2008. This shows the advancement and awareness of information security and particularly cryptography in just 10 years!

In 2010, NIST announced the five finalists for the SHA-3 competition. This was also known as round 3. The five finalists were BLAKE, Grøstle, JH, Keccak, and Skein.

On October 2, 2012, NIST announced Keccak as the winner. Keccak was adopted as a SHA-3 standard. In this section, we'll learn how to implement SHA-3.

Overview

Keccak was designed by Guido Bertoni, Joan Daemen, Michael Peeters, and Gilles Van Assche [Bertoni, 2012]. The Keccak algorithm can do a lot more than hashing. It can be used for random number generators (RNGs) and also as a stream cipher. However, Keccak was adopted for SHA-3 as a standard only for the hashing functionality, in a way that users can replace SHA-2 with SHA-3 without any major system changes.

Unlike previous hash algorithms, Keccak has a totally different internal structure. As mentioned previously, it's not based on the Merkle-Damgård principle; it's based on *sponge construction* and uses absorb and squeezing mechanisms. It absorbs data like a sponge absorbs water, and it releases data like a sponge releases water when squeezed. The sponge analogy indicates that the Keccak algorithm can process a lot more data than any other hash algorithm we've discussed.

Before understating the details of the algorithm, we must understand the parameters needed to define Keccak. The algorithm uses two important parameters, the number of bits to be absorbed, denoted by b, and the number of rounds needed to process the bits, denoted by n_r. The total numbers of bits, b, that can be absorb by the algorithm is calculated as follows:

$b = 25 \times 2^l$

Where l = 0,1,2,3,4,5,6. For SHA-3, l must be 6:

$b = 25 \times (2^6) = 25 \times 64 = 1,600$

b is comprised of r and c, as follows:

$b = r + c$

Here, r is the bit rate and defines the block length of the input message, and c is the capacity. The value of c is governed by the hash or the digest of SHA-3, as follows (where h is the hash or digest):

$c = 2 \times h$

For example, if the SHA-3 output (i.e., the hash or the digest) needs to be 256-bit, then the values of c must be 2 × 256 = 512-bit. That also gives us the value for r. In this example, r = 1,600 − 512. The block size for the input message, r, will be 1,088 bits.

The SHA-3 functionality of the Keccak algorithm only processes the r bits or the data equal to the input message block size. Table 4.1 shows the various values for r and c based on the digest values. The digest values used are the same as in SHA-2.

b Bits (Width or State)	r Bits	c Bits	Hash or Digest
1,600	1,152	448	224
1,600	1,088	512	256
1,600	832	768	384
1,600	576	1,024	512

Table 4.1 r and c Values for Various SHA-3 Hash Outputs

Lastly, we need to define n_r, as follows:

$n_r = 12 + 2 \times l$ (in case of SHA-3, l is always 6)
$n_r = 12 + 2 \times 6 = 24$

For SHA-3, there will be 24 rounds.

At a high level, the Keccak algorithm is divided into two functional areas: preprocessing and inner Keccak. *Preprocessing* is the functional part of the algorithm that prepares the input message for the actual hashing mechanism. While the *inner Keccak* functionality performs the actual sponge operation—absorbing and squeezing the data.

In preprocessing, the message is divided into the input block size and padded. The input block sizes are equal to the bit rate r. For example, if the hash output value is

512 bits, then the c value will be 512 × 2 = 1,024, and the r or the block size should be 576 (1,600 − 1,024 = 576). The number of blocks must be an exact multiple of 576. If that isn't the case, the last block is padded; however, padding in Keccak works a bit differently than the other algorithms. In Keccak, the first bit and the last bit are padded with 1, and everything between these two 1s are padded with 0s. It's denoted as *10*1*, where *0** represents the number of 0s between two 1s. These padding bits are appended to the predetermined bit string P. P is 01 for SHA-2-compatible or replacement hash values of SHA-3.

The top part of Figure 4.8 shows the high-level construction of Keccak. The fixed length b bits, also called the width of function f, are fed to function f. The b bits are first divided into r and c. The r bits are XORed with the input message block. The output of XOR and c bits are fed to the function. The output of the function again goes through the same process with the next input message block. The process continues until the last block of the message. This part of the process is an absorbing process.

After the last block is ingested, the function starts putting out the bits, and that is where the squeeze process starts. The first output is taken as the hash value or a digest for SHA-3. The remaining outputs aren't used for SHA-3 implementation. Going back to our example of 512-bit hash output, the SHA-3 squeezes out the first 512 bits as the hash value. However, the b or the total fixed length is 1,600 bits, and the remaining bits are squeezed out in the subsequent outputs, but not used by Keccak when used as a SHA-3.

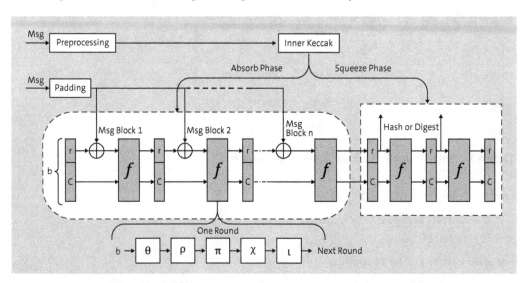

Figure 4.8 SHA-3 (Keccak) Overview

Implementation

The first values of b are all set to 0s. These can be compared to the initialized values of registers (or buffers) A through H in SHA-2. These initialized r values are then XORed with the first block of the message. The output of the function sets the computed values in r. For subsequent blocks, r takes the values from the previous functions.

Function f is at the heart of the functionality of Keccak. f processes the 1,600 bits as an array rather than a string. To understand this concept, we'll have to think of b as a three-dimensional variable. The 1,600 bits are arranged in a 5×5×64 array. For SHA-3, the x-axis and y-axis coordinates are 5 each, while the z-axis is 64. The values along the x-axis are known as *rows*, the values along the y-axis are known as *columns*, and the values along the z-axis are known as *lanes*.

The string is converted by first placing all bits along the z-axis for each (x, y) coordinate. The entire string, b, of 1,600 bits is converted into state array A. The function uses this array to perform the permutation and substitution in five stages. These stages are θ (theta), ρ (rho), π (pi), χ (chi), and ι (iota). The round variable, n_r, for SHA-3, is 24. That means the functions run through 24 times or rounds for each input message block. For each round, the function computes θ (theta), ρ (rho), π (pi), χ (chi), and ι (iota). Figure 4.9 shows the workings of various states in each round.

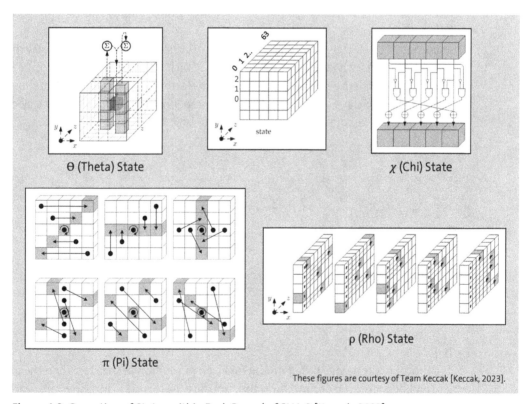

Figure 4.9 Operation of States within Each Round of SHA-3 [Keccak, 2023]

Let's take a closer look at each:

- **θ (Theta)**

  Theta performs the computation in three steps: the C function, the D function, and storing the final value. Let's walk through the mathematical calculation for each step.

  For x between 0 and 4 and z between 0 and 63, $C[x, z]$ can be computed as follows:

  $C[x, z] = A[x, 0, z] \oplus A[x, 1, z] \oplus A[x, 2, z] \oplus A[x, 3, z] \oplus A[x, 4, z]$

  The C function simply calculates the values of each of the five columns ($y = 0$ through 4 for $x = 0$ to 4 and $z = 0$ to 63) and performs the XOR operation on the columns. The D function performs the rotation by 1 bit along the z-axis:

  $D[x, z] = C[(x - 1) \bmod 5, z] \oplus C[(x + 1) \bmod 5, (z - 1) \bmod w]$

  Function C performs the bitwise XOR operation along the columns or y-axis, while D rotates these values by 1 bit along the x-axis.

  Finally, the θ step stores the computed values back in array A', as follows:

  $A'[x, y, z] = A[x, y, z] \oplus D[x, z]$

- **ρ (Rho)**

  The output of the θ step is used as an input for this step. This step is very easy. It rotates the bits in each lane by a fixed number of bits. This operation is performed for all 25 lanes (from the z-axis). The rotation is performed over *mod64*. The rotation is different for each lane and is predefined. The predefined rotation values are called rotation offsets. The rotation offsets are given in Table 4.2.

	x = 3	x = 4	x = 0	x = 1	x = 2
y = 2	25	39	3	10	43
y = 1	55	20	36	44	6
y = 0	28	27	0	1	62
y = 4	56	14	18	2	61
y = 3	21	8	41	45	15

Table 4.2 Rotation Offsets for the ρ Steps

- **π (Pi)**

  If the rho step was named based on the function of rotation in the step, then the pi step is named after the permutation functionality. In this step, the lanes are rearranged based on the following mathematical function:

  $A'[x, y, z] = A[(x + 3y) \bmod 5, x, z]$

  Here, A is an input state array (output of the ρ step), and A' is the output of the π step.

- χ (Chi)

  The output array from the previous step becomes the input for this step. This function operates on the lanes, that is, on the z-axis. It takes a lane from one location and XOR with the output of the logical AND of the other two lanes. One input of the AND is inverted. Refer to the sectional view of the process shown in Figure 4.9.

  Mathematically speaking, this looks like the following:

  $A[x,y] = A[x,y] \oplus ((\neg A[x + 1,y])$ && $A[x + 2,y])$

- ι (Iota)

  This is the easiest step of all five steps. This step adds a predefined constant to a lane at a specific location by XORing the constant to that bit. There is one constant for each round, and the constant changes with each round. There are 24 rounds in SHA-3, and the constants for each round are listed in Table 4.3.

RC[0]	0x0000000000000001	RC[8]	0x000000000000008A	RC[16]	0x8000000000008002
RC[1]	0x0000000000008082	RC[9]	0x0000000000000088	RC[17]	0x8000000000000080
RC[2]	0x800000000000808A	RC[10]	0x0000000080008009	RC[18]	0x000000000000800A
RC[3]	0x8000000080008000	RC[11]	0x000000008000000A	RC[19]	0x800000008000000A
RC[4]	0x000000000000808B	RC[12]	0x000000008000808B	RC[20]	0x8000000080008081
RC[5]	0x0000000080000001	RC[13]	0x800000000000008B	RC[21]	0x8000000000008080
RC[6]	0x8000000080008081	RC[14]	0x8000000000008089	RC[22]	0x0000000080000001
RC[7]	0x8000000000008009	RC[15]	0x8000000000008003	RC[23]	0x8000000080008008

Table 4.3 Round Constants in Hexadecimal Notation for $n_r = 24$

The function goes through these five steps, from θ to ι in each round. SHA-3 performs 24 such rounds for each input message block. The most significant bits (MSBs) from the *r* values of the first output are used as the hash or the digest value. If Keccak is used as a keystream generator in the stream cipher or as an RNG, then the remaining outputs are used.

One of the major advantages of the Keccak is its adoptability in implementations. The SHA-3 is suitable for software as well as hardware implementations. A complete overview of SHA-3 implementation is given by the inventors [Bertoni, 2012] (for another good reference, see Paar, 2017). The performance benchmark is published by eBACS: ECRYPT Benchmarking of Cryptographic Systems [Bernstein, 2019].

4 Cryptography Services

4.2 Message Authentication Codes

Message authentication codes (MACs) are used, as the name implies, to authenticate the message. MAC is based on symmetric algorithms. The idea is to use the shared symmetric key while generating and verifying the MAC. The trade-off is that MAC doesn't provide nonrepudiation—only integrity and authenticity. If the MAC is verified at the receiving end, then the receiver knows that the person who generated the MAC and sent it over is the person with the other symmetric key. The MACs are also known as keyed-hash functions or cryptographic checksums.

This is in contrast to digital signatures, which do provide nonrepudiation, in addition to integrity and authenticity. However, digital signatures are based on asymmetric cryptography algorithms, which are resource intensive and slow. We'll discuss digital signatures in more detail in Section 4.3.

Hashed-based message authentication codes (HMACs) use the hash functions in generating the MAC. This idea was proposed by Mihir Bellare [Bellare, 1996] in 1996. The implementation of HMAC using SHA-512 is shown in Figure 4.10. The secret key is added to the message before hashing. The HMAC function performs the hashing twice. These are called inner hash and outer hash.

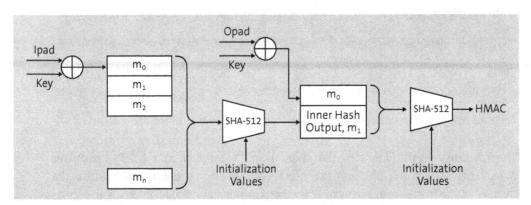

Figure 4.10 Implementation of HMAC Using SHA-512

In recent years, SHA-2 or SHA-3 hash functions have been used in HMAC. In our discussion, we've used SHA-512 as our hash function with HMAC. Before performing the inner hash, the input message is divided into 1,024-bit blocks—the message block size for SHA-512. The first block is generated using the key and the inner padding (*ipad*). The secret key is typically 128 bits long (or 256 bits at the most), and the remaining leading bits of the key block are filled with all 0s. The ipad has a fixed pattern of bits—0011 0110—that is repeated for the entire block. The key block and ipad blocks are bitwise XORed, and the output of the XOR is used as the first block of the message string.

The process continues with the outer hashing. The key is again XORed with the outer padding (*opad*). The pattern for opad is 0101 1100 and is repeated for the entire block.

The output of this XOR function and the output of the inner hash SHA-512 are again hashed using the same SHA-512 function.

In both inner hash and outer hash, the functions are fed the initialized values. These values initialize the buffer variables *A* through *H* in SHA-512. It's important to note that although the hashing happens twice, the message is hashed only once. The second hashing operation is performed only on two blocks: (1) the XOR output of the key and the opad and (2) the hash or the digest of the inner hash. Therefore, the overhead of using the second hash function is minimal.

Like hash functions, MACs can also be constructed using block ciphers. One of the common use cases of HMACs is with Transport Layer Security (TLS) protocols, which is explained in detail in Chapter 6, Section 6.2.1.

4.3 Digital Signature

So far, you've learned that encryption provides confidentiality, the hash function provides integrity, and the MAC provides authenticity. But we still need a way to hold the sender accountable. How do we prove that a message came from the particular sender? In the case of the MAC, the sender and receiver both have the same key, so in the event of a dispute, there's no way to say which of the two might have sent the message or carried out the transaction. Digital signature solves that problem, as you'll see in the following sections.

4.3.1 Primer on Digital Signatures

First let's take a look at the traditional transaction. In the world of paper and ink, a physical signature on paper is used as proof of transaction. (In this case, a signature is considered solid proof and assumed that no one else can duplicate the signature.) However, a signature in the digital world isn't considered a robust proof of transaction. This is because even if the digital transaction is signed, the signature is converted into binary, and anyone who can access the document can alter those binary digits. People can forge the signature or even change the transaction details before the document is encrypted. The risk is very high. For example, if account holder X sends a request to transfer $1,000 to account Y, the bank must confirm two things:

- Whether the amount that was asked to transfer was indeed $1,000 and not only $10 or $100, to ensure that the details of the transaction weren't compromised, and a hacker hasn't changed the amount or the recipient's name
- That this request did come from account X and not some hacker or even whether account Y is posing as account holder X

The first problem of data integrity can be solved using the hash function, and the second problem of authenticity, confirming that the transaction was indeed requested by

X, is solved using the MAC. But what if X denies requesting this transaction? The digital signature solves this problem by using the key to sign the message or the transaction.

The digital signature is generated using asymmetric cryptography. As you learned in Chapter 3, asymmetric cryptography has two keys—a private key and a public key. In asymmetric cryptography, the sender uses the receiver's public key to encrypt the message. The receiver is the only one with the corresponding private key and decrypts the message. In digital signature, it's reversed: the sender generates a signature using his private key, and the receiver verifies the signature using the sender's public key. Since there is only one private key for the corresponding public key, the owner of the private key, the sender, is the only person who could have signed the message or transaction. There is no way to deny or back out of a deal.

In a very simple form, the digital signature algorithm works like this: The algorithm ingests the message (or the document) and the private key of the sender (or the document owner) to generate the signature. This signature is then appended to the message. The appended message is transmitted to the receiver. The receiver extracts the signature and runs it through the verification algorithm using the public key of the sender. The output of the verification algorithm is Boolean—True or False. If the signature matches, the output will be True (else False). Because the signature was verified using the sender's public key, it confirms that the sender has signed and sent the document.

The process is very straightforward if we ignore the amount of time asymmetric cryptography takes in signing a huge document or a large message. This problem is partially solved using the hash function. The message or the document is hashed first. The hash value or the digest is then fed to the signature algorithm. The use of the hash function also provides another cryptography service—integrity.

The digital signature thus provides three of the four cryptographic properties—integrity, authenticity, and nonrepudiation.

4.3.2 Digital Signature Standard

Although RSA, Elgamal, or elliptical curve asymmetric or public key cryptography algorithms can be used to generate the digital signature, NIST has published a Digital Signature Standard (DSS) or Digital Signature Algorithm (DSA) in FIPS publication 186. The first adoption and standardization happened in the 1990s. An overview of how the DSA works is shown in Figure 4.11.

The first step in DSA is to prepare for the algorithm. This is also known as the initial setup. This is usually done by the sender of the message. The sender must first generate public parameters (sometimes also referred to as global parameters), create private and public keys, and select an RNG, K. The sender freely shares the public parameters and the public key with the receiver—technically these can be shared with everyone.

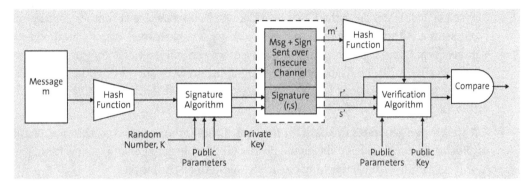

Figure 4.11 Overview of Digital Signal Algorithm (DSA)

The message or the document whose signature is to be generated is fed to the hash function. The output of the hash function—the hash or digest—is fed to the digital signature algorithm. The digital signature algorithm uses public (or global) parameters, a private key, and a random number K and generates the digital signature (r,s). This signature (r,s) is appended to the message m and then sent to the receiver over the public or insecure channel.

The receiver, on the other side, receives the message m' along with the signature (r',s'). The signature (r',s') and the message m' are fed to the verification algorithm. The verification algorithm also uses the public parameters and the public key of the sender. The output of the verification algorithm is compared with the r' component of the signature. If they match, the signature is a valid signature.

FIPS in its latest release of FIPS 186-5 in February 2023 [FIPS 186, 2023] discontinued the use of the DSA. NIST recommends the industry standard algorithms, such as RSA and other elliptic curve-based digital signature algorithms. The implementation of RSA and Elliptic Curve Digital Signature Algorithm (ECDSA) are extensions or small modifications of their original encryption implementations, which we've discussed in detail in Chapter 3.

ECDSA functionally works the same way as the DSA—the sender generates the signature, appends it to the message, and sends it over. The receiver compares the signature, and the verification algorithm provides a Boolean answer: True or False. However, what happens under the hood—the mathematical operations—are quite different. In ECDSA, the private key, the public key, and domain parameters are calculated using the elliptic curve. The ECDSA standard is based on the elliptic curve over prime fields Z_p. Refer to Chapter 3, Section 3.2.4, for more details on elliptic curve cryptography (ECC).

4.4 Merkle Trees

The concept of concatenating the hash values to create a tree-like structure was proposed by Ralph Merkle in the 1980s [Merkle, 1988]. A *Merkle tree* is another application of hashing. Although a Merkle tree (also known as a hash tree) is not classified as a direct cryptographic service, for our purposes we can call it a tier 2 cryptographic service because it can be used within cryptographic services such as digital signatures. Merkle trees are also used with cryptocurrency, which we'll discuss in Chapter 8, Section 8.2.2.

A Merkle tree generates the hash for the data blocks or the message portion. Each hash value of the data block is concatenated with another block to generate a new hash. The process continues until reaching one hash digest for all the blocks.

For example, take four data blocks, A, B, C and D. Use a hash algorithm such as SHA-512 to generate a hash digest for each block such that the output of each block is represented as follows:

- Data block A = HashA
- Data block B = HashB
- Data block C = HashC
- Data block D = HashD

These hashes, HashA, HashB, HashC, and HashD, are known as *leaves* of the Merkle tree.

Next, concatenate HashA and HashB as well as HashC and HashD to generate two new hash digests, HashAB and HashCD, as follows:

HashA + HashB = HashAB
HashC + HashC = HashCD

HashAB and HashCD are known as *nodes* of the tree.

Next, concatenate HashAB and HashCD to generate HashABCD. HashABCD is known as a *Merkle root* (or hash root).

You might be wondering why we need to do this. Merkle trees provide the following advantages:

- They save disk space. Instead of storing each hashed digest, the root can be saved for long-term archiving. This saves a considerable amount of disk space in the long run.
- They improve process efficiency since verification can be performed at the root level. If the value of any of the leaves or nodes is changed, the hash value of the root will change. By verifying the root, you can safely attest to the integrity of the entire tree.
- They are very useful in handling large datasets. These large datasets can be divided into small chunks, and a tree can be generated to improve efficiency. This also helps

with scalability. If another large dataset is added, then the tree can be expanded further without changing the existing tree.

Figure 4.12 shows the use of a Merkle tree in creating a digital signature. The large message or database is divided into small blocks. The blocks are independently hashed using SHA-512. Each block is used as a leaf. The process of concatenating and creating the tree is performed as explained previously. The root is generated as the final step.

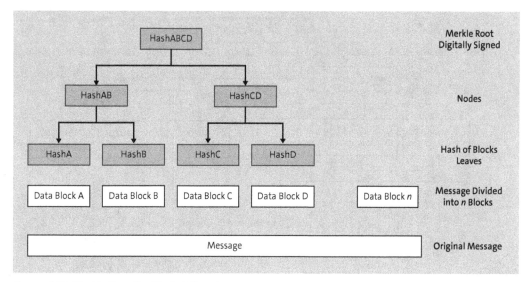

Figure 4.12 Merkle Tree for Digital Signatures

The hash root or Merkle root is digitally signed by one of the algorithms we studied in Section 4.3. This essentially signs the entire message. To verify any particular message block, you'll need the hash value of that block and all involved nodes. For example, to verify message block D, you'll need the hash value of message block D, HashCD, and the root. However, if any of the block is tampered with, the root hash value and the digital signature will change, indicating the compromise of integrity.

Since asymmetric cryptography takes a long time to digitally sign a large message, signing just a root significantly improves efficiency.

4.5 Summary

Historically, confidentiality was and has been the primary motive for using cryptography. However, with modern technologies and the heavy use of digital transactions in everyday life, other usages of cryptography have also become common in the past 30 years.

4   Cryptography Services

In this chapter, we studied the cryptography services—integrity, authenticity, and non-repudiation—and how to achieve these services using hash functions, MACs, and digital signatures. We also learned how to implement MD5, SHA-512, and SHA-3 algorithms as well as HMAC and DSA.

The cryptography tree shown in Figure 4.13 summarizes the various algorithms used under the keyless or one-way cryptography.

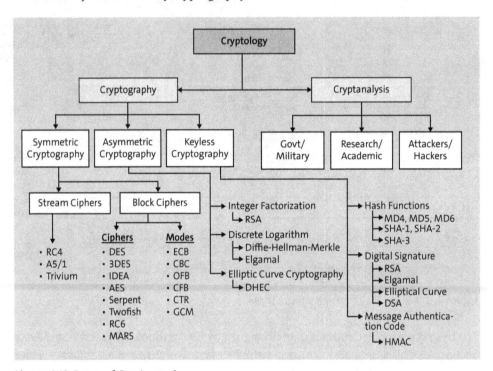

Figure 4.13 Types of Cryptography

The hash functions are known as keyless cryptography because they don't use the key in the hashing operation. The hash functions, digital signature, and MACs are also known as one-way cryptography because there is no way to get to the original or input message from the output values.

> **Key Takeaways from This Chapter**
> - Reviewed hash function, digital signature, and MAC
> - Studied three primary techniques of hashing—MD5, SHA-512, and SHA-3
> - Learned about the implementation of MAC, digital signatures, and Merkle trees

In Chapter 2, we learned about symmetric cryptography, and in the next chapter, we'll learn how to use symmetric cryptography to protect and encrypt the data at rest by exploring various techniques.

PART II
Modern Cryptography in Practice

Chapter 5
Storage Security: Data Encryption at Rest

One of the most basic applications of modern cryptography is to protect data at rest. In the digital world, encrypting data at rest is required not only to protect the data but also to meet various compliance and regulatory needs. This chapter focuses on understanding and implementing techniques to protect data at rest.

In the 21st-century economy, data plays as an important role as oil played in the 20th century. Data, intertwined with security measures, now stands as the cornerstone of modern business operations. Storage security is essentially focused on protecting data.

The invention of modern technology accelerated the use of computers in the business world in the 1960s, which generated lots of digital assets, or data. This situation posed another problem to computer engineers: How do we protect these digital assets, or data? Modern cryptography offers various ways to protect data using encryption at different levels in the system. Data encryption at rest is also known by the DARE acronym: data-at-rest encryption. Because data is kept in some form of a storage device while at "rest," data encryption at rest is also an important part of storage security strategies.

> **Chapter Highlights**
> - Fundamentals of data encryption at rest
> - Understanding various methods of data encryption at rest
> - Learning to implement encryption on data at rest at various levels in the system

So far in the book, we've covered details about symmetric, asymmetric, and one-way cryptography algorithms. Now it's time to put these algorithms into practice. In this chapter, we'll learn the fundamentals and various methods to protect data and storage using data-at-rest encryption.

The chapter will start with a primer on data security in Section 5.1. Then, the chapter will dive deeper into various methods to encrypt data at rest in Section 5.2. The chapter will be focused on the current trends and tools used in practice today. Although storage security includes many different techniques and tools to protect data in storage (aka at rest), the focus of this chapter is primarily on storage security through cryptography, in particular, encryption at rest.

5 Storage Security: Data Encryption at Rest

5.1 Primer on Data Security

The Open Worldwide Application Security Project (OWASP) releases Top 10 application security vulnerabilities every three years. In the OWASP Top 10 list released in 2021, Cryptographic Failures ranked at number 2 [OWASP, 2021]! This is concerning because, in previous years, there was no vulnerability category called Cryptographic Failures. The cryptographic attacks and failures were part of the Sensitive Data Exposure category. To demonstrate the significance of this change, in the list published in 2013, Sensitive Data Exposure was at number 6. This was bumped to number 3 in 2017. In 2021, unfortunately, cryptography established its own category of vulnerability and ended up as the runner-up on the list—second only to Broken Access Control.

The rise in cryptography vulnerabilities is partly because other application vulnerabilities, such as SQL injection or insufficient logging and monitoring, are understood well by security practitioners. This isn't the case with cryptography because it's difficult to understand and design cryptography solutions and even more difficult to implement correctly.

With the importance of the topic in mind, let's begin our discussion with the basics of data and data security. The first section is all about understanding the data—the asset we want to protect. In the second section, you'll learn about reasons and requirements to protect the data—why we want to use cryptography to protect the data and how we can implement it. If you're an experienced security professional, you can skip the first section if you like, but we recommend that you at least quickly skim it as this section sets the stage for the rest of the chapter.

5.1.1 Understanding the Data to Be Protected

Before we discuss how to protect our data, we need to understand it. In the following sections, we'll learn about the data lifecycle, data stages, and data classification.

Data Lifecycle

Like everything else in the world, data goes through a lifecycle. Following are the key stages of the data lifecycle:

- **Data creation**
  In this stage, data is created by entering data into the system, copying data from other sources, or collecting data from other tools. In this stage, data appears for the first time in the system.

- **Data storage**
  The newly created or entered data must be kept or stored somewhere. The data is stored when it's not being processed (or used) and not moving. Data in this stage is considered *data at rest* because it's not moving or being processed. Data is also

backed up regularly and, in this chapter, you'll also learn how to protect the data backups, which are also considered data at rest.

- **Data processing**
  Data is in the processing or usage stage when it's being updated or worked on—or it's not in storage. Typically, it's hard to implement security measures while data is in this stage.

- **Data archiving**
  Data is archived when not needed on a regular basis. Data is archived for a variety of reasons, for example, data isn't required in routine operations, the data storage is at or close to the capacity or even to save the storage cost. Usually, it costs less to archive data than to keep it in active storage. Archived data, in addition to long-term storage and data backups, is considered data at rest and must be protected.

- **Data destruction**
  The last stage in the data lifecycle is the destruction stage. Once the data is no longer needed, the data must be destroyed properly. Cryptography does come to our rescue in this stage also. The process of cryptographically deleting data is known as crypto-shredding.

> **Crypto-Shredding**
>
> *Crypto-shredding* is a technique for securely deleting data. The data to be deleted is encrypted, and then the encryption key is destroyed or deleted, making data inaccessible. In crypto-shredding, once the key is destroyed, then the data itself is also deleted, and the storage media is sanitized. Storage media can be sanitized using erasing, overwriting, degaussing, or purging. Later on, even if a hacker gets his hands on the storage media and is able to recover the data, he has no way of decrypting it because the key is deleted. This is the most secure way of deleting the data.
>
> Crypto-shredding involves deliberately deleting the encryption key. However, in the real world, many times, people lose the encryption key by accidentally deleting it. In this situation, data also becomes useless or inaccessible. Millions of dollars' worth of cryptocurrency is useless because people don't remember the key! You'll hear more about that in Chapter 8.

Data Stages

Data goes through different stages or states every time it changes the lifecycle phase. As shown in Figure 5.1, data can go through one of three stages during the data lifecycle:

- **Data in use**
  Data is said to be in use when data is being updated, created, or worked on. This doesn't necessarily mean that data is used or being worked on by a human, but it includes cases when automated tools and other applications are updating or

5   Storage Security: Data Encryption at Rest

consuming the data. This is the most vulnerable stage for the data because it's usually not encrypted while being used. In most cases, data is encrypted while at rest and in transit. In Chapter 12, we'll learn about how to protect the data in use.

- **Data in transit**
  Data is in transit when data is moving from one device or location to another. This doesn't necessarily mean when we copy or physically drag and drop data from one drive to another, but this includes situations such as sending an email, sending a message, performing online banking, or executing any e-commerce transaction. Hackers have many techniques and tools—including all the AI-based options—to get their hands on the data while the data is in transit, so it's crucial to protect the data while in transit. You'll learn all about protecting data in transit in Chapter 6.

- **Data at rest**
  This is the hero of our story in this chapter. We've already defined data at rest, so we won't repeat it here. Plus, this entire chapter is about protecting data at rest, so we'll discuss a lot more about how to protect data in this state.

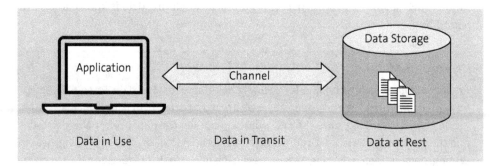

Figure 5.1 Data Stages

Data Classifications

Data is classified and protected based on the sensitivity of the information. In private industries, data is usually classified in the following manner:

- **Confidential**
  Proprietary information, personally identifiable information (PII), protected health information (PHI), and so on are classified as confidential. Confidential data is accessed only on a need-to-know basis. The confidential data, almost always, is encrypted. If confidential data is stolen or exposed, the company's reputation is damaged or even incurs financial loss.

- **Secret**
  Secret data (some businesses label this data as sensitive) is internal to the business, but most employees of the company can access the information. The control is a bit less strict compared to confidential data. This type of data is generally not encrypted.

- **Internal**

  Internal classification is used when data isn't made publicly available but can be shared as needed. Customer presentations and other client facing-documents are classified as internal or shared as needed. This type of data is generally not encrypted, but there are some exceptions.

- **Public**

  As the name indicates, this type of data is shared openly and is easily available to everyone. Public-facing websites and other media information is classified as public. This data isn't encrypted.

You might be wondering why we need to classify data. Why can't we encrypt all data? The reason is that data protection comes with a price tag in terms of cost and performance. Every time a security control is applied to protect the data, it costs to implement and manage the control, as well as some computational overhead, which impacts performance. Encryption can be resource intensive, so encrypting data is very expensive. It's essential to protect confidential data with the best and perhaps multiple protections because if this data is leaked, the business can suffer.

> **Data Classification 101**
>
> The data classifications given here are for private industries and aren't universal. Private businesses can choose the classification labels and categories that fit their business model. On the other hand, the US government and defense agencies have their own well-defined classifications: top secret, secret, confidential, and unclassified.

To protect the data at rest, we must understand three things about the data to be encrypted: the lifecycle stage must be at storage, backup, or archived, the data state must be at rest, and the classification must be confidential (or secret in some cases).

5.1.2 Understanding Data Security

There are many ways to protect or secure data and data storage. One of the most common security controls is *access control*. The idea behind access control is to make sure that only the authorized user has access to the data. Implementing access control doesn't really mean that the data is safe from every unauthorized user in the world. Hackers can use very advanced techniques to execute social engineering that can compromise access control very easily. Once the access is compromised, there is no stopping the hackers. Another scenario is the insider threat. If an insider gains access to the system that he isn't supposed to, the insider can see or use the data. This data can be mistakenly or maliciously shared or exposed to unauthorized users. Moreover, when the entire storage device is stolen or even somehow files are extracted, the data can be exposed. The robust logical access control won't protect data when the drive itself is physically stolen. Other controls such as firewall, intrusion detection system (IDS)/ intrusion prevention system (IPS), logging and monitoring, and so on can help protect the data up to a certain extent, but if that fails, the data can be exposed.

It's also important to consider ransomware attacks, where attackers encrypt the data and ask for the ransom amount. If the amount isn't paid, the attackers will expose the data or sell it on the dark web. In either case, access control or security control around storage won't help.

These are the reasons why security control is needed over the data itself. If you recall our discussion of the defense-in-depth strategy (see Chapter 1, Section 1.2.4, Figure 1.3), the data protection controls, such as data encryption, are at the core of the security controls.

Let's first discuss some reasons why we need to encrypt data at rest:

- When the security of data in transit versus data at rest is compared, it may seem obvious that the data is more likely to be attacked while in transit when it's out of the owner's control. During transit, the attacker has more opportunities to hack the data compared to data at rest. Data at rest is (supposedly) under the watch of security professionals all the time, and hackers have less opportunity. However, this is a misconception. Security of data in transit is well defined and generally well understood. Plus, everyone must meet security standards to communicate or interact with each other. This isn't the case with data at rest. Data at rest is usually within the control of the organization and doesn't impact anyone outside. Therefore, what level of security control is applied to the storage and to the data is completely up to the organization. The business may skip some security controls to save money or boost performance.

- The availability of data to the attacker is another reason for encrypting data at rest. The availability of data to the attacker in transit is very short, while data at rest is available for the attacker for a long time—as long as the data is in storage, backup, or archive. This provides the attackers enough time to plan and execute the attack. In a nutshell, although it appears that data at rest is more secure than data in transit, that isn't (always) the case.

- There are more reasons to secure storage at the data level than just to protect confidential information. Businesses need to protect/encrypt data to meet compliance requirements. For example, a business may be required to encrypt data as part of Payment Card Industry-Data Security Standard (PCI-DSS) compliance if it's processing credit card information. Similarly, if the business is dealing with health data, it may need to be compliant with the Health Insurance Portability and Accountability Act (HIPAA) requirements, or if the company is doing business in the European Union, it may need to be compliant with the General Data Protection Regulation (GDPR). In all of these examples, data encryption at rest is mandated by compliance requirements.

Based on the business needs and compliance requirements, data is stored in a variety of devices, such as tape drives, hard disks, data lakes, storage area network (SAN), network-attached storage (NAS), and so on. These storage devices can be secured with a variety of controls—physical, administrative, and technical. In this chapter, we'll only discuss securing data through a technical control, namely encryption.

Encrypting the data at rest is the best way to protect the data and helps even if all other controls fail. However, this doesn't mean that security professionals should take other controls lightly or skip them altogether. Data-at-rest encryption should be at the heart of the defense-in-depth strategy.

5.2 Data-at-Rest Encryption Methods

Encrypting data at rest is trickier than you might think. Every time a user wants to work with data, data must be available to the user in plain text—that means data needs to be decrypted. The process of encryption and decryption must be fast, affordable, and, most importantly, transparent to the user. That is why there is no one-size-fits-all solution for data-at-rest encryption.

Data at rest can be encrypted at various levels and using different methods depending on the performance, cost, and sensitivity requirements of the business. Figure 5.2 shows various data-at-rest encryption methods.

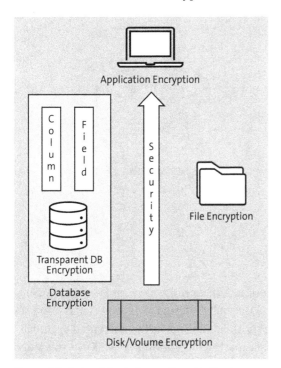

Figure 5.2 Data at Rest: Encryption Methods

Disk encryption is at the bottom of the stack with volume encryption. Database and file encryption are in the middle, while application-level encryption (ALE) is at the top. Security increases as you go from bottom to top but the complexity of implementing and managing the encryption also increases.

In this section, we'll discuss the implementation of each method of encrypting data at rest, starting at the bottom of the stack.

5.2.1 Disk Encryption

Disk encryption is the process of encrypting the entire disk, and is also known as *full-disk encryption (FDE)*. The disk encryption process encrypts all the data, including the OS, on the disk—unless the disk being encrypted is a boot disk or stores the code to decrypt the disk. In that case, a bootable portion of the code isn't encrypted. The unencrypted part of the code is responsible for processing the password and decrypting the rest of the disk. It's also on the hook for booting the system. Disk encryption works on client workstations (e.g., laptops), on servers, and on external drives. In the case of external drives, disk encryption encrypts the entire drive because there is no boot code or decryption code stored on the external drives.

In the public cloud, disk encryption uses *envelope encryption* to solve the problem of securing the encryption key for disk encryption. Figure 5.3 shows the concept of envelope encryption.

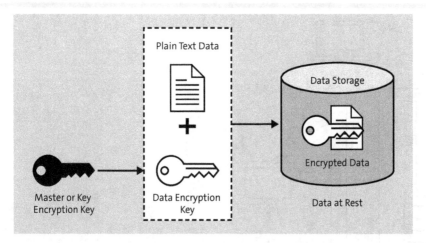

Figure 5.3 Overview of Envelope Encryption

Disk encryption uses symmetric encryption, Advanced Encryption Standard (AES), to encrypt the data. This AES encryption key is usually referred to as a data encryption key (DEK)—because the key is used to encrypt the data. This key is encrypted again with another key—key encryption key (KEK) or master encryption key (MEK). At this stage, it's safe to store the DEK with data storage while the MEK or KEK is stored in the key vault or dedicated hardware module. The idea is to use the KEK/MEK as an envelope and to put the DEK safely in that envelope.

It's important to understand the concept of envelope encryption because various methods to encrypt data at rest use envelope encryption in one form or another. In the public cloud environment, envelope encryption uses a hybrid approach (AES and RSA). DEK is usually encrypted with a symmetric encryption algorithm, such as AES, while KEK is encrypted with a public key algorithm, such as RSA. We'll discuss more about data security in the cloud in Chapter 7.

In client-side devices, such as laptops and workstations, there is no key vault or dedicated hardware module to store the key. The disk encryption process varies slightly based on the manufacturer, for example, Microsoft and Apple. Although both use symmetric key encryption to encrypt data, the mechanism to store the key is different.

Windows uses a hardware chip called a Trusted Platform Module (TPM). The TPM is a powerful cryptography chip on the motherboard and can handle a variety of tasks, but our focus is on cryptography-related tasks: key generation, hash algorithm, and hardware authentication or integrity. The TPM generates the key for the hardware or laptop and encrypts this key. This process is called wrapping or binding. The key can be decrypted and made available by TPM when a correct PIN is entered. This is where the disk decryption process starts. TPM also provides device health attestation. This is done by hashing the boot and other initialization software. The integrity is checked at the start of the device. The device starts the decryption process only if the correct PIN is entered and the integrity is confirmed. This feature prevents the system from booting if the software is corrupted. TPM is also used for hardware authentication. If the hardware or the disk is swapped or the disk is connected to a different device, TPM won't allow disk decryption. (Note that this is a general description of how TPM is implemented; the actual implementation could vary as each manufacturer decides which features to offer or enable.)

MacOS, on the other hand, doesn't use TPM. It uses the Apple T2 chip. Unlike TPM, T2 doesn't offer a wide range of features and depends on external sources to provide various features. The end result is very similar, but the implementation is different. In MacOS, T2 and various other components help perform the tasks, whereas Windows uses one chip, TPM, to perform all tasks.

Windows BitLocker is also available in server configurations. Windows servers used in the cloud and in the data center can also use the BitLocker feature to encrypt the entire server disk. This helps protect data when the various storage volumes are attached and detached to the server. We'll learn more about it in Chapter 7, Section 7.1.1.

Disk encryption is one of the most basic forms of data-at-rest encryption methods and offers several advantages and disadvantages, as shown in Table 5.1.

Advantages of Disk Encryption	Disadvantages of Disk Encryption
▪ Disk encryption is very inexpensive for the user, and, for the most part, it's available as the default or out-of-the-box functionality of the disk/hardware. ▪ Disk encryption is transparent to the user and has little performance overheads. Once the correct PIN is entered, the user doesn't even notice that the decryption process is happening in the background. ▪ Disk encryption offers a seamless user experience, as the PIN is required only once at the beginning, and afterward, data on the entire disk can be accessed without any additional steps. ▪ Disk encryption provides protection from the physical theft of the device or the drive. If the laptop or the drive is physically stolen, disk encryption protects data. ▪ It meets the minimum compliance requirements—if you're looking to meet a checkbox in the list of compliance requirements, disk encryption can meet that at a very low cost.	▪ Once the disk is decrypted, the entire disk data is available to grab. The advantage of a seamless user experience goes away if the hacker takes control after it was decrypted. ▪ If it's a bootable disk, a small portion of the disk isn't encrypted. This bootable portion can be altered by malicious attacks ▪ Depending on the size of the data, it may take a long time the first time the user encrypts the disk. Depending on the system, the disk may not be usable during this time. ▪ In a data center/cloud environment, several people can have access to the disk, and it's difficult to protect sensitive information.

Table 5.1 Disk Encryption: Advantages and Disadvantages

In a nutshell, disk encryption is a low-cost data-at-rest encryption method but doesn't offer robust data protection. To optimize the benefits, it's best to use this in the form of data-at-rest encryption.

5.2.2 Volume Encryption

At a high level, volume encryption is very similar but a slightly better version of disk encryption. First, let's define volume. *Volume* is a unit of the storage or the disk. The storage volume can be a physical volume or a logical volume. Volume can easily be explained in the cloud or data center environment where multiple users need to access data from a very large storage unit. The large storage disk or drive is divided into logical units—based on the users, type of data, and so on. Compare this to a file cabinet in your workspace. The disk is like the entire file cabinet, and the volume is each drawer. If you have only one drawer in the cabinet, then the cabinet and drawer are of the same size. Similarly, if we have only one volume, then the volume and disk are the same size, although we wouldn't have created just one volume.

> **Volume versus Partition**
>
> Don't confuse volume with partition. Disk *partition* is a created logical section or part (hence the name partition). A partition is almost like physically dividing the disk, while volume is an area logically defined for the user that hosts the user's file system.

The concept of volume is very useful in data centers, public clouds, or virtual environments because the volume can be mounted when needed and unmounted when not in use. In a cloud or data center environment, a large storage device, usually a disk, is shared by many users and for a variety of purposes. This is usually done through NAS. The disk or even multiple disks are mapped into volumes. Users then use the volumes to work on the data. It's possible to create a task-specific volume and assign the security based on the type of data it stores. For example, a small volume can be created just to save PII and other confidential data only. Access to this volume can be highly restricted. That way, chances of accidentally exposing or misusing the confidential data is reduced.

The process of encrypting a volume is very similar to disk encryption. Volume encryption also uses symmetric encryption (typically AES) and envelope encryption. It uses the DEK or the volume key (VK) to encrypt/decrypt volume and the MEK to encrypt/decrypt the DEK/VK. The MEK is stored in the key vault or some kind of key management system.

Volume encryption is slightly above the disk encryption on the data-at-rest encryption stack because it does offer some advantages over disk encryption (but it also has similar disadvantages), as listed in Table 5.2.

Advantages of Volume Encryption	Disadvantages of Volume Encryption
■ The first and foremost advantage is the performance. Because only the volume is encrypted instead of the entire disk, the volume encryption is significantly faster. ■ Only the volume that is attached to the server (or the volume being actively accessed) is decrypted and not the entire disk. This way, if the encryption key or password is compromised, only that volume is at risk and not the entire disk. ■ Each volume has its own encryption key and password. This provides slightly better control and improved security.	■ If the encryption key or password is compromised, the data of the entire volume is at risk. ■ Once the volume is decrypted, the entire volume is available to the attacker. If the attacker somehow manages to bypass the access control, all bets are off. ■ Like disk encryption, it offers protection primarily against physical theft. It doesn't offer much in terms of granular data protection.

Table 5.2 Volume Encryption: Advantages and Disadvantages

Advantages of Volume Encryption	Disadvantages of Volume Encryption
- It's economical because only the volume with confidential information needs to be encrypted and not the entire disk. This saves costs, improves performance, and reduces key management overhead. - All unattached or unmounted volumes always remain encrypted.	

Table 5.2 Volume Encryption: Advantages and Disadvantages (Cont.)

We'll learn about volume encryption offered by various public cloud providers in Chapter 7.

5.2.3 File Encryption

Disk and volume encryption protects data from the physical theft of the hardware but doesn't help much when a skilled attacker bypasses the access control or when credentials are compromised. Once the attacker gets into the disk or volume, the attacker has complete access to all the data in the disk or in that volume—there is no stopping them. Worst of all, many data owners don't even realize right away that their data is compromised. How can data be protected in this situation? To solve this problem, we'll have to move on to file encryption in the data-at-rest encryption stack.

File encryption, also known as *file-based encryption (FBE)*, is the process of encrypting files, as shown in Figure 5.4. A user can encrypt individual files or folders to have granular control. File encryption uses a combination of symmetric and asymmetric cryptography. Each file or folder when encrypted uses a randomly generated encryption key. This key is known as the file encryption key (FEK). The file or folder is encrypted using symmetric encryption. The same FEK is used to encrypt and decrypt the file or folder. An asymmetric key pair is used to encrypt or wrap this FEK. The FEK is encrypted using the user's public key and kept with the encrypted file. The FEK can only be decrypted with the user's private key. The asymmetric key pair is generated by the system for each user.

> **Let's Be Objective about the File**
> Don't get confused about file-level storage versus object-level storage. File storage is used for structured data. It stores files and folders in a hierarchical fashion. The metadata is stored with the file. On the other hand, objects store data with a flat namespace. Metadata is attached to each object as a key-value pair. The objects are accessed using application programming interfaces (APIs), whereas files are accessed using traditional file systems.

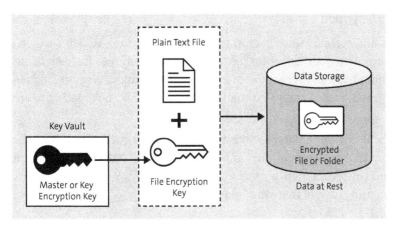

Figure 5.4 Overview of File-Based Encryption

The user's key pair is generated when the user first logs in and will remain on the local system or on the domain controller. In the public cloud, the key pair can also be kept with the key management system.

File-based encryption offers advantages and disadvantages, as shown in Table 5.3.

Advantages of File-Based Encryption	Disadvantages of File-Based Encryption
■ Files are individually encrypted. Even if an attacker gets access to the runtime system, the attacker won't be able to read the data from the files. ■ Every file is encrypted with its own FEK, so even if a key is compromised, data from only one file is exposed. (On some systems, this is a configurable feature, which means the administrator can decide to have one FEK for all files or have one key per file.) ■ Because only one file is encrypted or decrypted, there's almost no impact on performance. ■ The system can also be used to encrypt the folders, instead of files. The mechanism works the same way, and all the files under the folder become unreadable when encrypted.	■ File-based encryption doesn't encrypt the metadata of the file. That means if an attacker gets access to the system, the attacker can see when the file was last used, the file size, the timestamp, and even file names (in some cases). Attacker gets enough information to figure out which file is worth targeting and plans an attack. ■ If the PIN or password is compromised, then the attacker can get access to the key pair, which can be used to encrypt/decrypt all the FEKs.

Table 5.3 File-Based Encryption: Advantages and Disadvantages

Using file-based encryption with disk or volume encryption provides good protection to data at rest.

5.2.4 Database Encryption

Most applications and systems store data in some sort of database, which is why it's important to secure databases. Securing the database is largely vendor dependent. Each database vendor has its own method of encrypting the database. In this section, we've explained the database encryption methods in general and not specific to any vendor.

The primary ways to encrypt a database are transparent data encryption (TDE; also called transparent database encryption), column encryption, and field encryption. Each database encryption method protects data at rest only. You'll learn how to encrypt a database using each of these methods in the following sections.

Transparent Data Encryption

TDE encrypts the entire database. Conceptually, it works similar to the file encryption technique or disk encryption technique we discussed earlier. The database is encrypted using symmetric encryption, such as AES. The database is encrypted with a DEK, which resides within the database. When the user queries the database, the database decrypts the data and provides a response in plain text. The user can also save or write data into a database using plain text.

The data is encrypted just before writing into the database and decrypted just before sending it to the user. The entire process of encrypting data before storing it in the database and decrypting it before retrieving is transparent to the user, hence the "transparent" part of the name. The user and the application interacting with the database has absolutely no idea what is happening under the hood. Some database vendors use this technique to encrypt the tablespace, that is, the entire logical storage container. You can compare this with volume encryption.

Tablespace versus Table: Recapping the Database Terminology

Data in the database is stored in a structured format. The data is in relation to another piece of data. The system that manages the data and their relationship is known as a relational database management system (RDBMS). Let's recap some of the basic terminology of the relational database:

- **Tablespace**
  Think of this as a logical storage container that keeps the actual data files. The tablespace allows the database administrator to manage the storage allocation for the database within the physical storage.
- **Table**
  Think of the table as a spreadsheet in the database. The tables have the actual data to be stored in the database. Tables provide the structure to the structured database. The table stores the data in rows and columns.

- **Column/row**
  The use of columns and rows is very similar to the use of columns and rows in spreadsheets.
- **Field**
  Field is defined in a couple of different ways, depending on the context. The term is used interchangeably with columns and in some cases, like in some unstructured databases, it's used to describe one piece of information, a data field, such as a name or phone number.

Although it depends on the type of database and the vendor, in many cases, TDE usually encrypts the tablespace.

TDE methods have advantages and disadvantages, as shown in Table 5.4.

Advantages of TDE	Disadvantages of TDE
<ul><li>It's transparent to users.</li><li>There's low impact on performance as the entire tablespace is decrypted at the first interaction. Once the user session is authorized and authenticated, the user can keep working, and the tablespace is decrypted for the session.</li><li>TDE can provide security against physical theft.</li><li>Each tablespace is encrypted separately. If user credentials are compromised, the data of one tablespace is at risk.</li><li>TDE is easy to implement and maintain in large data center settings.</li></ul>	<ul><li>The entire tablespace is encrypted with one key. If the user credential or the key is compromised, data in every table within that tablespace is compromised.</li><li>Primarily, TDE provides security against physical theft. However, it offers very little in terms of data protection.</li><li>If hackers somehow bypass the authentication, the entire tablespace data can be compromised.</li><li>Server administrators, database administrators, and everyone else who has access to the database has access to the data. This makes data more vulnerable.</li><li>The TDE implementation varies depending on the type and vendor of the database. In addition, database vendors could and do change the implementation as technology evolves, and this makes it difficult at times to manage.</li></ul>

Table 5.4 TDE: Advantages and Disadvantages

Column-Level Encryption

Data in the database is stored in the table, and tables store data in columns. Some columns have different classifications of data. In the case of TDE, the entire tablespace is encrypted—that means every table in that tablespace and every column in each table is encrypted. However, that isn't the case with column-level encryption. In column-level encryption, every column is individually encrypted. The column-level encryption provides more security to the data because each column (or set of columns, in some cases) has an encryption key. The column encryption key is stored in the database. These column encryption keys are then encrypted with the MEK. The MEK is then stored in the key vault or key management system. The process of encryption and decryption is still transparent to users as with TDE.

The column encryption keys are usually symmetric keys. Most databases and vendors use AES-128 or AES-256 symmetric encryption algorithms to encrypt the data at the column level. However, the master key is encrypted with either an asymmetric algorithm or a symmetric algorithm. This largely depends on factors such as where the database is hosted, the database vendor, and the type of database in use. In general, column-level encryption is specific to the type of database and the vendor; as a result, most of the information and literature available is specific to a database or vendor.

The advantages and disadvantages of column-level encryption are shown in Table 5.5.

Advantages of Column-Level Encryption	Disadvantages of Column-Level Encryption
- The primary advantage of column-level encryption compared to TDE is the fact that even if the user credentials or key is compromised, the entire database isn't compromised—only columns accessed by that user may get impacted. - Moreover, unlike a transparent database, not all server administrators and database administrators can have access to the data. This is because the administrator may have access to tablespace or tables, but the data is protected by individual column encryption. This also gives some protection from the malicious insider threat.	- This additional layer of security comes with a price tag—it costs more to implement and manage the column-level encryption. - It slows down the performance by a bit because using data may require decrypting multiple columns. - Managing the keys adds some overhead and cost.

Table 5.5 Column-Level Encryption: Advantages and Disadvantages

Field-Level Encryption

As mentioned earlier, the term *field* is used interchangeably with *column*. So, what is the difference between column-level encryption and field-level encryption? At a very high level, there is no difference. Field-level encryption works just like column-level encryption. However, field-level encryption is used to provide very granular control. In column-level encryption, all columns are encrypted, but in field-level encryption, only selected columns or fields are encrypted. For example, if a database table has seven columns, but only three of those seven columns have sensitive information that needs to be protected, why encrypt all seven columns and even marginally degrade the performance? (We use *marginally* here because, for the most part, the impact of encryption and decryption on database performance is very low.) This is where field encryption

5.2 Data-at-Rest Encryption Methods

comes into play. Field encryption protects the data by encrypting the columns that contain sensitive information, while other columns can remain in clear text or can be secured by other measures.

Access to these encrypted columns is granted by using *role-based access control (RBAC)*. Figure 5.5 shows how role-based access works in conjunction with field encryption. For example, User_Payroll and User_Benefit have access to the employee's payroll-related columns or fields. In addition, the group named Group HR and all the users in that group have access to the employee's payroll-related fields. However, other users, such as User_Engineering, wouldn't have access to the encrypted fields and wouldn't be able to decrypt the columns (or wouldn't even see the columns) related to employee payroll information.

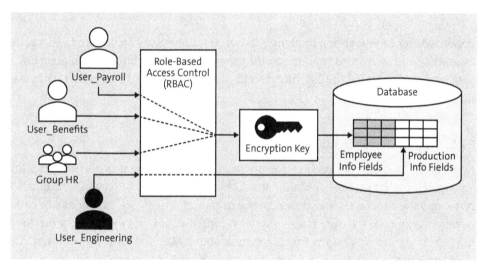

Figure 5.5 Field-Level Encryption with RBAC

Some database vendors or cloud providers also offer the ability to encrypt the fields (or columns) with sensitive data on the client side and then send the encrypted data to the database. Encryption key management for field-level encryption works exactly the same as column-level encryption. The field encryption key is stored with the database, and the MEK is stored with the key management service.

The advantages and disadvantages of field-level encryption are shown in Table 5.6.

Advantages of Field-Level Encryption	Disadvantages of Field-Level Encryption
■ The major advantage of field-level encryption is the ability to provide finer access control to sensitive data.	■ The key management for the field-level encryption can be a nightmare. It can be difficult and expensive to manage individual keys for each column.

Table 5.6 Field-Level Encryption: Advantages and Disadvantages

5 Storage Security: Data Encryption at Rest

Advantages of Field-Level Encryption	Disadvantages of Field-Level Encryption
■ It improves the performance compared to the transparent database and the column-level database, as field-level encryption encrypts only selected columns and doesn't encrypt all the columns within a table or the entire tablespace. ■ If the user's access is compromised or the system is hacked, only partial data (the one to which the user has access) is exposed, not entire table or database.	

Table 5.6 Field-Level Encryption: Advantages and Disadvantages (Cont.)

Because database encryption implementation techniques vary by type and vendor, we encourage you to refer to vendor-specific literature for more information on specific databases. At the end of the day, incorrect implementation of an encryption scheme is no encryption at all!

5.2.5 Application Encryption

Application-level encryption (ALE), as the name implies, is the process of encrypting and decrypting data at the application level. This is at the top of the stack in the data-at-rest encryption methods and closest to the users (refer to Figure 5.2). Mobile devices, remote workforces, software as a service (SaaS) applications, and dependency on cloud infrastructure are all reasons for us to think about ALE. The data is encrypted and decrypted by the application software and stays encrypted through transit and at rest. Figure 5.6 shows the overview of ALE.

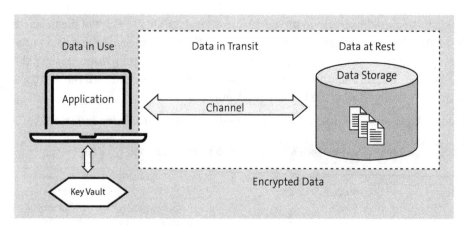

Figure 5.6 Overview of Application-Level Encryption

5.2 Data-at-Rest Encryption Methods

When we studied disk encryption, volume encryption, file encryption, and various methods of database encryption, we considered how the encryption and decryption process is transparent to the users or applications. So why do we want another type of encryption on the application level?

The data at rest encryption techniques discussed earlier are all used to protect data primarily when the hardware is stolen. However, these techniques fall short in certain scenarios, like when data isn't protected during runtime, from insider threats, and when attackers bypass the access control. However, ALE has some serious drawbacks as well. Both advantages and disadvantages are shown in Table 5.7.

Advantages of ALE	Disadvantages of ALE
■ The primary reason for using ALE is to provide layered data protection or defense-in-depth strategy. Over the past few years, the workplace landscape has changed dramatically—more people are working remotely, and more companies and people are using cloud-based infrastructures. This makes security professionals nervous. Encrypting data at the application level provides an additional layer of protection. ■ Another reason to use ALE is to perform client-side encryption. This is sometimes required by cloud providers. ■ In most data-at-rest techniques, administrators, privileged users, and some other users can access the data. This also opens the door for privileged creeps and malicious insiders. If the data is encrypted at the application level before storing it in the disk or database, there's no way for the system administrator, database administrator, or any other unauthorized users to access the data. ■ Some cloud providers don't use encrypted communication within the data center. The exchanges between servers, between servers and storage, and within a virtual private cloud (VPC) are often in plain text. Since the data is in plain text within a controlled environment and for only a fraction of a second, the likelihood of potential attack is very low; however, ALE protects data in this situation as well.	■ Depending on the application and software, implementing ALE can be complicated and difficult. ■ ALE needs a higher-than-average skill level. This could create a potential problem while upgrading or changing the application software. It's critical to get it right because making a mistake in implementing or using the incorrect/incompatible library can be a big problem. ■ ALE could also impact performance. Encrypting and decrypting at runtime could take a small toll on overall performance. ■ One of the major issues is the encryption key. It's complicated to design and implement proper key management. The key should be stored away from the software and the data itself and should be rotated regularly. Care must be taken to decrypt and re-encrypt the data every time the key is rotated. Lastly, the architects must plan a proper access control for the key. One way to minimize the key vault interaction is to use envelope encryption. Getting any step wrong in the key management process could jeopardize the security of the data.

Table 5.7 ALE: Advantages and Disadvantages

Advantages of ALE	Disadvantages of ALE
■ ALE is also used to protect the data from outdated hardware and security standards. For example, if the hard disk is old or the database version is at the end of its life (and not upgraded), then ALE can be used.	

Table 5.7 ALE: Advantages and Disadvantages (Cont.)

Encrypting data closer to the client or application always provides more security and end-to-end protection. ALE can be made part of the software development lifecycle to align with the shift-left philosophy.

Although ALE is categorized under data-at-rest protection, it also provides protection during the data in transit. We'll learn more about data in transit encryption in Chapter 6, but it's important to note at this stage that data in transit mechanisms such as Transport Layer Security (TLS) don't encrypt data at all points during the transmission. ALE protects data in this situation.

Disk encryption, volume encryption, file encryption, TDE, column-level encryption, and field-level encryption are all considered infrastructure-level encryption methods as opposed to ALE. Infrastructure-level encryption is done, as the name indicates, at the hardware level, while ALE is performed by the application software.

5.3 Summary

Data encryption at rest is one of the most important layers in data protection and in the defense-in-depth strategy, but its importance is always underestimated. Many assume that data encryption at rest is the hardware provider's responsibility. While that's partly true, the data owner is on the hook for protecting data. Hardware providers, such as data center owners or cloud providers do provide infrastructure-level encryption, but it doesn't tell the whole story. The other part of the story is the lack of awareness and available literature on the topic.

In this chapter, we've explained the various techniques for data encryption at rest at various levels, starting from the lowest level of disk encryption to the highest level of ALE. Data encryption at rest uses the symmetric encryption algorithms, AES, for the most part. We haven't discussed the process of encryption in this chapter because AES and other symmetric and asymmetric encryption algorithms are discussed in detail in Chapter 2 and Chapter 3, respectively.

Key Takeaways from This Chapter
- Reviewed fundamentals of data encryption at rest
- Learned various techniques for encrypting data at rest
- Studied the difference between infrastructure-level encryption and ALE

In the next chapter, you'll learn about data encryption in transit. Data encryption in transit uses asymmetric encryption algorithms. Get ready to implement what you learned in Chapter 3.

Chapter 6
Web Security: Data Encryption in Transit

In the digital world, people need to share all sorts of information, from personally identifiable information to business secrets in real time over the internet. This information, while traveling through the internet, is considered data in transit and must be protected from malicious users and hackers. This chapter explains how communication happens over the internet and how to secure data in transit using various protocols.

For thousands of years, the primary goal of cryptography was to protect the messages (or data) in communication or in transit—that was the goal behind the Caesar cipher, and it's the goal behind modern communication protocols. If you look at the history of cryptography, you'll find that it was invented, evolved, and improved for only one purpose—to provide confidentiality to the data in transit. Other applications of cryptography, such as data encryption at rest, hashing, digital signatures, and so on, made their way in only after the 1970s. The longevity of the concept of data protection in transit alone is a good reason for us to learn more about the subject.

It wouldn't be a complete overstatement to say that information security professionals have a better grasp of the situation when implementing data in transit protection, compared to data-at-rest security. There are several reasons for this statement. First, data in transit, in most cases, involves more than one party or organization. To make communication successful, both ends must come to some sort of agreement on what protocols to use. This forces more and more organizations to implement secure protocols. Second, because the issue of data security in transit impacts all organizations and society as a whole, government agencies have stepped in, and the protocols have been standardized. There are no such strict requirements for data-at-rest protection. In the case of data-at-rest protection, as we saw in Chapter 5, every organization acts based on its own budget and requirements. There are some compliance and regulatory requirements for encryption at rest but that is largely achieved by disk encryption, volume encryption, or transparent data encryption (TDE). However, this is just an exercise to mark the checkbox. Third, hackers always have their eyes on the data in the wild and it's relatively easier for hackers to tap into communication than to hack into the data center. In fact, this is another reason why businesses are a little relaxed about enforcing data-at-rest encryption.

6 Web Security: Data Encryption in Transit

Although cryptography has evolved significantly over the years, our data, information, and privacy are more vulnerable today than ever before. We hear news of data breaches and cyberattacks every day, making protecting data in transit absolutely necessary.

> **Chapter Highlights**
> - Overview and necessity of data encryption in transit
> - Understand and implement Transport Layer Security (TLS) and Internet Protocol Security (IPsec) protocols for web security
> - Learn about public key infrastructure (PKI) and how to use PKI in web security

This chapter is divided into four parts. Section 6.1 starts with an overview of web security where we'll briefly look at the history and the necessity of web security. Section 6.2 forms the core and will focus on various protocols and techniques used in securing communication or data in transit. The chapter will also show you how cryptography is used in daily life and some practical applications in Section 6.3. Section 6.4 will explain how PKI works and how encryption keys should be managed. Of course, web security and online communication include many different techniques and tools to protect data in transit, but the focus of this chapter is primarily on data security through cryptography.

6.1 Primer on Web Security

Security of data in transit was a concern of only the military and other handfuls of people until late in the 20th century. Even after computers made their way into businesses, the security of data at rest was the primary concern at the time. Throughout the late 1980s and even the first couple of years into the 1990s, the internet was used by academia and tech wizards. However, things were slowly turning around in the early 1990s, and businesses started using the internet for communications and information. At the time, there was no fear of data loss or hacking because businesses still weren't using the internet for e-commerce or sensitive transactions. However, it was inevitable that the use of the internet (or web) would explode eventually. It was necessary to plan and develop the technology to protect data and information when people started using the web.

Ideally, there shouldn't be any reason to worry as researchers and developers have techniques like symmetric encryption, asymmetric encryption, hashing, message authentication codes (MAC), and digital signature in their arsenal to protect the data at each and every stage throughout the data lifecycle. The availability of these techniques made cryptography an obvious choice for data protection in transit.

There was one problem, though: the man-in-the-middle attack. There is no way to confirm if the public key belongs to the person the receiver thinks it belongs to—it could be any "man in the middle."

To understand the man-in-the-middle attack, let's first recap the asymmetric or public key cryptography. In asymmetric key cryptography, there are two keys—a public key and a private key. These keys are mathematically matched. The owner of the key pair keeps the private key confidential and shares the public key freely with the sender of the message. The public key is used to encrypt the message, and the private key is used to decrypt the message. The sender of the message uses the owner or receiver's public key and encrypts the message. The receiver (or the owner of the keys) can decrypt the message with the private key. We studied several algorithms of public key cryptography in Chapter 3. The asymmetric public key system is robust and can't easily be cracked. It's almost impossible for any hacker to derive a private key from a public key. The system is rock solid against eavesdropping and passive listeners. So, if a hacker in the middle can't eavesdrop or read the data, then what is the problem?

The problem comes from an active attacker. The system is vulnerable when a hacker has access to the communication link, such as the public internet, and the hacker can replace or modify the message. Figure 6.1 shows how asymmetric cryptography can be vulnerable to an active attacker using the Diffie–Hellman key exchange.

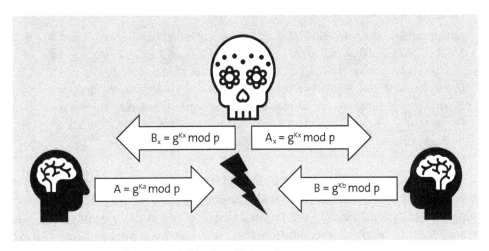

Figure 6.1 Overview of the Man-in-the-Middle Attack

Person A chooses the private key K_a and calculates the public key $A = g^{ka} \bmod p$. Person A shares this public key A with Person B. Person B, on the other hand, follows the same process by choosing a private key K_b and calculating the public key $B = g^{kb} \bmod p$. However, an active hacker, Person X, in the middle can intercept this communication and replace the public keys A and B with A_x and B_x. Person A and Person B will receive A_x and

B_x as the public keys and establish the communication without realizing the public key received is from the hacker and that they are communicating with a "man in the middle." Refer to Chapter 3, Section 3.2, for more information about the Diffie–Hellman key exchange.

Something must be done to stop man-in-the-middle attacks. In the early 1990s, people in the industry started thinking about how to prevent this type of attack and how to provide some sort of authentication so that all parties involved in communication know that the public key used is the authentic key. Secure communication for sensitive data and e-commerce transactions was needed.

In the early days of web browsing, Netscape's browser Netscape Navigator was very popular and widely used. To secure communication, Netscape developed a protocol called *Secure Socket Layer (SSL)* that not only provided confidentiality but also attempted to provide integrity and authentication. In 1993, Netscape tested the SSLv1.0 internally and found that there were many issues. As a result, SSLv1.0 was never released for public use. Netscape improved SSLv1.0 and, in February 1995, released SSLv2.0 [Davies, 2011]. SSLv2.0 was a step in the right direction but wasn't enough. It has some significant issues, such as the use of the same key for encryption and authentication, the use of a weak MD5-based hash function in the MAC, session caching issues, and so on.

Father of SSL

Among others, Horstel Feistel, Martin Hellman, Whitfield Diffie, Ron Rivest, Ralph Merkle, Taher Elgamal, Adi Shamir, and Paul Adleman are considered pioneers of modern cryptography. However, Taher Elgamal is also famous as the "Father of SSL" [Messmer, 2012]. Dr. Elgamal is not only known for the Elgamal cryptosystem and Elgamal digital signature but also for his contribution to developing the SSL protocol. Dr. Elgamal significantly contributed to developing an early version of the SSL protocol during the early to mid-1990s at Netscape Communications.

It's interesting to note that Microsoft released its own browser, Internet Explorer, in 1995 [Benaloh, 1995]. It turned out that Microsoft was also working in parallel on securing communication. Internet Explorer used Microsoft's *Private Communications Technology (PCT)* to secure communication. Fortunately, PCT was very similar to SSL, so servers were able to communicate with Netscape Navigator or Internet Explorer.

Meanwhile, given all these issues with SSLv2.0, Netscape released an improved version of the protocol as SSLv3.0 in November 1996. SSLv3.0 was a much-improved version and widely used in the mid- to late 1990s, primarily because that was the only choice (Internet Explorer was still catching up). However, there was a growing concern about every browser company developing its own protocol. This wasn't going to work, and it had to be standardized.

The Internet Engineering Task Force (IETF) formed a workgroup to establish a standard protocol to be used by everyone. Meanwhile, SSLv3.0 encountered some vulnerabilities, including the famous POODLE attack. (You'll learn more about this attack in Chapter 12.) This gave even more reasons for the workgroup to come up with a better, standardized protocol. The outcome of the IETF workgroup was the *Transport Layer Security (TLS)* protocol. The workgroup released TLSv1.0 in January 1999. TLSv1.0 has fixed all the vulnerabilities of SSLv3.0. It was basically a slightly improved version of SSLv3.0. However, the name was changed so that it was accepted by all parties as a new standard. Officially, SSLv2.0 was deprecated in 2011, and SSLv3.0 was deprecated in 2015.

> **What Is TLSv1.0?**
>
> TLSv1.0 simply fixed all the SSLv3.0 vulnerabilities. In fact, RFC 2246 noted the following:
>
> *This document describes TLS Version 1.0, which uses the version { 3, 1 }. The version value 3.1 is historical: TLS version 1.0 is a minor modification to the SSL 3.0 protocol, which bears the version value 3.0 [RFC 2246, 1999].*

TLSv1.0 was improved over time, and the newer versions TLSv1.1, TLSv1.2, and TLSv1.3 were released. Table 6.1 shows the timeline for each SSL/TLS version. In Section 1.2, you'll learn in detail how TLS works and how to implement it. In this section, we'll get a quick overview of the networking basics.

Protocol	Year of Release	Remarks
SSLv1.0 was internal only and not released for public use.		
SSLv2.0	1995	First public release of the protocol, which had many security flaws. Officially deprecated in 2011.
SSLv3.0	1996	Enhancement over SSLv2.0 that was widely used with Netscape Navigator. Officially deprecated in 2015.
TLSv1.0	1999	SSLv3.0 rebranded as TLSv1.0 after improvements. Officially deprecated in 2021.
TLSv1.1	2006	Officially deprecated in 2021.
TLSv1.2	2008	In use as of the time of writing.
TLSv1.3	2018	In use as of the time of writing.

Table 6.1 TLS Protocol Timeline

Everything in networking starts with the *Open Systems Interconnect (OSI) model*. The OSI model is a reference or conceptual model and isn't used in implementation. The

OSI model was first conceptualized in the 1970s and refined over the years. The model that is most popular and that we're familiar with now was established in the early to mid-1980s.

The network model is used to connect two systems. The OSI model is divided into seven layers, and each layer performs a specific task. Figure 6.2 shows the seven layers of the OSI model. The model is always represented with layer 7 at the top and layer 1 at the bottom because the model is presented from the destination or receiving end. Each layer communicates with the layer above and the layer below. Data starts traveling from the application layer at the source or client end and passes through all the layers from the application to the physical layer. During this time, data goes through the encapsulation process. *Encapsulation* is the process of creating a formatted message by adding the header and other necessary information to data at each layer. At the receiving or the server end, the process is reversed. Data goes from the physical layer to the application layer, and while passing through each layer, decapsulation happens. *Decapsulation*, you guessed it, is the reverse process of encapsulation. In decapsulation, the header and other information are removed at each layer as data moves up from the physical layer to the application layer.

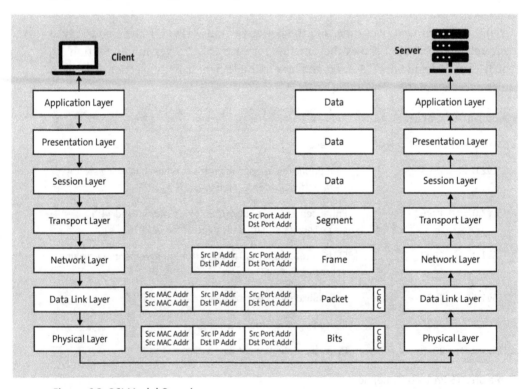

Figure 6.2 OSI Model Overview

Let's briefly explain the functionality of each layer:

- **Application layer (layer 7)**

  The application layer is the closest to the end user or the client. The application itself doesn't reside in this layer but the interface to the application and user, such as application programing interfaces (APIs) and other services, resides here. The application layer passes on the data to the presentation layer. Following are some examples of protocols operated at the application layer:

  – Electronic Data Interchange (EDI)
  – File Transfer Protocol (FTP)
  – Hypertext Transfer Protocol (HTTP)
  – Internet Message Access Protocol (IMAP)
  – Post Office Protocol version 3 (POP3)
  – Simple Mail Transfer Protocol (SMTP)
  – Simple Network Message Protocol (SNMP)
  – Telnet

  The application layer is also very important in terms of filtering and managing traffic. The web application firewall (WAF) operates at this layer and filters out unwanted traffic. In addition, the application gateway works at this layer. Based on the protocols that reside in the application layer, it provides services such as file transfer, file/directory services, mail services, and so on.

- **Presentation layer (layer 6)**

  As the name indicates, the presentation layer "presents" data to the next level. On the client or the source side, the presentation layer presents the application layer data to the session layer in the network-ready format; on the server side, the presentation layer presents the session layer data as the application-ready data. Encoding/decoding, encryption/decryption, and compression/decompression happen at this layer. Just to be clear—encoding, encryption, and compression happen on the client or the source side, while decoding, decryption, and decompression take place on the server side. This layer essentially works as an interface between the application and the network. Following are some examples of the formats that operate on the presentation layer:

  – American Standard Code for Information Interchange (ASCII)
  – Extended Binary-Coded Decimal Interchange Mode (EBCDICM)
  – Joint Photographic Experts Group (JPEG)
  – Moving Picture Experts Group (MPEG)
  – Musical Instrument Digital Interface (MIDI)
  – Tagged Image File Format (TIFF)

- **Session layer (layer 5)**

    People sometimes get confused between session and connection. The session isn't the same thing as the connection. *Connection* is establishing a physical link between the two computers or network points. The *session*, on the other hand, is establishing communication over the existing connected points. We need a connection to establish a session, but we don't have to have a session to connect two computers. In the OSI model, the session layer is responsible for establishing, maintaining, and terminating the session. The session layer also offers a message or data synchronization between the source and destination. This is done by using checkpoints or acknowledgement messages at a regular interval. The destination server sends an acknowledgement message at every regular interval to the source or clients informing if the data is received or not. The third functionality of the session layer is a dialogue control. The sessions can communicate (dialogue) in three ways: simplex, half-duplex, and duplex.

    In *simplex* mode, communication takes place in one direction only. Most broadcast messages are an example of a simple communication mode. In *half-duplex* mode, communication can happen in both directions, but messages or data can be sent in only one direction at a time. A walkie-talkie is an example of a half-duplex communication mode. Lastly, *duplex* (sometimes referred to as a full duplex) mode of communication is a normal communication where both parties can send and receive messages at the same time. Phones are an example of duplex communication.

    Following are two examples of protocols operated at the session layer:

    - Network File System (NFS)
    - Structured Query Language (SQL)

- **Transport layer (layer 4)**

    The transport layer is responsible for establishing a logical connection between two applications. The session layer passes on the data to the transport layer, and the transport layer creates data segments. Because the transport layer establishes logical connections between the applications that run on the ports, the transport layer requires adding the port addresses or port numbers to the data segment. The transport layer also adds the sequence control number to the data segment. *Sequence control* provides reliable data delivery. A sequence control number is used to make sure that all the data segments are received, and the message is recreated in the proper order at the destination. The port address and sequence control number are added as a header to the data segment as part of the encapsulation process. The transport control layer also ensures flow control, error control, data loss control, and data duplication control. Following are two examples of protocols operated at the transport layer:

    - Transport Control Protocol (TCP)
    - User Datagram Protocol (UDP)

TCP is a connection-oriented protocol. TCP uses a three-way handshake process between source and destination to confirm that a connection is established to begin the communication. We'll discuss this handshake in more detail in Section 6.2.1. TCP also confirms that the communication is terminated gracefully (using the FIN flag). UDP, on the other hand, is a connectionless protocol. It doesn't guarantee an established connection and doesn't have any flow or sequence control either. Because of less overhead, UDP is faster than TCP. When we visit a YouTube channel, the connection is established using TCP, but the video is streamed using UDP.

> **Not in the Transport Layer**
>
> The TLS protocol isn't part of the transport layer protocols. However, in the TCP/IP model, it resides between the application layer and the transport layer. This is because it works with the application layer protocol, such as HTTP and FTP, as well as the transport layer protocols, such as TCP and UDP.

- **Network layer (layer 3)**

  Data segments are converted into packets at the network layer. The primary functionality of the network layer is routing. The network layer achieves this functionality by adding source and destination Internet Protocol (IP) addresses to the packet. The IP addresses are added as headers to the data packets as part of the encapsulation process. The IP addresses are logical addresses and are used by routers in the forwarding tables. Routers are operated at the network layer.

  Following are some examples of protocols operated at the network layer:
    - Internet Control Message Protocol (ICMP)
    - Internet Group Management Protocol (IGMP)
    - Internet Protocol (IP)
    - Internet Protocol Security (IPsec)
    - Network Address Translation (NAT)
    - Routing Information Protocol (RIP)

- **Data link layer (layer 2)**

  The data link layer receives the packets from the network layer. The packets, however, can't be sent to the physical layer (e.g., Ethernet) for transmission without proper formatting. The primary responsibility of the data link layer is to convert the packets into frames that can be transmitted using ethernet (or the appropriate medium). The data link layer also adds the media access control (MAC) address as the header and cyclic redundancy checksum (CRC) as the footer of the message. The MAC address is the physical address of the device. It's 48 bits (6 bytes) long and represented in a six-digit hexadecimal number. The CRC is added for integrity.

  The data link layer is divided into two sublayers based on the functionality: the logical link control (LLC) layer and the MAC layer. The LLC layer interacts with the network layer while the MAC layer interacts with the physical layer. Hardware, such as switches and bridges, are operated at the data link layer. The data link layer at the receiving end confirms if the MAC address is correct for the receiving device and then removes the header.

Following are some examples of protocols operated at the data link layer:
- Address Resolution Protocol (ARP)
- Integrated Services Digital Network (ISDN)
- Point-to-Point Protocol (PPP)
- Point-to-Point Tunneling Protocol (PPTP)
- Reverse Address Resolution Protocol (RARP)
- Serial Line Internet Protocol (SLIP)

- **Physical layer (layer 1)**

The primary function of the physical layer is to provide a physical connection between two devices. The data can only be sent if there is a connection between two devices. The physical layer is also responsible for receiving frames from the data link layer and converting them to a series of binary digits because messages can't be transmitted in the frame format. These binary bits are then transmitted over an ethernet or other medium connecting two devices. The physical link layer also synchronizes and controls the message rate or throughput between the devices based on the medium. Network Interface Cards (NICs), hubs, and repeaters work on the physical layer.

At the receiving end, the physical layer converts the binary bit stream into frames and forwards the frames to the data link layer on the destination side.

Following are some examples of protocols operated at the physical layer:
- Bluetooth
- Ethernet
- Integrated Services Digital Network (ISDN)
- Universal Serial Bus (USB)

As mentioned earlier, the OSI model is a reference model for how the message should be structured and isn't implemented in practice. The TCP/IP networking model is the actual implementation model used in the real world. As shown in Figure 6.3, the TCP/IP model comprises only four layers: application layer, transport layer, internet layer, and host-to-network layer. The application layer, presentation layer, and session layer of the OSI model are consolidated in one layer, the *application layer*, in the TCP/IP model. The transport layer and network layer are the same in both architectures. However, the network layer is also referred to as the *internet layer* in the TCP/IP model. The functionality of the last two layers of the OSI model, the data link layer and the physical layer, are consolidated in the *host-to-network* layer in the TCP/IP model. This layer is also known as simply the network layer or the link layer of the TCP/IP model. The naming of the TCP/IP layers is somewhat flexible, partly because people are trying to align or map the functionality of the two models and partly to avoid any confusion or misunderstanding because of naming. The functionality of TCP/IP layers is exactly the same as the OSI model, so we won't describe the TCP/IP layers here.

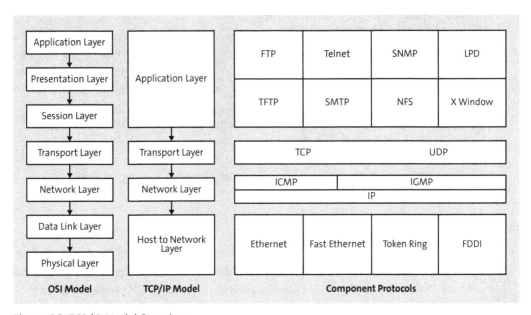

Figure 6.3 TCP/IP Model Overview

6.2 Web Security Protocols

When using web applications, such as email systems, e-commerce sites, and so on, we want to make sure that the data moving between client and server is secured. Data in transit security is implemented using various protocols. You'll be surprised to know that in implementing data in transit security in the real world, we'll use everything we've learned so far about cryptography—symmetric encryption, asymmetric encryption, hash functions, message authentication codes (MACs), and digital signatures.

Two primary or most commonly used protocols for securing data in transit are the TLS protocol and the IPsec protocol. The TLS protocol is used with the public internet, while the IPsec protocol is used primarily in virtual private networks (VPNs). In this section, we'll discuss both the TLS and IPsec protocols.

6.2.1 Implementing the TLS Protocol

TLS is the most widely used protocol for securing online communication. The latest version of TLS is version 1.3. As we saw earlier in Section 6.1, previous versions of SSL and TLS were found to be vulnerable and were eventually deprecated. TLSv1.2 and TLSv1.3 are the active versions. In this section, you'll learn how TLS works and how to implement TLS. The description and steps will be as close to TLSv1.2 as possible. Then, you'll learn more about TLSv1.3 and the benefits it offers over the previous versions.

> **TLS Protocol: A Workhorse of Web Security**
>
> The TLS protocol is used to secure the connection between two computers—server and client, email exchanges, file transfers—almost everything that we do online is secured using the TLS protocol. It's almost everywhere and works with almost everything, primarily because the protocol is designed to be independent of the applications, software, and the type of the system (hardware). For example, a client might be connecting to a server using a web browser while another client may be connecting to the same server from a mobile app, and the TLS protocol works. Or one application may be coded in Java, and another may have the protocol developed in Python, and the TLS protocol will still work. Lastly, it doesn't matter to the protocol what is being encrypted.

TLS Handshake

The TLS protocol works in two parts: TLS handshake and TLS record. The *TLS handshake* protocol, as shown in Figure 6.4, provides authentication by establishing the identity of the server (also of the client when needed) and also establishes the key agreement to be used in encrypting the actual data. The TLS record protocol guarantees the confidentiality and integrity of the data exchanged using the protocol.

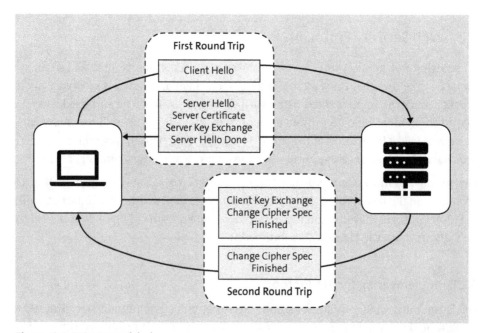

Figure 6.4 TLS 1.2 Handshake

The TLS handshake starts when the client reaches out to the server. (In an ideal world, the server can also initiate the handshake, but, in reality, this doesn't happen for the most part—after all, why would *www.amazon.com* or *www.google.com* reach out to your laptop? But technically, this is feasible.)

Communication starts by establishing a connection between the client and the server using the three-way handshake of TCP. The client first sends a SYN flag to the server. This is a "starts synchronizing" message from the client to the server. The server responds by saying "I acknowledge your synchronizing request" with an SYN/ACK flag. The client then acknowledges with an ACK flag that they received the SYN/ACK flag from the server. Now the connection is established between the client and the server. Next, the TLS protocol exchange begins.

The structure of the TLS handshake message is defined as shown in Listing 6.1 [RFC 5246, 2008]:

```
struct {
    HandshakeType msg_type;
    uint24 length;
    select (HandshakeType) {
        case hello_request:       HelloRequest;
        case client_hello:        ClientHello;
        case server_hello:        ServerHello;
        case certificate:         Certificate;
        case server_key_exchange: ServerKeyExchange;
        case certificate_request: CertificateRequest;
        case server_hello_done:   ServerHelloDone;
        case certificate_verify:  CertificateVerify;
        case client_key_exchange: ClientKeyExchange;
        case finished:            Finished;
    } body;
} Handshake;
```

Listing 6.1 TLS Handshake Message

The client first sends the ClientHello message to the server. The client can send the client hello request to the server at any time to initiate a new session, when it receives a HelloRequest from the server, or at any time the client sees a need to renegotiate the security parameters. The ClientHello message contains a lot of information, including the latest version of the TLS that is supported by the client, available cipher suites, and a random number to be used in generating a shared secret. These three pieces of information are important requirements to establish secure communication between the client and the server.

The ClientHello message can be up to 512 bytes. Some servers have a very strict requirement that the ClientHello message must be exactly 512 bytes. This type of server requires padding to meet the size requirements. Some servers are flexible and accept the `ClientHello` message without padding.

Following are some of the key parameters included in the `ClientHello` message:

- **Handshake type**
  This byte indicates the type of the handshake message. 0x01 represents the client hello message. This byte is always 0x01 for the `ClientHello` message.

- **Message size**
  The next 3 bytes represent the message size; that is, the number of bytes to be followed. The length excludes the first byte, the handshake type, and the three bytes reserved for the length. Everything after the length is counted and included in the length. The maximum length is 508 bytes (512 – 4).

- **Version**
  The next 2 bytes in the client hello message indicate the highest TLS version supported by the client. These two bytes are generally 0x0303 for TLSv1.2 and 0x0304 for TLSv1.3. This allows the server to pick the latest version available. If the client and the server both support the latest versions, then the server will pick that version; otherwise, the latest version supported by both the client and the server is chosen by the server.

- **Client random**
  The client random is a 32-byte long random number generated and sent by the client to the server as part of the client hello message. Of the 32-byte random number, 28 bytes are later used as a part of the Diffie–Hellman key exchange to generate the shared secret. The first 4 bytes represent the client's timestamp.

> **Timing of the Timestamp**
>
> The first 4 bytes of the random number are used to indicate the time in seconds since midnight January 1st, 1970, GMT. This was originally introduced to preserve the randomness of the number. The fear was if the random number generator (RNG) somehow ended up generating the same number more than once, then any eavesdropper could eventually calculate the shared secret. This makes TLS very vulnerable. To avoid this situation, a time in seconds was introduced in the first 4 bytes of the random number. However, over time, it was noticed that not all computers are synchronized, and the timings are different. This allows an attacker to identify or fingerprint the client. For this reason, it's recommended that the use of timestamps should be discontinued in the random number [Matthewson, 2013].

- **Session ID**
  The session ID is associated with the session and remains the same for the session. This is primarily used by the server to figure out if this is a part of the existing session that needs to be resumed or if a new session needs to be established. The server

checks the cache, and if the session ID is in the cache, then the server knows it's an existing session. Otherwise, the server starts working on the handshake. The session ID allows the client to continue with the existing session and save time starting over a new session. The value for the session ID is always 0x00 if it's a new session.

- **Cipher suite**

  The next 2 bytes represent the number of cipher suites supported and then the list of the supported cipher suites. Each cipher suite is represented by 2 bytes. For example, if there is only one cipher suite supported, then the length is 2 bytes. It will be 0x0002 and 0x0033 (code for the tone cipher suite). Usually, it's a list of all the cipher options supported by the client sent, such as key exchange primitive, symmetric algorithm for data encryption, hashing algorithm for the MAC, and so on. The most desirable or the client's favorite cipher suites are listed first.

- **Compression method**

  Each compression method is represented by 1 byte. If there is a nonzero value, then it indicates a compression method will be used during the TLS record protocol. The compression method negotiated in the TLS handshake protocol will be used during the actual data exchange. This byte is 0x00 if no compression method is used. By default, the TLS protocol uses the compression method as null. If compression is used, it must be lossless compression. The compression method is discontinued in TLSv1.3. For more information, refer to RFC 3749: Transport Layer Security Protocol Compression Methods.

- **Extensions**

  The extension field allows the client to share optional or additional information. This was introduced in TLSv1.2. Some examples of extensions are signature algorithms, Server Name Indication (SNI), renegotiation information, elliptic curve type supported (if elliptic curve cryptography [ECC] is used) and so on—plus, of course, the padding! The extension field starts with the total length of the extension bytes and ends with the length of the padding bytes and the padding bytes themselves. Each extension type starts with the length and is followed by the extension.

Putting these pieces together, the structure of the `ClientHello` message is shown in Listing 6.2 [RFC 5246, 2008].

```
struct {
    ProtocolVersion client_version;
    ClientRandom random;
    SessionID session_id;
    CipherSuiteLng cipher_suites_lng;
    CipherSuite cipher_suites;
    CompressionMethodLng compression_methods_lng;
    CompressionMethod compression_methods;
    select (extensions_present) {
            case false:
```

```
                struct {};
                case true:
                Extension extensions;
            };
        } ClientHello;
```

Listing 6.2 ClientHello Message

The client sends the `ClientHello` message to initiate secure communication. The client then waits for the server to respond with a `ServerHello` message. Any message other than a `ServerHello` from the server will end up in a fatal error and terminate the connection. The server picks the latest version of the protocol and responds with a `ServerHello` message. If, however, the server doesn't support the versions offered by the client, then the server sends a failure message, and the connection is closed. If the server supports TLSv1.3, but the `ClientHello` message indicates that the client can support only up to TLSv1.2, then the server accepts this and responds with TLSv1.2 as the protocol version to be used. In the `ServerHello` message, the server also sends the session ID, a random number to be used in generating a shared secret and selected cipher suite.

Following are some of the key parameters included in the `ServerHello` message:

- **Handshake type**
  This byte indicates the type of the handshake message. 0x02 represents the `ServerHello` message. This byte is always 0x02 for the `ServerHello` message.

- **Message size**
  The next 3 bytes represent the message size—the number of bytes to be followed. The length excludes the first byte, the handshake type, and the 3 bytes reserved for the length. Everything after the length is counted and included in the length.

- **Version**
  The next 2 bytes in the server hello message indicate the highest TLS version supported by the client. These 2 bytes are generally 0x0303 for TLSv1.2 and 0x0304 for TLSv1.3. Based on the version offered by the client in the client hello message, the server picks the TLS version and responds with the latest version possible. If the client and the server both support the latest versions, then the server will pick that version; otherwise, the server chooses the latest version supported by both the client and the server.

TLS Versioning

Do you wonder why TLS version bytes in the protocol for TLSv1.2 and TLSv1.3 are 0x0303 and 0x0304, respectively? Why aren't the values for the TLS version bytes 0x0102 and 0x0103? The answer lies in the SSL versioning. As we discussed earlier, TLSv1.0 was technically an improvement on SSLv3.0. The IETF decided to keep TLSv1.0 as version 3.1

> (in continuation of the SSL versioning scheme) in the protocol and indicated it by 0x0301. When TLSv1.1 was released, this number in the protocol was changed to 0x0302. TLSv1.2 is represented by 0x0303, and TLSv1.3 is represented by 0x0304.

- **Server random**
  The server random is a 32-byte long random number generated and sent by the client to the server as part of the client hello message. Of the 32-byte random number, 28 bytes are eventually used as a part of the Diffie–Hellman key exchange to generate the shared secret. The first 4 bytes represent the server timestamp.

- **Session ID**
  If the client hello message has session ID as 0x00 that means it's a new session. This tells the server that there is no previous session ID for this session and a new session ID needs to be assigned. The new session ID number will be saved by the server and the client to avoid the handshake process for the same session again. This could be very helpful on the server side and save a lot of time if it's a busy site.

- **Cipher suite**
  The server selects a cipher suite from the list shared by the client via the `ClientHello` message. The server informs the client about the choice of the cipher suite using this field.

- **Compression method**
  The server selects one compression method from the list offered by the client. If the value of this field is 0x00, there is no compression.

- **Extensions**
  The extension field allows the client to share optional or additional information. The server can return an extension that was sent by the client—nothing outside of what was sent in the `ClientHello` message.

The structure of the `ServerHello` message can be defined as shown in Listing 6.3 [RFC 5246, 2008].

```
struct {
    ProtocolVersion server_version;
    serverRandom random;
    SessionID session_id;
    CipherSuiteLng cipher_suites_lng;
    CipherSuite cipher_suites;
    CompressionMethodLng compression_methods_lng;
    CompressionMethod compression_methods;
    select (extensions_present) {
            case false:
            struct {};
            case true:
```

```
            Extension extensions;
        };
} ServerHello;
```

Listing 6.3 ServerHello Message

Once the `ServerHello` message is sent over to the client, the server must follow with a `ServerCertificate` message if the agreed key exchange method uses a certificate for authentication. This message is used to send the server's digital *certificate* to the client to prove the server's identity. The `ServerCertificate` message includes a certificate that contains a public key of the server and a digital signature (signed by the server's private key). The certificate tells the client that the public key belongs to the server and proves that the key really belongs to that particular server. This action takes care of the authentication piece of the cryptography services. (The other two services—confidentiality and integrity—will be done during the actual data exchange.)

The server certificate also includes a signature by the certificate authority (CA). The `ServerCertificate` message includes all the certificates signed by leaf or intermediate CAs as well as the root certificate. The `ServerCertificate` message includes the following key parameters:

- **Handshake type**
  This byte indicates the type of the handshake message. 0x0B represents the `ServerCertificate` message. This byte is always 0x0B for the `ServerCertificate` message.

- **Message size**
  This field indicates the size of the message or the length of the message.

- **Certificate length**
  This field indicates the length of all the certificates to follow after this field.

- **Certificate**
  This field has the actual certificate or certificates. In cases where there is more than one certificate in the message, the server's certificate must come first. All other certificates can follow after the server's certificate. Each certificate should start with a length of the certificate to be followed. To clarify, the certificate length field indicates the total length of all certificates. In this field each certificate starts with its own length, for example, LNGTH1, CERT1, LNGTH2, CERT2., LGNTHn, CERTn.

It's the client's responsibility to validate each certificate received from the server. The client is also on the hook to check for the revocation list using either the certificate revocation list (CRL) or Online Certificate Status Protocol (OCSP). For more information, see Section 6.4.

The client can also send a similar message, `ClientCertificate`, to the server if needed. However, in most cases, the client is initiating the connection, so there is no need to prove the client's identity.

So far, the client and server have completed the following:

1. Established a secure connection using a three-way TCP handshake
2. Agreed on the protocol version to be used for the data exchange, decided which cipher suite to use to provide confidentiality and integrity to the payload, and shared the random numbers that can be used in generating the shared secret
3. Authenticated the server

The next step is to make sure that both the client and the server have the same (symmetric) key to encrypt and hash the data or the payload. To achieve this, the server sends the `ServerKeyExchange` message right after the `ServerCertificate` message. This message is optional because this information is largely dependent on the type of cipher suite that was used. If the message is sent, the information sent within the message can also vary based on the type of algorithm or the cipher suite used. We'll continue with our Diffie–Hellman key exchange algorithm example from Section 6.1. In the Diffie–Hellman algorithm, public parameters—generator g, prime number p, and the public key of the server—are shared with the client.

The server chooses the private key K_s and calculates the public key $Y = g^{Ks} \mod p$. The server shares this public key, the generator g, and the prime number p with the client using the `ServerKeyExchange` message. The client, on the other hand, chooses the private key K_c and calculates the public key $X = g^K \mod p$. The client shares this public key X with the server using the `ClientKeyExchange` message (explained next).

The goal of the `ServerKeyExchange` message is to send these public parameters to the client. The `ServerKeyExchange` message looks like this:

- **Handshake type**
  This byte indicates the type of the handshake message. 0x0C represents the `ServerKeyExchange` message. This byte is always 0x0c for the `ServerKeyExchange` message.
- **Message size**
  This field indicates the size of the message or the length of the message.
- **Prime p**
  The server shares the length of public parameter p and the number or the value of p.
- **Generator g**
  The server provides the length of the generator followed by generator g.
- **Public key**
  The server sends the length of the public key and the Diffie–Hellman public key, Y.
- **Hash algorithms and signature**
  The server sends the hash algorithm being used for the signature generation and the digital signature parameters to be used with the Diffie–Hellman digital signature. This signature uses the private key from the server's certificate. The same key is used in the `ServerHello` message for the authentication.

The structure of the `ServerKeyExchange` message is defined as shown in Listing 6.4 [RFC 5246, 2008].

```
struct {
      select (KeyExchangeAlgorithm) {
          case dh_anon:
              ServerDHParams params;
          case dhe_dss:
          case dhe_rsa:
              ServerDHParams params;
              digitally-signed struct {
                  opaque client_random[32];
                  opaque server_random[32];
                  ServerDHParams params;
              } signed_params;
          case rsa:
          case dh_dss:
          case dh_rsa:
              struct {} ;
              /* message is omitted for rsa, dh_dss, and dh_rsa */
          /* may be extended, e.g., for ECDH -- see [TLSECC] */
    } ServerKeyExchange;
```

Listing 6.4 ServerKeyExchange Message

After the `ServerKeyExchange` message is sent, the server can send a request for the client's certificate if the server needs to verify the client's identity. However, this message is optional and isn't sent for most communication. If this message is sent, then the client is required to send the certificate chain. The client's certificate must be signed by the algorithm agreed upon earlier. If the server chooses not to request the client's certificate, then the server sends `ServerHelloDone`.

The `ServerHelloDone` message is sent to indicate to the client that the `ServerHello` message is over, and the server has sent everything required for the client to calculate the shared secret or the parameters needed to calculate the encryption key. The `ServerHelloDone` message has the record header and handshake type field:

- **Record header**
  The record header includes the type of the message, the protocol version, and the number of bytes to follow. The type of the message is the handshake record, which is 0x16; the version is TLSv1.2, which is 0x0303; and 4 bytes of the handshake message to follow, which is 0x04.

- **Handshake type**
  This byte indicates the type of the handshake message. 0x0E represents the `ServerHelloDone` message. This byte is always 0x0E for the `ServerHelloDone` message.

Now the ball is in the client's court. Upon receiving the `ServerHelloDone` message, the client must send the certificate if the server has requested it. Otherwise, the client can proceed with the `ClientKeyExchange` message.

The server and the client have already agreed to use the Diffie–Hellman key exchange. The server has already shared the public parameters, g and p. The only information the client needs to share with the server is the client's public key. The client sends the public key in the `ClientKeyExchange` message to the server using the following parameters:

- **Handshake type**
  This byte indicates the type of the handshake message. 0x10 represents the `ClientKeyExchange` message. This byte is always 0x10 for the `ClientKeyExchange` message.
- **Public key**
  The client sends the length of the public key and the Diffie–Hellman public key, X.

The structure of the `ClientKeyExchange` message is defined as shown in Listing 6.5 [RFC 5246, 2008].

```
struct {
        select (KeyExchangeAlgorithm) {
            case rsa:
                EncryptedPreMasterSecret;
            case dhe_dss:
            case dhe_rsa:
            case dh_dss:
            case dh_rsa:
            case dh_anon:
                ClientDiffieHellmanPublic;
        } exchange_keys;
    } ClientKeyExchange;
```

Listing 6.5 ClientKeyExchange Message

Once the client sends the public key, the next message is the `ChangeCipherSpec` message. The client sends this to indicate to the server that it has completed all the required calculations to generate the shared encryption key and the next message sent by the client will be the encrypted handshake message. The `ChangeCipherSpec` message has the following key parameters:

- **Handshake type**
  This byte indicates the type of handshake message. 0x14 represents the `ChangeCipherSpec` message. This byte is always 0x14 for the `ChangeCipherSpec` message.
- **Message**
  The message is always 1 byte long and has the value of 0x01.

Lastly, the client sends the Finished message right after sending the ChangeCipherSpec message to indicate that the key exchange and the authentication process have been completed successfully. The Finished message is the first message to be encrypted by the newly agreed upon symmetric key. The client performs the following steps to generate this message:

1. Concatenate all the messages exchanged so far with the server—ClientHello, ServerHello, ServerCertificate, ServerKeyExchaange, ServerHelloDone, ClientKeyExchange, and ChangeCipherSpec. This doesn't include the server request if the communication was initiated by the server and the handshake record header.
2. Generate the hash of this concatenated message.
3. Encrypt the hash using the agreed encryption key and send the cipher text to the server.

The server also sends the ChangeCipherSpec message to the client to indicate that it has all the required information and completed the calculations successfully. The Finished message is followed immediately after the ChangeCipherSpec message. The server needs to generate the Finished message in exactly the same way and should have all the messages like the client. However, there's a small twist here. The server must decrypt the Finished message sent by the client, append that to the rest of the messages, and then generate the hash and concatenate.

The purpose of the Finished messages is to verify that the messages haven't been tampered with during the negotiation process. This is to protect the communication from man-in-the-middle and downgrade attacks. The server sends a fatal error alert and aborts the connection if the Finished message isn't received right after the ChangeCipherSpec message. This concludes the TLS handshake protocol.

TLS Record

Once this is successfully completed, the client and the server send the data securely over the public internet using the TLS record protocol. The data sent is encrypted and is appended with a MAC to provide confidentiality and integrity using the encryption key agreed upon by the client and the server during the handshake protocol.

The MAC keys are derived from the master secret. The master secret is derived from the premaster secret and the random numbers exchanged during the hello messages. The premaster secret is nothing but the common values derived by the server and the client after the Diffie–Hellman public parameters and public key exchanges.

TLSv1.3

Although TLSv1.2 is still active and being used, TLSv1.3 was introduced in 2018 as a more robust and faster version. Under the hood, the functionality of TLSv1.3 is very similar to the functionality of TLSv1.2, such as a handshake, key exchange, and so on. However,

TLSv1.3 is a much better version because of all the improvements. Following are some of the key improvements of TLSv1.3 over TLSv1.2:

- TLSv1.3 introduces *forward secrecy*. This means if the server's private key is compromised, the session keys (and previous data) aren't compromised. (The session key is the symmetric key, which isn't compromised in this situation, so the data encrypted before isn't compromised either.)
- TLSv1.3 doesn't support any short 64-bit or 80-bit encryption key lengths. This automatically removes support for weaker ciphers, such as Data Encryption Standard (DES). In fact, TLSv1.3 supports only two block ciphers, AES-GCM and AES-CCM, and one stream cipher, ChaCha20. TLSv1.3 doesn't support other weak ciphers like RC4 or Advanced Encryption Standard (AES) in cipher block chaining (CBC) mode.
- TLSv1.3 also dropped support for the weak hash functions, such as MD5 and SHA-1. TLSv1.3 supports only SHA-256 or better for hashing used in the MAC.
- TLSv1.3 stopped supporting compression methods. This move protects the data from the CRIME attack.
- TLSv1.2 relies on SessionID to resume the session. If the SessionID isn't available or it expired, a new handshake needs to be performed. TLSv1.3 improves this by using a *session resumption* technique. With session resumption, TLSv1.3 sends a preshared key and a new public key in the ClientHello message. The server is able to use this information to generate a shared secret. The server creates a MAC of this data and sends it to the client in the ServerHello request. The client validates the MAC and knows that this is the right server.
- TLSv1.3 has introduced robust downgrade attack protection. In TLSv1.2, the attacker can intercept the ClientHello message and trick the server into using an older (and potentially weaker) version of the TLS protocol. In TLSv1.3, the server is using a 32-byte random number in the ServerHello message to protect from the downgrade attack. The server uses the first 8 bytes of the 32-byte random number to send a specific pattern to the client if the client has requested TLS version 1.1 or 1.2. If the client has requested TLSv1.3, these 8 bytes are random. In this case, if an attacker has tried to downgrade the version to 1.2, the client will know because the client isn't expecting a fixed pattern in the ServerHello.

Removing all the weaker options, TLSv1.3 has become a lot leaner and more efficient. Even the handshake process is simplified to be completed in only one round trip. As shown in Figure 6.5, there is only a ClientHello request going from the client to the server, which includes the possible cipher suites and a public key. The server then confirms the version and the cipher suite and also sends additional information such as key share (public parameters), certificate, finish message, and so on. The client then sends the Finish message. (The client also sends a certificate and other information as requested by the server.)

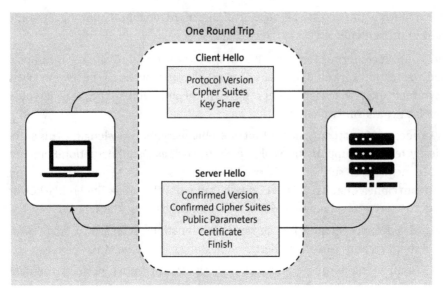

Figure 6.5 TLS 1.3 Handshake

> **TLS with UDP**
>
> The TLS protocol can work with TCP as well as with UDP. As we saw in the previous section, UDP is a connectionless protocol and works somewhat differently from TCP. The UDP version of the TLS protocol is different, and it's known as the Datagram Transport Layer Security (DTLS) protocol.

6.2.2 Implementing VPNs Using IPS

In recent years, the number of people working remotely has significantly increased. These remote workers access their company's network to perform their duties. However, this increases the risk for companies because an attacker can eavesdrop or even intercept while the employee is accessing the data on the company's network. To avoid this situation, organizations use a virtual private network (VPN). A VPN is basically a tunnel/private network in which the data traveling through this private network is concealed from the outside world. VPN is a virtual network established on the existing physical network. (It's like requesting a private meeting room within a public library.)

There are two primary ways to use a VPN:

- **Remote access**
  In a remote access VPN, a user connects his laptop to a remote network or site. This type of VPN is primarily used by employees to securely connect to the company's network while working off-site or traveling. Most people use remote access VPNs.

- **Site-to-site**

  The site-to-site VPN securely connects two different networks. The site-to-site VPN became more popular after the use of the public cloud exploded in recent years. When any organization heavily uses the public cloud, the organization's network is usually connected to the public cloud using a site-to-site VPN.

We learn more about the ways to connect to the public cloud in Chapter 7. Regardless of whether a remote access VPN or site-to-site VPN is used, the VPN secures the data by creating a virtual (or logical) tunnel. This tunnel or secure communication is created using a set of protocols. At a high level, these protocols work similarly to the TLS protocol you learned about in the previous section.

> **Virtual Private Network versus Software-Defined Network**
>
> In the cloud or dynamic environment, software-based VPN architectures are gaining popularity. This type of architecture is called a software-defined network (SDN). In an SDN, there are no VPN servers, but software controllers (SC) manage the authentication and connection parameters for the hosts. Hosts then negotiate the key exchange and set up the connection. SDN is a very attractive alternative to hardware-based VPNs. The SDN-based networks, including VPNs, can be set up easily. In the cloud and other shared/virtual environments, this helps because SDNs allow cloud providers to easily set up, manage, and tear down VPNs at any time.

Internet Protocol Security

The computers exchange traffic with each other on a network using logical addresses (IP addresses) and physical addresses (MAC addresses). When the networking models, such as the OSI model and TCP/IP, were designed, the goal was to move traffic from one network to another network in the most efficient way. All focus was on efficiency and accuracy. Security of the traffic probably didn't make the list of the design criteria or priorities, but rather was added later on. This resulted in a "bolt-on" security as opposed to a "built-in" security. *Internet Protocol Security (IPsec)* is a good example of bolt-on security.

IPsec was first published by IETF RFC 1825 in 1995 and later RFC 2401 in 1998. IPsec operates at the network layer (OSI model) or internet layer (TCP/IP model) [RFC 4301, 2005] [Dhall, 2012]. IPsec is a suite of protocols, primarily including the following:

- Authentication Header (AH) protocol
- Encapsulating Security Payload (ESP)
- Internet Key Exchange (IKE) protocol
- IP Payload Compression Protocol (IPComp)

Additionally, IPsec works in two modes—transport mode and tunnel mode. In *transport mode*, the data packet retains the original IP address, and only the data is encapsulated. On the other hand, in *tunnel mode*, the entire packet, including the original IP address, is encapsulated. The source router or the gateway assigns the new IP address of the destination router or the gateway. Once the messages reach the destination gateway or the VPN server, the packet is decrypted, and the data is routed to the original IP address. A quick overview of tunnel mode and transport mode VPN is given in Table 6.2.

Tunnel Mode VPN	Transport Mode VPN
▪ Entire packet encrypted ▪ Provides end-to-end security ▪ Used to connect two networks ▪ Additional overhead in encapsulating the entire packet	▪ Only data encapsulated ▪ Provides data security only ▪ Used to connect from host to another host or a network ▪ Less secure, but efficient

Table 6.2 Tunnel Mode versus Transport Mode

Under the hood, the IPsec protocol, like the TLS protocol, uses the fundamental components of cryptography—symmetric encryption, asymmetric encryption, and hashing. You should now be familiar with these algorithms, so we won't repeat the descriptions here.

Internet Key Exchange

The process of setting up a VPN using IPsec starts with the IKE protocol. IKE uses other key exchange protocols, such as the Internet Security Association Key Management Protocol (ISAKMP), Oakley protocol, and Security Key Exchange Mechanism (SKEME) protocol. The Oakley protocol is focused on the Diffie–Hellman key exchange while the SKEME is more versatile and offers more options for key exchanges. ISAKMP is a framework for the authentication and key exchange.

In this part of the process, both ends (i.e., the client and the server) work on authentication and negotiating the key parameters. It's in this phase that we confirm or authenticate the sender and negotiate the key parameters. Does this sound like a TLS handshake process? It probably does because it's a very similar implementation here; however, it operates on a different network layer. Like the TLS handshake, the authentication is achieved by a public key or a digital certificate. To generate the symmetric key or the shared secret—to be used later to encrypt the data or the payload—the Diffie–Hellman (or something similar, e.g., the ECC) method is used. The public parameters—prime number p, generator g, and the public key—are shared in this step of the process. If everything goes well, then at the end of this step, a *security association (SA)* is established; in other words, a connection is formed between the two parties.

The SA is a simplex communication link that sends data in only one direction. To establish a two-way communication, two SAs must be set up—one for each direction. The SAs have the following properties:

- Security Parameters Index (SPI), which is used in the AH or ESP header to associate a message with a particular SA.
- IP destination address, which is used to identify the destination for the SA link.
- The protocol identifier, which tells us which protocol is using this SA.
- The sequence number parameter of SA, which keeps track of the message packet and helps identify the repeats or missing packets.
- The Lifetime parameter, which keeps track of how long the SA should be kept alive. The SA connection is lost when it reaches the lifetime.
- The IPsec mode used, which is either the transport mode or the tunnel mode.

Authentication Header and Encapsulating Security Payload

Once the heavy lifting of key authentication and key exchange is completed, phase 2 of setting up of VPN begins. In this part, the other two protocols are used: Authentication Header (AH) and Encapsulating Security Payload (ESP).

The AH protocol adds the authentication and integrity data as the header of the packet. Figure 6.6 shows the AH protocol.

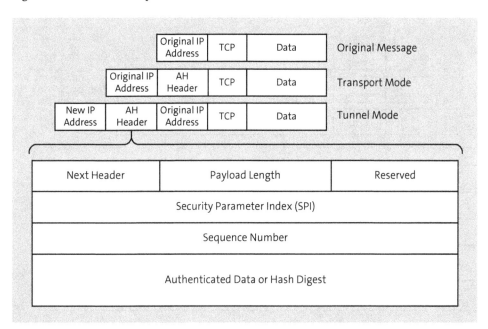

Figure 6.6 IPsec Authentication Header

The AH passes the data or the message through the hash function. The generated hash function is encrypted with the symmetric key, and the encrypted hash digest is appended as a header to the message. The data itself is still in the clear text. The receiver receives the message with the authentication header as part of the packet. The receiver decrypts the digest value with a symmetric key. The unencrypted hash value is sent through the same function and compares the generated hash value with the received hash values. If the values are the same, then the data isn't modified along the way. This provides data integrity. Additionally, the hash function is a keyed function, and the same symmetric key is used by the sender and the receiver. If the receiver can use the key to extract the hash digest, then it assures that the packet came from the right sender. This authenticates the sender. The symmetric key is generated in the previous phase using the IKE protocols.

The most important protocol of IPsec is ESP. The message sent using the AH protocol is in clear text and doesn't provide confidentiality. In fact, the National Institute and Standard Technology (NIST) discourages the use of the AH protocol in implementing IPsec [NIST 800-77, 2020]. The recommendation is to use ESP protocol, which encrypts the message and provides confidentiality. Figure 6.7 shows the header structure of the ESP protocol. The encapsulated data is sent either using the transport mode or the tunnel mode. In the transport mode, the original user IP address is visible, while in the tunnel mode, everything is encapsulated, and the new source and destination IP addresses are assigned.

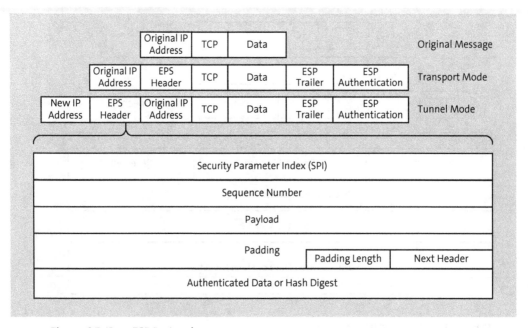

Figure 6.7 IPsec ESP Protocol

> **TLS versus IPsec**
>
> If TLS and IPsec both do almost the same things, then why do we have two protocols? This is a valid question. The TLS protocol operates at a higher level of the OSI and TCP/IP models—between the application and transport layers. On the other hand, the IPsec protocol operates at the network/internet layer. The higher the protocol, the less likely data is to be secured. The TLS protocol encrypts only the application data while the IPsec protocol, even in the transport mode, encrypts everything except the IP address and encrypts the entire packet, including the IP addresses in the tunnel mode. TLS is primarily used for point-to-point communication, such as from browser to server. IPsec is used to connect to an entire network.

IP Payload Compression Protocol

The last and probably the least used IPsec protocol is IPComp. This protocol is considered if the data needs to be compressed. It doesn't compress the packet if the message is small; it only tries to compress the larger packets. Many different compression algorithms are used, and if the compressed data is larger than the original data, then the compression is skipped even for a bigger packet. Data is encrypted after compression if data is compressed. IPComp adds overhead and can slow down the process. In most cases, when a very large amount of data is to be transmitted, it's efficient for the application to compress the data. Research efforts are underway to develop efficient compression algorithms that can be used with Internet of Things (IoT) devices.

6.3 Securing Web-Based Applications

In the previous section, we discussed securing data in transit while connecting to a web server or connecting to a work network from home. However, there are other important uses in daily life where we need to protect data in transit, such as emails, distributing software or music, and so on. In this section, you'll learn how to implement enterprise-wide email security and how to protect music or software from illegal distribution.

6.3.1 Securing Email Communication

Emails and mechanisms to send and receive email aren't new technology. In fact, the Simple Mail Transfer Protocol (SMTP) was proposed in 1982 by Jon Postel [RFC 821, 1982]. RFC 821, Simple Mail Transfer Protocol, states, "SMTP was developed to ensure a more reliable and efficient way to transport messages." In the early days, email was primarily used by academia, research, and tech enthusiasts only. At the time, it didn't even cross anyone's mind that the protocol developed for email delivery should not only include reliable and efficient email delivery but also secure delivery. When email became a main form of communication in the early 1990s, the security of our messages became a concern. The SMTP protocol was ASCII text messages, but the nature of email usage was changing. We needed something that could send pictures, audio, and even videos. To accommodate the new requirements of the changing times, SMTP was extended, and the new Multipurpose Internet Mail Extension (MIME) protocol was proposed in 1991—again, with no security controls in sight!

There are two protocols created to secure email communication: Pretty Good Privacy (PGP) and Secure Multipurpose Internet Mail Extension (S/MIME). Both use the same weapons from our cryptography arsenal: symmetric encryption, asymmetric encryption, hashing, and digital signature. We've extensively discussed these algorithms, so we won't repeat them here. You'll learn how to use or implement these algorithms in securing emails.

Pretty Good Privacy

PGP encrypts email messages with symmetric encryption. The symmetric key is known as a *session key*. This key is used to encrypt the message by the sender. The recipient of the email needs this key to decrypt the message. PGP uses asymmetric cryptography to achieve this. The sender must encrypt the session key with the public key of the recipient, which provides confidentiality for the email. This process is similar to the envelope encryption we discussed in Chapter 5. PGP also digitally signs the message. It generates the plain text message and then encrypts the hash with the private key of the sender. Once the message is encrypted and digitally signed, the email is sent over the public internet.

The receiver, upon receiving the email, uses the sender's public key to verify the signature. If the sender's public key works on the signature, then the receiver can conclude that the email was signed by the owner of the private key. This step authenticates the sender. Next, the receiver uses his own private key to decrypt the session key. Once the session key is decrypted, it can be used to decrypt the message. The early version of PGP used International Data Encryption Algorithm (IDEA) and then Triple DES for symmetric encryption while RSA was used for asymmetric encryption.

The process works very well on paper, but the million-dollar question is how the sender would know if the public key used for encrypting the session key really came from said receiver. To solve this problem, PGP proposed a *web of trust* mechanism. The web of trust is very similar to introducing people in a social setting. For example, if you move to a new town where you know only one family, then that family will introduce you to a couple of people. Those new contacts will introduce you to a few more, and slowly your web of trust will increase in a new town. The concept is exactly the same here—PGP assumes that over time, people will have public keys from many contacts. These contacts will have your public key. This is difficult to achieve in daily life when businesses need to send many emails to many new contacts every day.

In the late 1990s, PGP collaborated with IETF to create an open standard based on the proprietary PGP. The new standard was called *OpenPGP*. OpenPGP uses digital certificates to authenticate users [RFC 4880, 2007]. In addition, OpenPGP supported newer encryption algorithms. For symmetric keys, it offers the option to use AES, and for asymmetric algorithms, it offers RSA and public keys with certificates. For hashing, OpenPGP uses the SHA family algorithms.

There are security gateway servers available on both ends to encrypt, decrypt, sign, and verify the OpenPGP emails. The gateway servers make the process transparent to users. Many major mail clients—Mozilla Thunderbird, Apple Mail, Microsoft Outlook, and so on—have a plug-in available for OpenPGP.

Secure Multipurpose Internet Mail Extension

S/MIME is another method of encrypting and digitally signing emails. S/MIME has gained a lot of momentum in business use and is a popular choice for encrypting and digitally signing emails [NIST 800-45, 2007]. S/MIME was first developed by RSA Security Inc. in 1995 using RSA's proprietary standards and formats. RSA eventually collaborated with the IETF working group and, in 1999, the IETF working group released S/MIME version 3.

The concept of S/MIME is very similar to OpenPGP. The sender of the email encrypts the message using the recipient's public key. The email is signed using the sender's private key. The email also includes the digital certificate. The certificate has the email domain that confirms the domain of the sender. The certificate is usually signed by an independent third party (the CA, as we introduced in Section 6.2.1). Upon receiving the email, the message can be decrypted using the receiver's private key. The signature can be verified using the sender's public key.

The overall process of securing and authenticating email is very similar for PGP and S/MIME. The advantage of S/MIME is the use of a digital certificate for signing the email. On the other hand, PGP has the advantage that it can encrypt everything—the subject of the email as well as the metadata. S/MIME doesn't encrypt the subject and the metadata. This gives some idea of the email and the content to an eavesdropper.

S/MIME is the default encryption method in many major email clients. For example, the process of encrypting and digitally signing the message is almost silent or automatic in Microsoft email clients. All the user needs to do is download the certificate and assign it to the client.

Securing email using cryptography isn't very popular or widespread. If the receiver doesn't have a key pair, then the sender can't send the encrypted message. It also adds overhead to the system as encryption and decryption takes a lot of processing power. The security is compromised when both parties are using the email encryption but one is using a weaker algorithm. The weaker algorithm can easily be compromised. In this case, the security is only as strong as the weakest link.

> **Why Is Email Security Important?**
>
> Many successful cyberattacks start with emails. Spam emails lure users to click on the links that open up a vault of possibilities for an attacker. If the emails are encrypted and digitally signed, then it provides confidentiality and integrity. The receiver is confident when clicking the links in the digitally signed emails because the receiver is assured that the email did come from the said user.
>
> Proton Mail is known for the security it provides. Proton Mail uses asymmetric or public key cryptography to provide a way to encrypt the email using the password. This helps when the recipient doesn't have a key pair. The sender can send the email using password protection. The password must be communicated to the recipient outside the email—using a text message, phone call, or by providing a commonly known hint.

6.3.2 Securing Streaming and Downloading

Do you ever wonder how Netflix and Spotify protect video and audio content from illegal downloading and copying? Or how we can't read Amazon Kindle's books without a subscription? Well, you guessed it—the answer is encryption. So far, we learned that bulk data encryption is performed using symmetric key cryptography and then using public key cryptography to share the symmetric key—that is what we did in TLS handshake, IPsec IKE, and so on. However, the process of key exchange is slightly different with streaming and downloading.

Digital rights management (DRM) is the underlying technology used in protecting streaming media or downloading files. DRM uses symmetric cryptography, specifically, the AES algorithm. We've already learned about AES and various block cipher modes in previous chapters.

Figure 6.8 shows the DRM process flow. The media provider (Netflix, Spotify, Apple, Google, etc.) stores the encrypted content on its server. The content is encrypted using AES symmetric key encryption. The content is either encrypted by the streaming company or by the creator such as a music company. When a user clicks on the video or music to play, the browser or a dedicated player connects with the streaming server. The server sends the entire encrypted file plus a header to the browser or the player. The browser or the dedicated player (such as an app) has software (or hardware in some cases) called a Content Decryption Module (CDM). CDM extracts the header and validates that the request came from the rightful subscriber. Once the subscription is confirmed, the CDM reaches out to the key server. The key server responds with a session key to decrypt the media. The CDM then starts decrypting the media for playing. Optionally, the CDM also includes the session ID with the key request. The session ID is used by CDM and the key server to validate the request periodically.

6.3 Securing Web-Based Applications

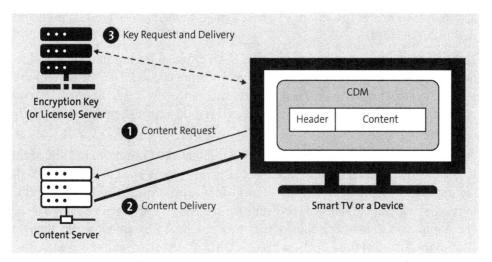

Figure 6.8 DRM Process Flow

The implementation is done using the Encrypted Media Extension (EME) specification. The EME specification is developed and maintained by the World Wide Web Consortium (W3C) [EME, 2024]. The EME specification describes the implementation of an API to control the playback of encrypted media.

This is a generalized overview of the process. The actual process slightly differs based on the content provider. Apple uses DRM technology called Apple FairPlay, Google has Widevine for DRM, and Microsoft enforces DRM using PlayReady. Apple uses CBC mode with AES while Google and Microsoft use counter (CTR) mode with AES. (We discussed CBC and CTR modes in detail in Chapter 1.) Another difference is the protocol—FairPlay uses HTTP Live Streaming (HLS); Google Widevine and Microsoft PlayReady, on the other hand, use Dynamic Adaptive Streaming over HTTP (MPEG-DASH). Most independent streaming devices, such as smart TVs, Roku, Android devices, and so on use Widevine.

Houston, we have a problem! How would Netflix handle a request if one user is using Apple TV and another user is using Chromecast? The problem is solved by the Common Encryption (CENC) scheme [CENC, 2023]. CENC uses the *encrypt by one, decrypt by many* philosophy, meaning the CENC keeps a key mapping that allows various systems to decrypt with different technologies. CENC isn't owned by any specific DRM system, so it works with all DRM systems. Remember, the security of the streaming content still depends on the encryption algorithm used; CENC only makes it easy for various encryption algorithms to decrypt content and stream using different streaming protocols.

Another major application of DRM is with e-books and documents. The overall concept is the same as with streaming media. However, it's easy to copy an e-book or a document and share it compared to a movie, so e-books and documents need more controls after they're downloaded.

The user connects to the server using a tablet, smartphone, or computer to download or read an e-book. The server verifies the login credentials, and if it's the rightful subscriber, the server sends the request to the license or the key server. Based on the type of subscription, the key server provides the key, and the server starts downloading the e-book or document.

The major difference between an e-book or document DRM compared to a streaming media DRM is the granular authorization. In the case of an e-book or document, DRM can implement various controls, such as read only, allow saving to the local disk, printing or copying, editing, and lastly revoking. The revoking is implemented using a timer. For example, if a subscriber is allowed to borrow an e-book for only two weeks, then the DRM makes sure that the e-book isn't accessible on the local drive after two weeks.

The use of cryptography is somewhat unconventional with DRM. Traditionally, cryptography is used to provide confidentiality; however, DRM technology is using it for authorization in a way. The decryption of content and the rights associated with the content are granted based on the subscription level of the user. The process is usually policy based. Each decryption key (the encryption key is referred to as a license in general use) checks the policy associated with the request, and it allows use of the content based on the policy. Even in streaming media, the CDM on the local device decides if the subscriber has the right to the content, and the CDM sends a request to the key server based on the permissions.

Another deviation from traditional cryptography use is the way the key is shared. Until now, we learned that public key cryptography is used to share the symmetric key or the session key. That isn't the case with DRM technology. Only a symmetric key is used with DRM. However, the key is sent to the device separately from the content itself (although this may be different depending on the type of DRM technology used as well as the content).

Last but not least, encryption is also used with software licensing. The installation of software unlocks only after the key is entered. This is a simple use case. The problem or the fatal flaw in the system is the fact that the key is usually distributed with the CD. However, it has improved slightly as most software is available through download now (instead of a CD), and the license key is sent via email. However, the key still stays with the software or the device after installation.

6.4 Public Key Infrastructure

Web security, data in transit security, email security, and many other real-world cryptography applications rely on one concept—we trust the public key. We believe that the public key we've used to encrypt data or email is from the person we believe is the owner. And that person has the corresponding private key to decrypt our message and read it. But if trusting the public key fails, the entire cryptography application structure collapses. How do we make sure that we trust the public key or the certificate?

The concept of *public key infrastructure (PKI)* is used by cryptography applications to create trust between communicating parties. In this section, you'll learn how PKI works and how you can get a certificate or even start signing a certificate. We'll discuss encryption key management in more detail in Chapter 7, Section 7.2.

The certificate or digital certificate is nothing but a binary file that contains the public key of the person or the entity being certified and the digital signature of the certifying entity. All certificates meet the X.509 standards. International Telecommunication Union (ITU) issued the first version of X.509 certificate in 1988, followed by version 2 in 1993. Version 3 was introduced in 2008.

The primary fields of a typical X.509 certificate are as follows:

- **Version**
  This is the version of the X.509 certificate.

- **Serial number**
  This is the serial number of the certificate. This number is usually issued by the certificate issuing authority. The number is very useful in revoking or searching for the certificate.

- **Signature**
  This is the digital signature of the issuing authority. Without this, there is no value of the certificate.

- **Signature algorithm**
  The CA must specify the hashing algorithm used in the digital signature. Without this, the user can't match and confirm the hash value and the signature.

- **Issuer**
  This is the distinguished name of the issuer (the name of the CA who issued and signed the certificate).

- **Validity**
  This is the validity period of the certificate. Every certificate has a predefined lifespan with an expiry date that usually indicates the issue and expiry date of the certificate.

- **Subject**
  This is the distinguished name of the subject or the owner of the public key. This name could be an individual, an organization, or a device. For example, if you have a company-issued laptop, the laptop may need a certificate to access some of the resources. Historically, this used to be a Common Name (CN), but it has been replaced with a Subject Alternative Name (SAN). The SAN is typically used by the server to list multiple domains, for example, one.xyz.com, two.xyz.com, and so on. This could also include a wildcard such as ?.xyz.com.

- **Subject's public key**

    This is the most important field of the certificate—the owner's public key. The certificate is nothing but a container of the public key and that is what's being certified by the CA in the certificate. This key will be used by the owner in communication with others.

- **Extensions**

    This field was introduced in version 3 and allows some customization to the CA. Extensions can be critical or noncritical. As you probably guessed, if the critical extension is missing or doesn't meet the condition, then the certificate won't work. On the other hand, access will still be granted even if the noncritical extensions don't meet the condition. The noncritical extensions are essentially optional or nice-to-have requirements.

The public key of the CA is easily available, and users use that public key to verify the digital signature of the CA in the certificate.

> **The Private Key**
>
> Note that when the CA generates the public key for any entity, the corresponding private key is also generated—otherwise, the public key is of no use. The private key, however, isn't part of the certificate. The private key is kept with the entity to whom the certificate and the key pair is issued. The private key (the file) is kept very confidential, and the file is usually password-protected.

If we can trust the CA, and the CA has a process of validating the identity of the subject, the process should work smoothly. The process of applying for a certificate is known as *registration*, and it's usually done by an independent party called the registration authority (RA). The admin or the owner of the domain can request a certificate from the CA. This is done using a process called a *certificate signing request (CSR)*. Once the admin fills out the CSR and provides proof of identity, the CA engages the RA. The RA then performs checks to make sure that the person listed in the CSR has the authority to request the certificate. The CSR should have all the necessary information—subject's name, domain name, public key, and so on—to populate the certificate if the request is approved. The role of the RA is kind of optional. It's up to the CA to decide if the validation of the requester's identity is performed in-house or if an RA will be engaged.

> **Let's Encrypt!**
>
> Getting a certificate and renewing it from the CA costs money; there is a fee associated with the certificate. To help people get certificates (and security) at no cost, Internet Security Research Group (ISRG) provides a free service called Let's Encrypt (*https://letsencrypt.org/about/*).

At the end or near the end of the certificate's validity period, the certificate must be renewed; otherwise, the certificate will expire. The process of renewal can be automated or manual.

Apart from the expired certificate, there are scenarios when the CA must revoke the use of certain certificates, for example, if the company is out of business, the domain isn't a valid domain, or the private key associated with the certificate is lost or compromised. In these situations, the CA must revoke the certificate and add the certificate serial number to the *certificate revocation list (CRL)*. The CA generates and publishes the CRL, and the client should validate that the certificate serial number isn't in the CRL before processing the request.

Alternatively, the client can simply query only one serial number and get a response if the number is revoked or not. This dynamic method is called the *Online Certificate Status Protocol (OSCP)*. The process of checking the CRL every time can take lots of CPU time and add stress on the web server. To avoid that, the web server can get a timestamped validation from the CA that the serial number assigned is a valid number and present when asked by a client. This process is called OSCP stapling.

Certificate pinning or *HTTP public key pinning* is the process of embedding the certificate in the code. This way, if there is a man-in-the-middle attack and an attacker has injected his own certificate, then it will be detected when the web server compares the certificate information with the information embedded in the code.

Whenever a key is compromised, the certificate must be revoked. But what if the CA's private key is compromised from the digital signature? To deal with this situation, the PKI doesn't depend on any one CA but has a chain of CAs or a hierarchy of CAs. The CA that signs the individual certificate is called a *leaf CA*. The leaf CAs are supported by intermediate CAs, and, finally, there is a root CA.

The information, fields, and format of the certificate make up a somewhat standardized process. There are two types of certificate standards: (1) X.509 by the IETF and (2) Public Key Cryptography Standards (PKCS) by the RSA. X.509 is the actual standard for the certificate, whereas PKCS is a storage format, such as binary, ASCII, and so on. The certificates are available in various formats:

- **Distinguished Encoded Rule (DER)**
  This is the binary format certificate, which is used by most browsers.
- **Privacy Enhanced Mail (PEM)**
  This is the ASCII format certificate that serves as the default for OpenSSL.
- **PKSC #12**
  This is a binary format used to ship the certificate or move from one server to another. The certificate stores both public and private keys. This type of certificate file is encrypted and must be accessed using a password.

- **PKSC #7**
  This is the same as PKCS #7 in ASCII format. It isn't encrypted.
- **P7B**
  This format stores the entire certificate chain, starting with the root CA certificate.

Lastly, there is no reason that you can't sign your own certificate—as long as someone can trust you. This is called a self-signing certificate. Self-signing is generally used by big corporations for internal use to keep the cost down as well as for testing and other development use cases.

6.5 Summary

Data in transit is the backbone of everything that we do in everyday life—from emailing to banking to shopping. It's crucial that we protect these transactions from any hacker. Securing data in transit provides security while we perform our daily business.

In this chapter, we explained the many techniques for data encryption in transit in various situations. We studied how to implement the TLS protocol to secure web browsing and how to implement IPsec to create a VPN tunnel. Lastly, we covered the backbone of the public key cryptography—PKI.

> **Key Takeaways from This Chapter**
> - Learned how to implement TLS and VPN
> - Understood the way to secure emails and streaming media
> - Reviewed how PKI works and how to use it

In the next chapter, you'll learn about securing data in the cloud using cryptography and how various cloud service providers use cryptography to secure data and information. The chapter will also discuss encryption key management, which is one of the most important aspects of using encryption in the cloud.

Chapter 7
Cloud and Connected Device Cryptography

This chapter focuses on protecting data in a place where the data owner has almost no control: the public cloud. The public cloud is probably the best use of cryptography. One of the "key" aspects of cloud cryptography is managing the encryption key. Cryptography is also used in your vehicle and in many other devices you use in daily life—without even realizing it! In this chapter, we'll discuss how cryptography is used to secure data in the cloud and in Internet of Things devices.

Although cloud computing has gained popularity only over the past 10 or 15 years, the concept of cloud computing dates back to the 1960s [White, 1971]. By the 1990s, with the easy availability of the internet, many companies started talking about cloud computing, and, in 1999, virtual machines (VMs), a backbone of cloud computing, were first released by VMware. In 2002, Amazon launched Amazon Web Services (AWS) as a free web service. The cloud computing that we know today was launched by AWS in 2006, followed by Google with Google Cloud Platform in 2008. Microsoft joined the cloud parade in 2010 with Microsoft Azure. But even then, people were a bit skeptical about using the cloud. The public cloud gained real traction only after 2014, and its use skyrocketed after the pandemic.

What took cloud computing 15 years to become popular? Economically, a pay-as-you-go model could save money for many businesses, so budget wasn't the reason for businesses. Instead, holding cloud growth back was security and availability of the data. Businesses weren't sure if their data was secure in the cloud and would be available when needed.

Chapter Highlights
- Implementing cryptography and managing encryption keys in the cloud
- Understanding the need for lightweight cryptography in protecting the Internet of Things (IoT)
- Securing vehicles with cryptography

Businesses started using cloud computing only after they were assured by the cloud service providers (CSPs) about the security and availability of their data. CSPs spent a lot of time and money educating potential users about data security and availability while using the cloud. But their efforts eventually paid off, and now the use of the public cloud has become an essential part of doing business.

This chapter is divided into four sections. Section 7.1 starts with an overview and implementation of cryptography in the cloud. This section will teach you how to secure your data, application, and infrastructure in the cloud. Section 7.2 discusses encryption key management strategies in the cloud. We'll discuss various strategies and show how to implement each strategy. You'll learn how cryptography is used by various CSPs and how you can implement cryptography with various CSPs to secure your data in Section 7.3. Finally, Section 7.4 teaches how cryptography works in various places that we don't even expect—in your car and with IoT. IoT is ingrained in our lifestyle, and securing the IoT data is crucial.

7.1 Primer on Cloud Cryptography

To understand why cryptography is such an important issue in the cloud environment (so much so that we must dedicate an entire chapter to the topic), we first need to understand the mindset of the users. As we mentioned earlier, when cloud computing was new, people were skeptical about using it. They have two primary concerns:

- **Security of the data**
  Users have no idea what kind of control CSPs could use and where the actual data will reside. Even after the active usage or the contract with the CSP ended, how would data be deleted, and how would CSPs destroy the media (and could they be trusted to do so)? There were many questions but no definite answers.

- **Availability of the data**
  How can you ensure that customers/employees or users will have access to applications and data in the cloud when needed?

There was no guarantee that the CSPs would live up to their promises. Of course, there was a legal contract between the providers and their customers, but that would be dealt with long after the actual impact on the end users. The end users would end up with a bad taste in their mouth regarding the user experience, which may cost businesses in the long run. The list of concerns was long, but for our purposes, we need to just worry about security and availability for now.

As you read this, you may be thinking: We have a solution to all these problems. Why do we need to discuss them? This is certainly the case right now. While AWS was first introduced in 2002 and offered as a free service to gauge the response, the paid cloud service of AWS was offered in 2006. After all, there must be a reason to offer AWS for free!

The point here is that people weren't ready to jump on the cloud computing bandwagon because, among many other issues, of their concerns about the security and availability of their data and applications. Luckily, there's a hero in this story—cryptography. Encryption of data made it possible for users to keep their data in the hands of unknown third parties (the CSPs) and sleep peacefully at night.

When we talk about data security, we're essentially talking about confidentiality and integrity. Technically, this means we're going to the basics of security—the security triad of confidentiality, integrity, and availability in the cloud. (We've discussed the security triad in greater detail in Chapter 1.) Our goal in this chapter is to learn to protect the security triad in the cloud by applying cryptography principles. Data in the cloud also needs to be protected at rest, in transit, and in process. Table 7.1 summarizes how cryptography protects each leg of the security triad in the cloud within each data stage. As you study the table, you may have a couple of questions.

	Data in Transit	Data at Rest	Data in Process
Confidentiality	Asymmetric encryption: Transport Layer Security (TLS), Internet Protocol Security (IPsec)/virtual private network (VPN)	Symmetric encryption: Disk, database, file encryption	Homomorphic encryption
Integrity	Hashing: Secure hash algorithm/message authentication code (SHA/MAC)	Hashing: SHA/MAC	Homomorphic encryption
Availability	Redundancy/encryption	Redundancy/encryption	Redundancy/encryption
	Infrastructure/Application	Infrastructure/Data	Application

Table 7.1 Cryptography Primitives in the Security Triad in the Cloud

You may be wondering what homomorphic encryption is. We'll study homomorphic encryption in Chapter 11, but in a nutshell, it allows data to be processed without decrypting. You may also be questioning how encryption helps with availability. The short answer is—it doesn't help directly; however, encryption does protect data when it's not available. We'll also discuss this concept later in Chapter 12.

In this section, you'll learn how to secure infrastructure, applications, and data in the cloud using the security triad principles.

> **What You Should Know**
>
> As you read through this chapter, we assume that you have a basic understanding and knowledge of cloud operations: the concept of VMs; various cloud offerings, such as IaaS, PaaS, and SaaS; cloud deployment models, such as public cloud, private cloud, hybrid cloud, and community cloud; and the workings of the cloud. This chapter has a very specific focus: using cryptography in the cloud.

7.1.1 Securing Infrastructure in the Cloud

You've already learned the fundamentals, various algorithms, and different applications of cryptography, and everything you've learned so far is applicable to software. However, cryptography can also be used to protect hardware or infrastructure. How can a mathematical algorithm protect something physical, like hardware? Let's find out in this section.

One approach to securing hardware infrastructure is to use a Trusted Platform Module (TPM). We briefly introduced TPM in Chapter 5, Section 5.2.1, with disk encryption. In this section, we'll go a little deeper and also learn how TPM is used with VMs [Sangster, 2011].

Hardware manufacturers have made significant advancements in chip development over the past 25 years. They strive for extremely fast performance, compact size, and economical integrated circuits or chips. One approach to meet all these goals was to create a single, efficient chip that can execute several functions effectively. These chips are known as *application-specific integrated circuits (ASICs)*. TPM is one such ASIC chip. TPM's specific application is to provide security to hardware that is attached to or interacting with the motherboard or computer. TPM is designed to provide security using cryptography services—encryption, hashing, and digital signature. TPM can provide the following cryptographic functions to protect hardware:

- **Key generation, storage, and management**
  TPM can use asymmetric algorithms and generate a key pair for RSA and elliptic curve cryptography (ECC). It can also generate a symmetric key for the Advanced Encryption Standard (AES), which is usually 128 bits in size. The generated keys are stored within the TPM in nonvolatile memory to save the keys even when the power is off. TPM also performs encryption key lifecycle management: generating, rotating, deactivating, and deleting the key.

- **Encryption, decryption, hashing, and digital signature**
  The keys wouldn't be much use if we didn't use them for encryption, decryption, and digital signatures. Although TPM doesn't encrypt and decrypt the data itself, it encrypts and decrypts the data encryption keys (DEKs).

- **Platform integrity**
  TPM uses hashing algorithms to provide platform integrity. Hash values of the important system components, such as the Basic Input/Output System (BIOS), bootloader, OS, and so on, are stored in the Platform Configuration Register (PCR). At the time of the system boot, the hash values are retrieved and compared with the newly calculated hash. If the hash doesn't match, the system won't boot. Any unauthorized change in the BIOS, bootloader, or any other system component will result in a new hash value and won't match with the values stored in the PCR.

- **Credential storage**
  TPM can also be used for storing other credential-related information, such as passwords, certificates, biometric data, multifactor authentication information storage, and so on. The advantages are the nonvolatile memory storage and that no one other than the authorized user can access the TPM data.

- **Hardware protection**
  The best part of using TPM is the tamperproof hardware security. It's almost impossible to destroy, damage, or bypass TPM. Moreover, TPM not only protects the BIOS, OS, and other system components but also protects peripherals such as the hard disk, RAM, and so on.

> **TPM: A Case for CaaS?**
>
> Infrastructure as a service (IaaS), platform as a service (PaaS), and software as a service (SaaS) are commonly used terminologies because each provides specific services in the cloud. TPM and virtual TPM (vTPM) provide all the cryptographic services to the servers in the cloud, which is a very specific service to the cloud users. It won't be an overstatement to recognize TPM as CaaS—cryptography as a service!

Figure 7.1 shows the basic components of TPM. The left side shows the cryptography functions, and the right side shows the TPM management functions. Previous chapters have described all the cryptographic functions, such as hashing, symmetric and asymmetric encryption, key generation, and so on, so we won't repeat them here. The only two terms that you must know are binding and sealing. *Binding* is a process of encrypting the DEK with the root key or the TPM key. The encrypted DEK can then be stored with the data in the hard drive. This process is the same as the envelope encryption described in Chapter 5. *Sealing* is a safe state of TPM. When TPM goes into a safe or sealed state, it records the conditions while encrypting. To unseal or decrypt, TPM must meet the same conditions; in other words, it has to be in the same state as the time of encryption to unseal or decrypt the data. To simplify, if anything changes while TPM is in a sealed or encrypted state, it won't decrypt.

7 Cloud and Connected Device Cryptography

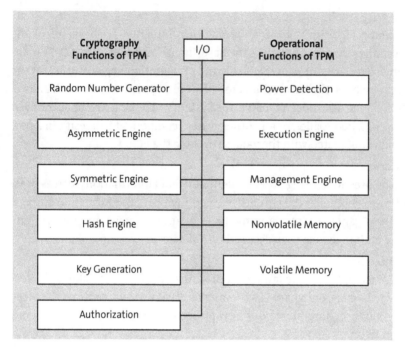

Figure 7.1 High-Level Architecture of TPM

TPM is very useful on a single computer or server and also in the cloud environment. Figure 7.2 shows how TPM interacts with boot and OS components on a single server and in the virtual environment.

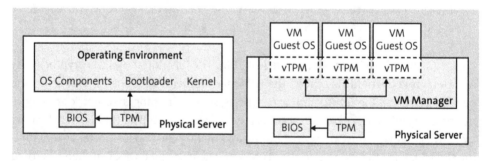

Figure 7.2 TPM on a Single Server and in a Virtual Environment

In a single server or computer, TPM resides on the motherboard and interacts with BIOS and the operating environment. This is the conventional use and a straightforward use case. In the cloud environment, the implementation is slightly different. The physical TPM resides on the physical server platform, and each VM hosted by that hypervisor uses a vTPM. vTPMs are derived from the physical TPM. This happens through the VM manager. Of course, the actual implementation of vTPM varies based on the type of VM manager and the vendor. No matter what the name and the processes are, the concept is the same—vTPMs are derived from the physical TPM, which is residing on the physical server.

vTPM can't be shared or used with more than one VM. vTPMs are software implementations or replications of the hardware TPM. vTPM is a very lightweight application and doesn't take any additional time or overhead when functioning. vTPM does increase the boot time of the VM slightly, but this isn't noticeable to the end user. The VM manager performs the following functions to manage the vTPMs:

- Creates a new vTPM every time a user creates a new VM
- Maintains the nonvolatile storage in the VM manager layer
- Must maintain the vTPM parameter state between the power cycles and reboots of the VM
- Helps establish the connection and the initialization routine—the VM attestation and other authentication take place at the beginning
- Makes sure that the vTPM and all the related parameters are deleted as soon as the data is deleted, which is also one of the main security features

TPM is very important in the cloud environment where users frequently detach and attach storage, disks, volumes, and so on. If TPM is configured correctly, then the data can't be read from the detached device (by connecting it to another VM). Moreover, vTPM works with disk encryption utilities such as BitLocker and DM-Crypt to encrypt the entire disk and the DEK and keep the key with the disk. When a user logs in, the user must enter a different password for the disk encryption utility. If the password is verified, then TPM decrypts the DEK, which then decrypts the disk.

vTPM also checks the boot sequence on the VMs, just like it does on the physical server. Although the process is much more involved in the cloud and virtual environment, vTPM performs the authentication, which means it compares the stored hash values of the boot and other OS components with the newly calculated values. The boot sequence will progress only if the hash values match, and the attestation is successful.

One of the important security concerns with vTPM is the nonvolatile memory. In TPM, the nonvolatile memory is stored locally; however, in the virtual environment, the nonvolatile storage is, usually, on the VM manager layer. This means, that if an attacker penetrates in the hypervisor layer and gains control over the VM manager, then the attacker can access the data stored by the vTPM.

7.1.2 Securing Data in the Cloud

Every security effort is centered around protecting data. Defining the strategy to protect data in the cloud is one of the most important decisions when planning to migrate to the cloud. Although the CSP is responsible for offering secure cloud operations and data protection, ultimately, it's up to the customer to decide the level of data protection required and the best security control. The data controller decides the data classification and respective security control. The decision is usually based on several factors, such as compliance requirements, competitive/market analysis, industry standards, and government regulations. But based on the direction from the data controller, the data processor (the CSP) implements the security control on the data. As we discussed in Chapter 5, protecting data costs money, and the level of security control should be decided based on the data classification only. This is very important in the cloud because the cloud works on a pay-per-use basis, and it costs extra to add any additional control.

> **Data Controller versus Data Processor**
>
> There are four major roles of data operators: data owner, data controller, data processor, and data custodian. Let's understand this with an example of a retail purchase. The sequence is the customer purchases from a retailer, and the retailer stores data in the public cloud:
>
> - **Data owner**
>   In our example, the customer is the data owner. The customer provides his data, such as name, address for delivery, credit card number, phone number, email address, and so on.
> - **Data controller**
>   The retailer is the data controller. The retailer stores data in the public cloud and decides what protection is needed for which data. This could be based on compliance requirements, industry standards, government regulations, and so on. In this case, for example, the retailer decides that the credit card number must always be encrypted.
> - **Data processor**
>   A data processor is a public CSP that keeps or stores data on behalf of a retailer. The processor follows the instructions of the data controller and provides security for the data.
> - **Data custodian**
>   The data custodian is part of the data processor or the IT department. The data custodian is the one who actually does the work, that is, implements the control.

As we've covered in Chapter 5 and Chapter 6, there are two types of data encryption strategies: data-at-rest encryption and data-in-transit encryption. The majority of the focus of data encryption in the cloud is on data-at-rest encryption. There are two primary strategies for encrypting data at rest in the cloud: client-side encryption and server-side encryption.

Client-Side Encryption

In *client-side encryption (CSE)*, data is encrypted on the client's device or location and then sent to the cloud for storage. When a user needs to access the data, the data must be retrieved from the cloud and then decrypted locally on the client's machine or location. CSE is also known as end-to-end (E2E) encryption because the data stays encrypted during the entire cycle when it leaves the client's machine and is stored in the cloud and then sent back to the client machine. The only time data is decrypted is while it's in the client's hand.

CSE is data-at-rest encryption and uses any of the symmetric encryption algorithms. The most popular algorithm recommends using an AES block cipher in Galois/Counter mode (GCM) mode with a 128-bit or 256-bit key size. However, this is a generic recommendation, and the application owner/customer can continue to use their own algorithm.

The CSE strategy is preferred by many cloud customers due to the following advantages:

- The first and foremost advantage of this strategy is that users don't have to change their encryption algorithms, key management strategy, or any other processes they have been following before the cloud. Users or cloud customers can use whatever algorithm they have been using to encrypt the data. There is minimal impact on operations.
- Cloud customers or users manage the encryption key. This gives users full control and flexibility over the key management strategy—when to rotate the key, when to deactivate the key, and when to delete the old key.
- Encryption and key management by the cloud service provider (CSP) usually cost users additional money. In CSE, however, the encryption and key management are performed by the client, saving these costs.
- In CSE, data is first encrypted by the client and then encrypted again when it travels from the client site to the provider's site. In addition, CSPs generally include some basic form of data-at-rest encryption. These encryptions in transit and at rest provide an additional layer of protection.
- CSE eliminates the worry of data-in-transit security because the data is already encrypted before it travels.

CSE has some disadvantages also:

- Data can't be accessed until retrieved. Even to view the data or make the simplest change in the data, the data must be retrieved and downloaded from the cloud to the local site and then decrypted. This adds additional overhead to the system.
- The cloud customer may not be able to use all the CSP's features because the CSP's tools can't access the data.
- The process is slow and time-consuming.

- Most cyberattacks happen or at least start with compromised credentials. If the access control isn't properly implemented or the admin credentials are compromised, the attacker can access the encryption key, and, at that point, all bets are off. Once the key is compromised, the data is vulnerable. Cloud customers are responsible for managing robust access control on their sites and machines.

One reason cloud customers opt for CSE is to meet very specific compliance requirements, industry standards, or regulatory requirements. This is the case when the data belongs to a niche industry or application, such as banking or healthcare.

The actual implementation of symmetric encryption depends on the encryption algorithm used. For more information on these algorithms, refer to Chapter 2.

When CSE is used, the CSP is (usually) used only for storage. All major CSPs work with CSE and provide flexible data storage plans and technology.

Server-Side Encryption

In *server-side encryption (SSE)*, unencrypted data arrives at the CSP. The CSP encrypts the data when it's written to the storage and decrypts it when it's accessed. SSE is also, like CSE, a form of data-at-rest encryption. A symmetric block cipher is used to perform encryption. Most CSPs use AES in GCM with the 128-bit or 256-bit encryption key. However, it's up to the CSPs to decide which encryption methods to use. The actual implementation of symmetric encryption depends on the encryption algorithm used. For more information on these algorithms, refer to Chapter 2.

The SSE strategy is preferred by many cloud customers due to the following advantages:

- The process is transparent to users or cloud customers. The CSP determines the encryption strategy and manages the entire process. The cloud customer or user has very little to worry about in the whole process.

    SSE is offered in various key configurations: the CSP-managed key, the customer-managed key (CMK), and the customer's own key (bring your own key [BYOK]). These configurations slightly change based on the provider, but customers can choose the configuration based on their risk appetite, business requirements, and cost. We'll discuss these options in detail in Section 7.2.

- The cloud customer only needs to manage access to the storage resource. The cloud customer doesn't have the additional overhead of managing the access control or the key (depending on the configuration).

- There is almost no additional cost for using SSE if the default key configuration, usually a CSP-managed key, is used.

- The biggest advantage is the ease of data access and processing. Whenever a user wants to access or process data, the user has easy access without going through any additional steps.

- Lastly, the cloud customer is able to use other security features offered by the CSP, such as logging, monitoring, alert notifications, and so on.

The SSE has some disadvantages also:

- The primary drawback is to encrypt the data in transit. In SSE, data is encrypted only when it's written to the storage device. The data is in plain text or unencrypted while traveling from the customer site to the CSP's site—unless a data-in-transit encryption mechanism is used.
- It's the cloud customer's responsibility to use TLS or VPN to protect the data while in transit. This usually ends up in additional costs and overhead.
- If the cloud customer chooses to use another encryption key configuration, such as a CMK or BYOK, then the customer will have additional responsibility to manage the access control to the key and additional cost.
- If the CSP's services are compromised, breached, or attacked, then all bets are off. The customer's data could be compromised. Even a malicious insider working at the CSP can pose a significant risk to the data.
- The customer has very little control over the encryption key management and overall encryption strategy.
- The security of the storage medium and your data is dependent on how effectively the CSP patches the vulnerable system and software in their environment.

One of the primary use cases for SSE is the decentralized workforce. If the data needs to be uploaded and accessed from various locations, then it becomes very difficult or almost impossible to perform and manage the CSE.

All major CSPs offer SSE with multiple options for managing the encryption key.

7.1.3 Securing Applications in the Cloud

We learned about application-level encryption (ALE) in Chapter 5, Section 5.2.5. The fundamental concept and operations remain the same—the application performs the process of encryption and decryption itself. However, the implementation considerations are slightly different depending on whether the application resides in the cloud or on the client's device.

If the application resides on the client's device, then ALE works more like CSE. The data in the application is encrypted right on the client's device and then sent to the CSP. It's always considered a better security practice if the data is encrypted closer to the end user because the data is decrypted just before the use and encrypted right after the use. This gives very little chance to any hacker to tap into the data. This is the best way to keep data secure in the cloud. However, this approach is extremely complicated and requires very skilled people to perform the task. A tiny mistake can cost lots of debugging time. This process must be repeated every time the data needs to be changed. In short, all the advantages and disadvantages we discussed about ALE in Chapter 5 and about CSE in the previous section are applicable here.

Again, one of the issues with CSE is its inability to perform operations on data unless you download and decrypt the data. Homomorphic encryption solves this problem by allowing the processing of the data while the data is encrypted, giving an edge to CSE. The data can be processed in the cloud while in the encrypted state without downloading or decrypting it. However, homomorphic encryption has its own problems and isn't easy to use—yet. We'll learn more about homomorphic encryption in Chapter 11.

If the application is architected in such a way that the application itself resides in the cloud, then ALE works more like SSE. The process of encrypting and decrypting in the application happens in the cloud. Everything we discussed about SSE in the previous section is applicable here. A disclaimer is necessary here—this is only applicable to truly cloud-based applications. Most cloud-based applications use an access layer, such as a thin client or mobile app, and ALE happens right at the access layer and not in the application.

One difficult task with ALE is key management. The responsibility of managing the encryption key lies on the cloud customer. This must be handled very carefully because most issues with encrypted data originate from mishandling the encryption key. Encryption key management will be covered in detail in the next section.

CSPs offer SSE, and we're not against using this option when appropriate. However, many (probably most!) server-side options offered by various CSPs are only good for meeting compliance and other requirement checkboxes. These SSE options don't provide meaningful protection to the data. In a true sense, they encrypt the entire storage because they are more like disk, volume, and file encryption. The protection offered by these encryption schemes is useful for physical theft of the data. This is where ALE or application layer encryption is useful.

This book isn't about advocating one CSP over another or one technology over another, but about giving you a good sense of how the underlying technology and algorithms work. Hopefully, this chapter will help you decide which encryption algorithm is the best for your use case in the cloud.

7.2 Encryption Key Management

Per Kerchoff's principle (see Chapter 1, Section 1.2.2), a cryptosystem should be secure even if everything about it is known except the encryption key. Modern cryptography is heavily dependent on the secrecy of the encryption key. Managing the encryption key is as important as or even more than the encryption itself. In this section, you'll learn why it's important to manage the key and what our options are when it comes to managing encryption options—primarily managing the key in the public cloud.

As most of you probably know, there are two main types of offerings for IT infrastructure and services, the official industry term being *deployment models*: on-premise and the cloud. Within the cloud, there are subcategories, such as public cloud, private cloud, hybrid cloud, and community cloud. In this section, we'll focus on key management, primarily in the public cloud.

In the case of an on-premise deployment or a local data center, key management is local to the organization. The security team can decide on the key management policy based on the business and compliance requirements. The situation is a bit complex in the cloud.

Encryption key management can be loosely defined as the administration of generating, activating, deactivating, rotating, storing, and deleting encryption keys; managing access control to the keys; and documenting and maintaining a key management policy.

There's a lot to unpack in this statement. Let's dissect each element of the definition to understand the importance of key management:

- **Key management policy**
  The first thing any organization must do is to create a formal key management policy. The document should spell out when a new key should be generated, how long it should be active, when it should be rotated, and when it needs to be deleted. The policy should also spell out the details of the access control to the key and the key vault. This should be a living document and should be changed to meet new requirements and to handle technological advancements and organizational changes.

- **Generating, activating/deactivating, rotating, and deleting the key**
  The keys should be used for a defined period of time and rotated at regular intervals. A new key must be generated every time to replace the existing key. The old keys should be deactivated first. The old key should be deleted only after the new key is tested.

  This is *very* important: never rotate a key and delete it right away. It's a good practice to deactivate the old key after rotating. Test the new key to make sure everything is working as expected, and then delete the old key. If anything goes wrong with the new key, you reactivate the old key. If you find out that the new key isn't working after the old key is deleted, then you don't have any way to recover it!

- **Storing the key**
  If the key is compromised, almost all data and the credibility of the organization are lost. Therefore, it's very important for the organization to keep the key in a very secure place—something like a hardware security module (HSM).

- **Key management approach**
  The organization must decide between centralized and decentralized key management. In centralized key management, one or very few master keys are maintained. The actual encryption/decryption process is handled by subkeys and sub-subkeys.

In a decentralized approach, there are several master keys, for example, one for HR data, one for financial data, and so on. Again, subkeys are also more decentralized.

Both approaches have advantages and disadvantages. The centralized approach has fewer keys to manage, but if the key is compromised, a lot will be at stake. On the other hand, in a decentralized approach, there are many keys to manage and keep track of, but if the key is compromised, very little is lost.

- **Access control**

  This is one of the most important items in key management. If the wrong person has access or the access isn't granular enough, everything else is useless. Care must be taken to restrict the control to the HSM or the key vault and the keys. All standard access control measures, such as least privilege, need-to-know, and so on, must be applied.

- **Managing the key vault**

  Care must be taken to manage the HSM or the key vault itself. This includes physical and technical measures. The HSM must be in a very secure place, and only authorized people should be able to access it. The HSM must be patched and updated regularly. If any vulnerability is left unpatched, an attacker can exploit it and gain access to the HSM.

> **What Are Keys, Key Vaults, and the HSM?**
>
> Most literature on cryptography and key management shows the key as a traditional key and the hardware security module (HSM) or the key vault as a box. It makes intuitive sense to visualize it this way. However, the reality is different.
>
> As discussed in Chapter 2 and Chapter 3, the encryption key is nothing more than a series of binary bits. The HSM is just like a Peripheral Connector Interface (PCI) card inserted in a slot on the server. The HSM itself is an ASIC chip. The chip and the card come with a layer of a foil-like material, something that looks like a cover or a protective sleeve, that provides protection from electromagnetic interference (EMI) on the side-channel attacks.
>
> The HSM is a physical key vault. Most CSPs also offer a virtual key management service (KMS) to customers. In most cases, the cloud customers end up using this virtual KMS. However, CSPs do store their master keys in the HSM.

In an on-premise environment, the organization has a lot of flexibility and control over the key management policy and strategy, but the flexibility and control come with a lot of work. On the other hand, in the cloud environment, the cloud customer has a lot less control and flexibility over the key management policy and strategy, but less work is required. Many organizations also use a hybrid approach where they maintain the key on-site, but this adds complexity in transferring the key to the cloud when it's necessary to decrypt or encrypt the data.

Figure 7.3 shows the key management options in the public cloud environment: CSP-managed key, customer-managed key (CMK), and bring your own key (BYOK). We'll explore each option in detail in the next sections.

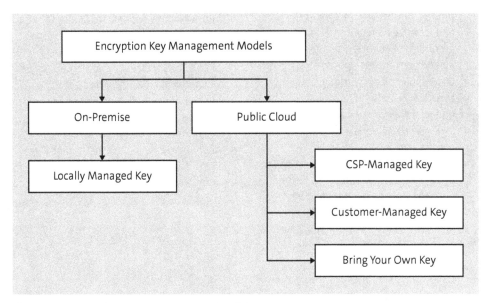

Figure 7.3 Encryption Key Management Models

7.2.1 Cloud Service Provider-Managed Key

A CSP-managed key is the default encryption option for most CSPs. This key management option is a default option in the sense that cloud customers don't have to sign up for the service or pay anything extra for it. Data will always be encrypted using this encryption key option.

In the CSP-managed key option, the key is provided and managed by the CSP. The CSP manages the key throughout its lifecycle, from generation to deletion. The CSP decides the rotation frequency and controls everything about the key and the process. The secure storage of the key is the CSP's responsibility.

The CSP controls the type of encryption algorithm used and the algorithm's key size. The CSP keeps administrative access to the key and key storage. The customer has no insight into the process, which is totally transparent to them. The data is encrypted as soon as it's written into the storage and decrypted as soon as it's accessed or read from the storage.

Table 7.2 lists the advantages and disadvantages of CSP-managed keys.

Advantages of the CSP-Managed Key	Disadvantages of the CSP-Managed Key
■ The cloud customer or the user doesn't have to design or set up the encryption process. Almost zero investment of time and resources is required. No skilled programmers or engineers are needed to set up and perform encryption/decryption. ■ Encryption and decryption of data is almost automatic, and the process has no impact on performance. ■ There is no additional charge to set up or use the service. ■ There is no need to worry about the HSM's physical security or about regularly patching and updating the HSM and the key server. ■ The CSP-managed key simplifies the IT operations and reduces the capital investment. ■ CSPs usually stay ahead of the new technologies, and customers can benefit from the latest tools and technology. ■ CSPs benefit from the scale of the economy, which they can pass on to customers.	■ Cloud customers have no idea who has access to their data. ■ The customers depend on the CSP to ensure proper access control to the HSM and the key server. ■ The CSP's processes and procedures dictate the patching and security of the HSM. ■ The CSP controls the end-to-end key management process. The customer has no insight into the process at all. ■ Anyone who can access the storage can see the data in the plain text. There is no granular control over the decryption of the data.

Table 7.2 CSP-Managed Keys: Advantages and Disadvantages

7.2.2 Customer-Managed Key

The customer-managed key (CMK) gives customers much more control over managing the encryption key. The CSP provides the key material (binary bit string), but it's up to the customer to generate the key and manage the key lifecycle. The CMKs stay in the CSP's KMS. The key material can't ever be exported out of the KMS. The only exception is the public key of the asymmetric key pair. In fact, cloud customers can't even have access to the key material.

The CMK allows customers to enable their own key access policy and identity and access management (IAM) policy, providing granular control over who can access and view the data.

Table 7.3 lists the advantages and disadvantages of CMKs.

Advantages of the Customer-Managed Key	Disadvantages of the Customer-Managed Key
■ The KMS allows users to store symmetric and asymmetric keys. The symmetric key and the private key of the asymmetric key pair never leave the KMS. These keys are used to wrap the data keys (envelope encryption discussed in Chapter 5). ■ In the event of suspicious activities, the cloud customers can access and view the complete audit trail of the key. ■ Customers have control over the key lifecycle with only a little bit of additional work. ■ Customers can create, rotate, or delete the keys in the KMS per their schedule. ■ Customers can customize the key and the IAM policies to manage the access control. ■ Very few specialized skills are needed to manage the KMS and the CMK.	■ The control and flexibility over the key come with a cost. There is a monthly and per-use cost associated with the CMK. ■ The cloud customer is responsible for creating a key management procedure and policy. This adds extra work for cloud customers. ■ The CMK doesn't provide full control. The customer still must use the KMS CMK within the CSP's framework. ■ Attaching the default key policy or making any mistake in the key policy or IAM policy defeats the purpose. ■ Potentially, the CSP could have access to the KMS and the key. ■ Although using CMK doesn't need a very high skill level, to maximize the benefits of KMS, the customer should have the ability to program and use the KMS application programming interface (API).

Table 7.3 Customer-Managed Keys: Advantages and Disadvantages

In a nutshell, this option provides a good balance between lots of work managing the keys versus absolutely no control over them.

7.2.3 Bring Your Own Key

If the CSP-managed key is at one end of the spectrum, then *bring your own key (BYOK)* is at the extreme other end. As the name implies, BYOK allows cloud customers to bring their own key material. This option gives cloud customers maximum control over the keys. The customer can import symmetric and asymmetric key pairs to the KMS.

Customers opt for this option in a number of situations, such as when the key material needs to meet certain compliance or regulatory requirements, the customer is using the key in a hybrid situation, or the customer simply wants to own the original key material.

Table 7.4 lists the advantages and disadvantages of BYOK.

Advantages of Bring Your Own Key	Disadvantages of Bring Your Own Key
■ Complete control, flexibility, and independence are available in managing the key lifecycle. This is almost like having a key in the on-premise deployment. ■ Total visibility into the key activities is provided. Customers can monitor and log all the access and activities. ■ Customers have the flexibility of importing, exporting, or deleting the key material. ■ Users can import all types of keys. ■ The customer, when in doubt, deletes the key or exports the key material, and the CSP won't have access to any data encrypted with that key.	■ There is no such thing as a free lunch. The flexibility and control offered by the BYOK encryption key model have a high price tag from the CSP. ■ This model also requires a high skill level on the cloud customer's team in implementing and managing the key lifecycle. ■ There's a high probability of introducing human error. ■ This is a very complicated process and adds a lot in terms of indirect workload and cost to the cloud customer.

Table 7.4 Bring Your Own Key: Advantages and Disadvantages

Table 7.5 shows the comparison of encryption key management models.

	CSP-Managed Key	Customer-Managed Key	Bring Your Own Key
Customer Control	No control	Partial control	Full control
Cost	No additional cost	Monthly and per usage cost	Very high cost
Audit Trail	No access	Yes, can view logged activities	Can monitor and log activities
Skill Level Needed to Use/Manage the Key	Low	Medium	High

Table 7.5 Comparison of Encryption Key Management Models

7.3 Cryptography as a Service by Major Cloud Service Providers

All major CSPs have one goal—to provide reliable and secure services to cloud users or customers. Providing reliable services is relatively simple (but very expensive). All major CSPs have multiple regions and availability zones as well as various options for data backups and quick data recovery. These concepts aren't new to the IT industry and have been implemented well by all major CSPs. Providing secure services, on the other hand, is more complex to implement because it's very hard to predict from where and how the data will be attacked. It could happen while data is in transit, in use, or at rest.

It could happen by a totally unknown person or by a malicious insider. Data can be hacked for espionage, money, or just for bragging rights. It's hard to predict and prepare for the attack vectors. This is why protecting data is complicated and the defense-in-depth concept is implemented to protect the data (see Chapter 1, Section 1.2.4).

However, when everything else fails to protect data, cryptography prevails! That is why all major CSPs (aka public cloud providers) offer a wide range of cryptography services and tools. In this section, we'll discuss, at a very high level, the cryptography services and tools offered by AWS, Microsoft Azure, and Google Cloud Platform.

7.3.1 Amazon Web Services

AWS relies, for the most part, on open-source cryptography algorithms. These algorithms have a proven track record, are well-studied and vetted by the community, and are researched by academia. As we use the cryptography services and tools offered by AWS, we notice that AWS offers a variety of algorithms and key sizes but recommends a default or most appropriate choice for any given task. (There are some exceptions where AWS strictly enforces the use of a certain algorithm only.)

AWS offers AES and Triple Data Encryption Standard (Triple DES) algorithms for symmetric cryptography. AES can be used with 128-, 192-, and 256-bit key sizes. For asymmetric algorithms, AWS chooses RSA and elliptic curve cryptography (ECC) algorithms.

Following is a brief overview of some of the commonly used cryptography services offered by AWS to protect the confidentiality and integrity of customer data:

- **AWS CloudHSM**

  AWS CloudHSM offers a HSM to cloud customers for securely storing and generating cryptographic keys. The HSM offered by AWS can perform most of the cryptographic services:

  – Generate, store, and manage cryptography keys.
  – Work with symmetric and asymmetric keys and perform both types of encryption.
  – Perform hashing as well as create and verify digital signatures.
  – Generate random data.

  This list probably sounds very familiar because TPM offers similar cryptographic services. AWS CloudHSM provides flexibility, reliability, and total control over cryptographic key management. However, all of these good features come with a hefty price tag.

- **AWS Key Management Service (AWS KMS)**

  AWS KMS can do almost everything for you that AWS CloudHSM can do but takes away the overhead and hassle of management. AWS KMS is a full-service key manager that can create, activate, deactivate, rotate, delete, and store keys, as well as work with symmetric and asymmetric cryptography. Most AWS services are integrated

and work seamlessly with AWS KMS, including AWS CloudTrail, which provides detailed logging of the cryptographic activities.

AWS KMS is a pay-as-you-go service and can cost much less compared to HSM.

> **AWS CloudHSM versus AWS KMS**
>
> It seems that both AWS CloudHSM and AWS KMS provide similar services, so why are two different services offered?
>
> CloudHSM is used when the customer must meet certain compliance, regulator, and industry requirements, such as using a certain entropy to generate the key. The CloudHSM option also works better when the customer is already using HSM and cryptography services on their data center and wants to move the key management to the cloud. Using CloudHSM will avoid re-encrypting the data and allow them to use the existing key.
>
> Conversely, AWS KMS is a full-service key management system that can be used for almost all other scenarios.

- **AWS PKI**

  In Chapter 6, Section 6.4, we discussed the public key infrastructure (PKI). AWS is a recognized certificate authority (CA) and can issue public and private certificates. AWS provides several PKI-related services, but we'll briefly mention the following two services:
  - AWS Certificate Manager can generate, issue, manage, renew, and delete private and public certificates for use with AWS websites and applications.
  - AWS Private Certificate Authority, as the name implies, provides services related to private certificates. Public certificates cost more money, and the idea of the private certificate is to keep the cost down by using internally generated certificates. The private certificates are only valid within the organization.

- **AWS Secrets Manager**

  AWS Secrets Manager allows customers to store and manage all other secrets, such as passwords, database credentials, third-party API keys, and more. The secrets manager replaces the hard-coded credentials in the code and also replaces the credentials at regular intervals.

  AWS Secrets Manager encrypts all the secrets stored using AWS KMS. You can store, retrieve, and manage secrets using the AWS Secrets Manager console, APIs, or the command-line interface (CLI).

- **AWS Encryption SDK**

  AWS Encryption SDK is a set of predefined or pre-coded software utilities for encrypting and decrypting software. It's more like a software library that lets people easily encrypt data with a minimum knowledge of cryptography. The software development kit (SDK) can be used on client-side applications.

AWS also offers AWS Database Encryption SDK, a software library that allows CSE within the database.

For data-in-transit security, AWS uses Secure Socket Layer (SSL)/Transport Layer Security (TLS), as well as AWS Direct Connect and VPN, to connect from an on-premise data center to AWS resources in the cloud. AWS also offers SSE for most storage and database services. This service is offered to the customer at no additional cost, and it uses the AWS default key or the CSP-managed key.

7.3.2 Microsoft Azure

Like AWS, Microsoft Azure also offers a variety of cryptography services, with a twist in the names. The following summarizes the similar services and highlights the differences:

- **Platform-managed key**
  The CSP-managed key, or the default encryption key, is known as the *platform-managed key* in Microsoft Azure, and is used for SSE. There is no additional cost to users, and the entire process is completely transparent.

- **Key vault and HSM services**
  Key management in Microsoft Azure is somewhat different as it breaks down into more specific scenarios (e.g., for payments). Microsoft Azure offers the following options:
  - Key management: The baseline Microsoft Azure key management offering is the Azure Key Vault Standard. This is used with CMKs in situations where Federal Information Processing Standards (FIPS) hardware or other strict requirements aren't applicable. Otherwise, the customer can choose Azure Key Vault Premium services.
  - HSM services: For various key usages, Microsoft Azure offers HSM services: Azure Payment HSM, Azure Dedicated HSM, and Azure Managed HSM. As you can guess from the names, the Azure Payment HSM is used when the key is used for payment-related processing. Azure Dedicated HSM is used for the single tenancy, and Azure Managed HSM is used for all other use cases.

- **Certificate services**
  As a CA, Microsoft Azure offers various certificate-related services.

- **Data-in-transit protection**
  For data-in-transit protection, in addition to using SSL/TLS protocols, Microsoft Azure offers the Azure VPN Gateway for site-to-site connectivity between the data center and the Microsoft Azure resources in the cloud. For securing web traffic, Microsoft provides the Azure Application Gateway.

7.3.3 Google Cloud Platform

Like AWS and Microsoft Azure, Google Cloud Platform offers a wide range of cryptography services. Most of the offerings are very similar to AWS and Microsoft Azure, with its own twist in the names.

Google offers a variety of encryption KMSs. The following is a summary of each:

- **Google-managed keys**
  Google-managed keys encrypt customer data in Google Cloud storage. This service is free. This key and service are used for SSE. The user doesn't need to configure anything additional, and the process is transparent to users.

- **Cloud Key Management Service**
  The next level up is the CMKs. The keys are managed in Cloud KMS, which is very similar to AWS KMS. There is a cost to use this service, but this is the best trade-off between convenience and cost.

  CSE is also possible with Cloud KMS. This is basic CSE that we discussed earlier but with a twist. The Cloud KMS key is used to encrypt the client's encryption key. This is nothing but envelope encryption, where the DEK is the key generated at the client side, and the data is encrypted with the DEK. The DEK is then wrapped or encrypted with the key encryption key (KEK), which is a key from Cloud KMS.

- **Google Cloud HSM**
  Cloud HSM is a key management service using an HSM. This option, as we know, provides the most control but adds management overhead and cost.

- **Cloud External Key Manager (Cloud EKM)**
  Google Cloud Platform offers an external CMK called Cloud EKM. This option allows users to store data with a partner, approved key service provider, or even on-premise. The key is always off-site. Google Cloud Platform services access the keys over the internet or via VPC. The customer bears all the responsibilities and costs.

- **Certificate services**
  Google also offers certificate services for public and private services. The functionality is very similar to that of AWS and Microsoft Azure.

- **Data-in-transit protection**
  For data-in-transit protection, in addition to using SSL/TLS protocols, Google Cloud Platform offers dedicated interconnections and VPNs to easily and securely access Google Cloud resources from the on-premise data center.

The three major CSPs—AWS, Microsoft Azure, and Google Cloud Platform—are at the top of their game when it comes to safeguarding customers' data.

7.3.4 Cryptography Services by Other Cloud Service Providers

Many other CSPs offer various cloud-based services. It's beyond the scope of this book to discuss the cryptography services offered by each CSP. However, IBM, Oracle, SAP, and Salesforce have a large enough footprint in the cloud market that we must briefly cover them while discussing the cloud.

IBM is a full-service CSP, which means it offers almost all the services a CSP could offer. However, IBM isn't as big in terms of market share as AWS, Microsoft Azure, and Google Cloud Platform. IBM Cloud provides SSE and encrypts data at rest in all storage devices and databases. For data-in-transit protection, IBM Cloud uses SSL/TLS protocols. For point-to-point connections, such as data centers to IBM Cloud, IBM uses a VPN gateway. IBM Cloud offers a Cloud Foundry Application Gateway to secure web traffic. For the key management part, IBM Cloud Key Protect offers KMSs that work very similarly to other KMSs we've discussed earlier. IBM Cloud Hyper Protect Crypto Services provides a full KMS, including an HSM.

Oracle and SAP are also CSPs that offer IaaS, PaaS, and SaaS services, while Salesforce mainly offers SaaS only. However, Oracle and SAP are primarily focused on cloud offerings of their own services. Oracle provides various options to encrypt Oracle databases, such as transparent data encryption (TDE), column encryption and, in some cases, even field encryption. For encryption key management, Oracle offers Oracle Cloud Infrastructure Vault (OCI Vault). OCI Vault offers cloud-based, full-service encryption key management.

SAP, a major player in the enterprise software business, offers most of its solutions on the cloud and collaborates with AWS, Microsoft Azure, and Google Cloud Platform to host them. SAP also offers some solutions as private cloud editions, where the solution is deployed in a dedicated VPC for a given customer. SAP Data Custodian offers KMSs.

All the CSPs offer robust cryptography solutions to protect customers' data and information. It doesn't really matter much whether your data is with AWS, Oracle, or any other provider. Most of the time, the issue is with the implementation and integration of the services offered. At times, the problem arises because cloud customers don't have people with the required skill level to implement and integrate the cryptography services offered by the cloud or the customer wants to save money and choose the lower level of security.

7.4 Lightweight Cryptography and the Internet of Things

Usage of the internet increased significantly after Wi-Fi came to every home. The use of Wi-Fi enables us to connect various devices or "things" to the internet easily. By connecting unconventional devices or things to the internet, people are able to do almost anything from almost anywhere! (By unconventional devices, we mean the devices that you didn't expect 15 years ago to be connected to the internet for use, such as smart baby diapers!) The concept of connecting devices or things to the internet for unconventional applications is known as the *Internet of Things (IoT)*. Because these devices are connected to the internet and are on local area networks (LANs) or wide area networks (WANs), NIST uses the term Network of Things (NoT). In this book, we'll stick with IoT.

In this section, we'll learn how the IoT works, its various applications, the risks associated with using it, and how cryptography can help protect it.

7.4.1 The IoT Concept

Computers, networks, routers, modems, and all other digital devices work on binary bit streams that are converted to electrical signals. The problem is that most devices (things) we interact with in our daily lives don't have digital output. Temperature, humidity, weight, sound, vehicle speed, coffee makers, and garage door openers are all producing analog signals. To use our analog devices with the internet, the first thing we need to do is to generate an electrical signal and then convert the signal into a digital signal. We achieve this by using *sensors*. Although we commonly refer to them as sensors, technically speaking, many of these sensors are *transducers*. At a high level, sensors and transducers do the same thing—they detect the change in the surroundings or the device they are attached to and generate an output. However, the sensors generate output in the same format while transducers convert the change into an electrical signal and generate an electrical signal as an output. For convenience, we'll use only one term and refer to them as sensors.

Figure 7.4 shows the IoT concept. The sensors are connected to whatever parameters we're measuring, and their output, in the electrical signal, is connected to the controller or the processor. The controller collects the data from the sensors, processes the data, and generates an output or the action. The action can go back to the device or the "thing" that the sensor is attached to, like a closed loop system, or can go to another device and act as an open loop system. Let's consider an example.

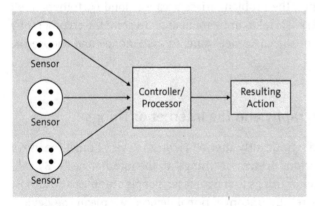

Figure 7.4 IoT Concept

When a baby is wearing a sensor-connected diaper, the sensor in the diaper will sense the wetness (probably through a litmus paper) when the diaper is wet and send a signal to the parent's phone. Assuming the parents have an app on their phone, the parents will be notified. This is an example of an *open-loop system*; someone else takes the corrective action.

But when a smart car like Tesla "sees" a person or an object in front of it, the sensor sends a signal to the controller. The controller processes the signal and takes action for the car to stop. The controller sends that signal to the brake module, which ultimately generates the signal and stops the car. There is no human intervention. This is a *closed-loop system*; the controller takes the corrective action. The signal came from the car and went back to the car.

A few years ago, our break room at work got a new coffee maker. In addition to all the bells and whistles, it has one feature that caught everyone's attention—an app. The coffee maker would notify you when your coffee was ready! How do you think that is happening? The coffee maker was connected to Wi-Fi. When the brewing is done, the sensor notifies the app.

We see so many examples of IoT around us but never really think about how they work.

> **Little Known History of IoT**
>
> The first known connected device was a Coca-Cola vending machine at Carnegie Mellon University, Pittsburgh (CMU). The machine was first operational in 1982! Like most inventions, this invention was also due to necessity. The machine was somewhat far away, and it was inconvenient for students to walk all that way to find that it was either empty or just refilled with Cokes that were still warm. To solve this problem, David Nichols and friends decided to monitor the status of the machine and share the status on the Advanced Research Projects Agency Network (ARPANET) and CMU's local network [Teicher, 2018].

7.4.2 Risks and Attacks Associated with IoT

Our society is fascinated by the hype of IoT. Every device around us is connected to the internet. The concept is very simple, and it works fine for everyone, right? Well, unfortunately, many things can go wrong!

The fact that IoT has a simple operating model goes against it because it's also relatively easy to manipulate or hack into IoT devices. Following are some common risks to IoT devices:

- **Insecure communication**
  The first and foremost risk of IoT devices is insecure communication. The data in transit between two devices or between a device and the controller isn't encrypted.

Most smart devices are used in homes and personal environments, and encrypting the data was never really a requirement. After all, why would someone want to encrypt the data from a coffee maker and a baby diaper? This is probably true. However, if the data is unencrypted, it could provide a hacker with valuable information as part of the reconnaissance phase. By tapping into the unencrypted communication between the coffee maker and the router, a hacker may be able to get an SSID for the business, IP address, and so on, and if the hacker is lucky, he may be able to access the password.

- **Lack of proper authentication and authorization**
  Most IoT devices don't have a proper authentication process established. After all, why would someone set up two-factor authentication for a coffee maker app? This is a valid reason for not being security conscious, but it can cost the business. If hackers can easily access the coffee maker account, eventually they can escalate.

- **Vulnerable software and/or firmware**
  It's very difficult for a manufacturer to track and update IoT devices regularly as they are sold as retail items directly to the consumer. If the consumer doesn't register the product with the manufacturer, then the manufacturer has no information about the device and can't upgrade it. Vulnerable software can be exploited.

- **Storage protection**
  Have you ever checked if your doorbell database storage was secured? Or do you know if your thermostat data is encrypted? Most of us don't worry about these kinds of "things," but they are important when things are connected to the internet. Again, this data alone probably can't do much harm, but this does give a starting point to the hacker. The data can be used as part of the reconnaissance.

- **Poor physical security**
  Last but not least, poor physical security is another risk factor associated with IoT devices. Physical access to IoT devices and even Industrial Internet of Things (IIoT) devices isn't secured very well. Most devices can easily be accessed by anyone. If a hacker can physically reach the device, the hacker can potentially manipulate it.

These risks aren't specific to IoT devices; they are general security risks that are applicable to any software, system, or device. However, they become critical in IoT simply because it's hard to track and manage them. These unmanaged risks ultimately lead to some major attacks on IoT devices. Following are some of the potential attacks on IoT devices:

- **Man-in-the-middle attack**
  In this attack, there is an active attacker on the network or has access to the network. In the best-case scenario, the attacker can simply listen to the conversation between the devices or between the device and the controller. But the attacker can go a step further and do much harder tasks. For example, if the attack is "sitting" on the network, he can drop the messages, modify or change the content of the message,

replay the message at a later date, or even inject some new information. In all of these situations, damage could be significant if the attacker has access to IoT device communication on a manufacturing plant, office buildings, or other important locations.

- **Denial of service attack**
  The denial of service (DoS) or distributed denial of service (DDoS) attack happens when the attacker sends too many false requests to the device on the internet or the router. These requests ultimately overload the system, and the system will deny the service to the legitimate user. In some situations, the whole system may crash.

- **Botnet attacks**
  In botnet attacks, the hacker introduces botnets on the network in the hope of gaining complete control of the network or most of the connected devices using botnets. Once the hacker has control of the devices or the network, he can manipulate or use the device or the network information.

- **Escalation of privileges and privilege creep**
  Poor authentication practices lead to this attack. Once the attacker gets his foot in the door through one of the connected devices, no matter how unimportant the device is, the attacker can slowly gather the information and try to get more access or even try to escalate the privilege and become an admin of the network. Once the attacker becomes an admin, all bets are off. The attacker has full control of the network.

7.4.3 Securing Connected Devices with Lightweight Cryptography

What is lightweight cryptography (LWC), and why do we need it when we've learned so many cryptography algorithms in Chapter 2 and Chapter 3? Can't we use one of those? Let's discuss LWC in the following sections.

Primer on Lightweight Cryptography

Algorithms that we learned and have been using, such as AES and RSA, are secure and almost impossible to break, but this robustness comes with a lot of baggage—it requires a lot of computational power and memory. IoT devices can't handle these algorithms. *Lightweight cryptography (LWC)*, on the other hand, is very efficient. The LWC algorithms consume less power, use minimum memory, provide high throughput, and can operate on a bare minimum hardware. NIST uses the phrase "cryptography for a constrained environment" for LWC. The constrained environment does need a lean algorithm to perform well.

Following are the reasons why IoT devices need LWC:

- IoT devices are very small, and they can't run traditional algorithms. They need algorithms with a very small hardware footprint. The implementation of the cryptosystem must fit within the tiny sensor and other "things."

- IoT devices need authenticated encryption (AE), which provides confidentiality and authenticity. The former is provided by encrypting the data while the latter is provided by using a MAC. The MAC can be produced before or after encrypting the plain text. Because AE kills two birds with one stone, it's preferrable to use it with smaller IoT devices.
- Most IoT devices run on batteries, and it's important to have a lightweight algorithm that consumes little power.
- There is hardly enough memory for the sensor to run, and additional features, such as cryptography, are left with little memory. The lightweight algorithms must perform with little memory consumption.
- Sensors in IoT devices send updates to the controller or processor frequently, sometimes the frequency is less than 1 second. If the algorithm takes closer to a second, it can't provide the output or the reliable output. The lightweight algorithms must have very low latency and high throughput.

It's important to note that the "light" features and simple design work against the lightweight algorithms in terms of security. LWC algorithms are more susceptible (because they are lightweight!) to side-channel attacks. A *side-channel attack* is an attack on a cryptosystem by analyzing secondary parameters, such as timing analysis, power consumption charts, noise vibration study, and so on, rather than an attack because of the design flaw. Because the algorithm is lightweight and is running, usually, on a smaller device, it's easier to measure the secondary parameters, and this makes them a target for side-channel attacks.

The requirements are clear. If we want to secure IoT devices, we need LWC.

Implementing Lightweight Cryptography Algorithms

The field of LWC is relatively new. NIST started investigating "cryptography for a constrained environment" in 2013. They began a standardization process for cryptography algorithms that provide authenticated encryption with associated data in a constrained environment and then announced a call for papers in late 2018. The first round began in March 2019, and NIST received 57 algorithms. By August 2019, NIST vetted and selected 32 algorithms for the second round, and then 10 algorithms were selected for the third and final round. The 10 finalists entered the third round in March 2021, and Ascon, an LWC algorithm, was declared the winner in February 2023 [Dobraunig, 2019].

The Ascon algorithm uses the *authenticated encryption with associated data (AEAD)* concept. In other words, Ascon provides both confidentiality and integrity. This is an essential property for a lightweight algorithm as it can't accommodate two different algorithms for encryption and hashing. In conventional cryptography algorithms, it takes two passes of data, first for encryption and then for hashing, but in AEAD, the idea is to achieve both in one pass.

7.4 Lightweight Cryptography and the Internet of Things

Ascon has two block sizes: 64-bit and 128-bit. The state is 320 bits in size, while the nonce and tag are both 128 bits. The key size for the Ascon cipher is 128 bits. The cipher uses sponge construction and a substitution-permutation network (SP-network). This is very similar to the SHA-3 algorithm (refer to Chapter 4 for more details on the SHA-3 sponge construction). The difference, however, is that it doesn't use the table lookup; instead, it relies on the XOR function. Figure 7.5 shows the construction of the Ascon cipher. The process is divided into three phases or stages: initialization, encryption, and finalization. The encryption stage is divided into two states—associated data and plain text—if the associated data is used. The permutation is represented by p and the number of times the permutation is performed is denoted by a and b, where a is 12 and b is 6. This essentially means that initialization and finalization go through 12 rounds of permutation while associated data and encryption have 6 rounds of permutation. In other words, in p^a, the permutation is performed 12 times.

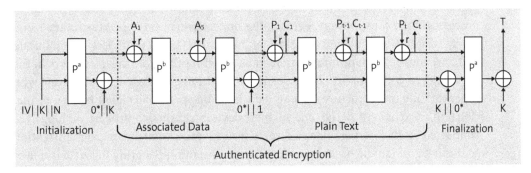

Figure 7.5 Ascon Lightweight Cipher

The initialization takes 320 bits of state, which is equal to 128 bits of key K + 128 bits of nonce N + 64 bits of IV. These 320 bits go through the initialization permutation p^a times (or 12 times). The output is XORed with the encryption key of 128 bits. At this point, the initialization is complete. Now we're ready to perform the encryption.

Next, we add the associated data to the mix. The 320-bit initialized state is XORed with associated data A. The length of associated data is r bits or the block size. (Recall from Chapter 4 the total value of state $b = r + c$, where r is the rate or the block size and c is the capacity.)

After adding the associated data, we're ready to encrypt the plain text by adding or XORing the plain text data with the 320-bit state. The output of the XOR is the cipher text C. The process is repeated for the entire plain text message. The permutation is performed in 6 rounds ($b = 6$) for associated data and plain text.

Once the cipher runs out of the entire plain text message, the permutation is performed in 12 rounds on the internal state (or the remaining 320 bits) and the key is added (XORed) to generate the final 128-bit tag, T. The tag provides authenticity to the associated data as well as to the cipher text.

In Chapter 4, we saw that the SHA-3 function represents the state of 1600 bits in a 5x5x64-bit 3D state. Ascon represents its state of 320 bits with a 5x64-bit 2D state.

The Ascon algorithm can process the data before the complete input or length is known. That means it can begin decrypting as soon as data arrives and doesn't have to receive the complete message. Ascon is inverse-free because p^a and p^b are only evaluated in one direction. No inverse operation is required even while decrypting, reducing the overhead. The Ascon algorithm is very efficient in hardware and software implementation.

7.4.4 Cryptography in Cars

It wouldn't be a stretch to say cryptography is at the heart of data security. Or, let's use a more appropriate analogy for this section: if information security is a car, then cryptography is the engine! You would be surprised to know that today's cars have more processors than some powerful VMs in the cloud. Controllers or processors are everywhere in cars, from engine control to infotainment. These controllers communicate and exchange important information and data with other controllers, sensors in the car, and even with support and dealerships. Self-driving cars even communicate with other cars and infrastructure. Various data, including personally identifiable information (PII), is exchanged all the time. All these devices (sensors, controllers, or things) are connected to the internet. Whether we, as security professionals, are paying attention to this data or not, the hackers are paying close attention. Securing this data is one of the important jobs for cryptographers. However, the most important security task for the automotive industry is to protect the most expensive asset—the car itself.

In the good old days, cars were easy to steal by simply cutting the wire from the ignition and connecting them directly; this process was commonly referred to as "hot wiring." In the mid- to late 1980s, automakers started introducing various mechanisms to make it difficult for thieves to steal the vehicle. By the 1990s, vehicles were equipped with Vehicle Anti-Theft System (VATS). This was a very simple system: a transponder built into the key with a fixed code. When the key was inserted into the ignition, the transporter was energized and read the code from the key. The code was compared in the module near the ignition cylinder. If the code matches, the starting sequence begins. This implementation of VATS was known as a Passkey system. By today's standards, this wasn't security. The code was fixed and can be read by using any tool that is used for vehicle diagnostics, but surprisingly enough, the overall vehicle theft numbers were significantly down. After all, thieves also have a learning curve!

Eventually, thieves started to catch up, but the automakers were a step ahead. They introduced a PassLock system. In the PassLock system, a similar concept was used, but the code was also compared with the engine controller—stored in the Engine Control Unit or Engine Control Module (ECU/ECM). This was slightly better than the PassKey system because it first validated the key and then compared the code with the ECU.

However, it was using fixed code and could be read with very little effort. It was just a matter of time before the thieves would catch up to the PassLock system.

So, automakers made it even harder! This time, a new vehicle anti-theft system was introduced and is commonly referred to as an "immobilizer." The immobilizer used a PassLock 3+ and a cryptographic chip to generate a random challenge response. It was a bit complicated for the manufacturing plants to "learn" the keys. The immobilizer worked in the following three stages:

1. **Learning phase**
   This is a one-time activity and usually happens in the assembly plant when the vehicle is built. This step is repeated only if the key, the PassKey 3+, or the ECM is replaced (and, in some cases, if the ECU is reflashed). In this step, the symmetric key is shared between the ECU and the PassKey 3+ module. The learning process is time-sensitive, and if the keys aren't learned within the given times, which is usually from 200–300 ms range, the process fails, and the ECU sets the diagnostic code. Once the keys are learned, the vehicle goes through the entire starting sequence.

2. **Initialization phase**
   This step is executed every time the driver starts the car. As soon as the driver enters the key in the ignition, the key is validated against the PassKey 3+ module. If that is the correct key, then the PassKey 3+ sends a fixed-value code to the ECU. The ECU matches this fix code and releases the engine for a short period (usually only 200–300 msecs). This step is very similar to the older PassLock system, but it allows the engine to crank for only about 200 msecs. The idea is to buy some time to perform the full challenge-response in the background, but the driver shouldn't feel that the vehicle didn't crank for a long time. At the same time, the time should be short enough so that the thief doesn't have time to move the vehicle out of sight, to the back alley or a less busy area.

3. **Engine release**
   Once the successful initialization happens, the ECU sends a cryptographic challenge to the PassLock 3+ system. The PassLock 3+ module takes the challenge, combined with the encryption key (learned in the learning phase in assembly) and generates the response. The response is sent back to the ECU. ECU also generates the response internally and compares it with the response received. If they are matched, full engine release is granted, and the starting sequence is completed. If the response isn't matched or not received, the engine stalls. The whole process should be completed in about 300 msecs.

What if a thief replaces the ECU? During the initialization phase, the ECU checks and validates the vehicle identification number (VIN). If the ECU is learned in one vehicle and installed in the other vehicle, or the ECU is replaced, it won't work.

This was developed in the early to mid-2000s. The system worked very well for the time. The cryptographic algorithm used was proprietary, and the key size was very small. Since then, technology has improved a lot. Now, LWC algorithms are used in vehicles to execute many tasks, including the immobilizer. Although the implementation and execution are much leaner and faster, the concept remains the same.

> **A Personal Note about Immobilizer**
>
> The topic of immobilizers is very close to my heart. In the early 2000s, I worked on developing the architecture and software for the Immobilizer system. I have a US Patent and Defensive Publication on the subject. This was very early in my career; I've spent numerous hours debugging the issues and have spent many, many nights teaching how to learn the keys at the assembly plants. Since then, I have always been drawn to cryptography.

Cryptography is used a lot more nowadays in connected vehicles and for many more purposes. The connected vehicles use asymmetric or public key cryptography. The key pairs are generated and assigned when the vehicle is built.

It's interesting to note that as technology has advanced over the years, more sophisticated security controls have been introduced to protect cars and data. However, that hasn't deterred thieves. In 2018, researchers were able to gain remote control over the infotainment system and controller area network (CAN) buses of more than 10 automotive models [Burkacky, 2020]. Apart from the vehicle itself and PII data, GPS locations have been another data set of interest to hackers. The hackers extract GPS location data to locate when people are home and at work to find the ideal time to enter the vehicle or even the home.

The need for robust security control in vehicles is growing very fast, and cryptography is equipped to provide the best security controls.

7.5 Summary

Confidentiality, integrity, and availability of a data are at the core of information security. No matter if the data is on-premise, in the cloud, in the IoT, or in a vehicle, the three sides of the security triad must be secured. Regardless of the location of the data, who is processing the data, and how the data is processed, the data must be protected.

In this chapter, we explored how cryptography is used in the cloud to protect infrastructure, data, and applications. Managing encryption keys in the cloud is even more important than encryption itself. We learned how to implement the various methods of managing encryption keys in the cloud. We also surveyed the cryptography services offered by various CSPs.

The second part of the chapter was focused on the IoT. We studied how IoT works and how we can secure the IoT network using LWC. LWC is a relatively new and developing field. We learned why LWC is needed and how we can implement it. Lastly, we discussed the use of cryptography in vehicles.

> **Key Takeaways from This Chapter**
> - Learned the role of cryptography and encryption key management in the cloud
> - Reviewed how IoT works and how LWC can help protect IoT
> - Importance and need of cryptography in the automotive industry

Cryptocurrency has attracted the attention of everyone, from technologists to investors. In the next chapter, we'll learn why and how cryptography is used with currency.

Chapter 8
Cryptography in Cryptocurrency

People have always been interested in money, and cryptocurrency is no exception. In fact, cryptocurrency has taken people's interest in money to a new height. People from all walks of life are taking a deep interest in cryptocurrency—for reasons ranging from exciting technology to the potential to become rich overnight. The focus of this chapter is on the technology aspect of cryptocurrency. We'll discuss what makes cryptocurrencies so sought after and what role cryptography services play in cryptocurrency.

Currency or money has been a catalyst for anything and everything we do in our lives. To make more money, people have worked harder, invented new things, studied and specialized in professions, fought wars—the list goes on. You name anything that is happening around us and 9 out of 10 things are probably happening for or because of money.

People's interest in money isn't new, but cryptography in money—now that certainly is a headline! As we've learned in earlier chapters, cryptography has been used for many centuries. The initial or primary purpose of cryptography has always been to provide confidentiality. Over time, the use of cryptography has been extended to provide integrity, authenticity, and nonrepudiation. Since the late 1990s, the use of cryptography in the technology world has exploded. The use of cryptography services has been extended to many applications from securing streaming data to protecting Internet of Things (IoT) devices to signing emails. Cryptography has become an integral part of our lives. Since 2009, cryptography has, officially, entered the financial world in the form of *cryptocurrency*. Since 2014, cryptocurrency has taken the financial world by storm.

> **Chapter Highlights**
> - Overview and fundamentals of cryptocurrency
> - Understand the use of cryptography in cryptocurrency
> - Learn about challenges, risks, and the future of cryptocurrency

So, what prompted the interest of financers and investors in cryptography, a side of technology that is heavily mathematical and difficult to deal with even for the nerdiest among us? What is crypto(graphy) doing in cryptocurrency? We'll attempt to answer these questions in this chapter.

This chapter is divided into three sections. Section 8.1 will start with an overview and fundamentals of cryptocurrency. We'll briefly look at the history of money and what led us to the use of cryptocurrency. The heart of this chapter, Section 8.2, is focused on how cryptography is used with cryptocurrency. The section will also discuss cryptographic algorithms used in cryptocurrency, and the working and use of crypto wallets. Section 8.3 will cover the outlook of cryptocurrency. We'll discuss the challenges of cryptocurrency and the risks faced by cryptocurrency. The chapter will end with the future of cryptocurrency and money in general.

8.1 Primer on Cryptocurrency

In this section, we'll provide an overview of the cryptocurrency fundamentals, including the underlying blockchain technology. But first, we'll take a look back at the history of currency to set the stage for where we are today.

8.1.1 History of Money

It's almost impossible to trace the origin of money because money or currency has been around in one form or another since the dawn of civilization. The history of money probably started at the same time people started cooking food or developing other necessities. Money has taken many forms over the years.

The first form of money is the barter system. Bartering is nothing but exchanging one thing for another. At times, bartering worked more on necessity than on value. For example, if a family is hungry and needs meat and eggs, then the family may end up trading a sheep for meat and eggs. At this time, hunger prevails over the value of the sheep. This worked well within small communities but wasn't a workable solution as civilization started expanding.

The barter system was eventually replaced with commodity money. In a commodity money system, one type of commodity works as an instrument or money. This could be salt, grain, or livestock. At a very high level, barter systems and commodity money are very similar—they both use some form of a commodity for an exchange, however, in a commodity money system, everyone uses only one item—salt, a specific type of grain, or a kind of livestock. The commodity money system established two things in society—*standardization* and *centralization*. First, it standardized the value for an item or labor. There was a fixed quantity of exchange, unlike the barter system where necessity ruled over value. Second, it centralized the control. Now one person, the king or the head of the community, decided the commodity money. For example, say the society decided that the commodity money would be wheat (grain). This increases the value of wheat, so people would be working on preserving wheat. But there was a fundamental problem with the system. The commodity itself wasn't in anyone's control. There could be a drought or flood, and wheat production could be impacted, or it could swing another way, and every would produce lots of wheat. The point is any of these factors could impact the economy and people's lives. Despite all the challenges, commodity money stayed around for a long time.

> **We Still Use Other Forms of Money Today**
>
> Kids in elementary school often trade baseball cards or other similar items as needed. This is one form of money. My kid was one of the youngest in the class and probably the last one to drive in his class during his high school years. He managed his rides by offering other drivers to print their homework! My other kid traded video games all the time with his friends. What do we call it—convenience or bartering?

Commodity money was slowly replaced with representation money. In this form of money, people were required to deposit valuables, usually precious metals, with a third party. This third party then issues a receipt or proof stating how much the deposit was worth. The person's buying power was based on the deposit receipt. This form of money was a bit more stable than commodity money because it assured that a person had a certain amount of assets, and the seller would know the exact value of that asset. In commodity money, there was no guarantee as to how much wheat crop would be available in a given year. Basically, the buying power could change for any unknown and uncontrollable reason, and the seller may not get the promised money. Representation money was more reliable or stable, but it involved a third party—usually a goldsmith or some sort of investor. But what if the third party defaults, cheats or lies? There was always a risk because the seller was relying on an unknown third party.

> **History of Money: What's in It for Me?**
>
> You may wonder why we're talking about the history of money and how it's relevant to cryptography. The goal here is to identify the weaknesses in various forms of money over time. Digital currency isn't immune to these weaknesses. The difference is that cryptography came to our rescue in securing digital money from these weaknesses.
>
> Note that the history of money outlined here isn't in perfect chronological order nor does it cover the entire history end to end. The purpose here is to help you understand the relevance of that history to cryptocurrency. For a complete history of money, refer to [Lewis, 2018] or watch the documentary *Bitcoin: The End of Money as We Know It!* [Hoffmann, 2015].

Eventually, kings and administrators introduced money in other forms, such as coins, leather, and paper. Many kings used precious metal coins, and the value was based on the type of the metal and the size of the coin. People started to cut the coin for smaller denominations and even used various shapes and sizes to fit their needs. The kings were supposed to have central control over the monetary system, but people started minting the coins at home to suit their needs. Similar issues were faced with leather money. Paper money is also traced way back in history. However, the origin of paper money was mainly attributed to bank notes, which derived from representation money.

Going back to using precious metal as a form of currency, the United States has a very interesting history of gold and money. In the late 1800s, the US government started minting money against the gold reserve. (In fact, it was silver and gold to start with, but, eventually, silver was dropped.) During the Great Depression of the early 1930s, the US government was facing a short supply of dollars, and the government needed to print more. In 1934, President Roosevelt introduced a Gold Reserve Act that required people to surrender all their gold to the government unless it was in the form of jewelry. Failing to do so could result in a fine or imprisonment. However, even after this, there wasn't enough gold with the US government to print more dollars, and the dollar to gold reserve was devalued.

> **Gold: Timeless Value**
>
> Gold has retained its value over centuries. Gold was precious and was valuable a thousand years ago, and, even today, it holds its value. One reason is the strength of gold over time—it doesn't tarnish or lose its shine. The second (economic) reason is the difficulty of mining and producing usable gold. The combination of these factors made gold very valuable. Gold is considered an inflation-free metal. The buying power of gold was the same 100 years ago as it is now—after adding inflation. (Although, post-pandemic inflation has slightly changed this philosophy.)

Near the end of the Second World War, the United States emerged as the most influential and powerful nation. In July 1944, the United States, Canada, Australia, most Western European countries, and 44 other countries met in the United States to negotiate and establish the Bretton Woods Monetary system. The system required every country to guarantee the convertibility of their currency to the US dollar. The US dollar was pegged to the price of gold: US $35 per one troy ounce.

The idea behind the Bretton Woods agreement was to develop economic stability around the world after the Second World War. By the 1960s, it was evident that the global economy was stabilizing, and, at the same time, the US government started going into a deficit due to overspending and inflation. This ultimately made it impossible for the US government to back the dollar with the gold reserve. In 1971, President Richard Nixon decided that the US dollar was no longer pegged to gold and that essentially ended the Bretton Woods agreement.

Since 1971, money or currency worldwide has been solely backed by the guarantee of each country's government and protected by their military power. (Conceptually, this is no different than the representation money.) Figure 8.1 shows currency over time, also known as *fiat money*. The term *fiat* in Latin means "let it be." Since the early 1970s, we've been using this fiat currency. With the onset of internet technology, the way we transfer/exchange currency from one hand to another has changed, but not the actual currency. In other words, only the mode of payment/exchange has changed from cash to electronic. This brings us to cryptocurrency, which we'll introduce next.

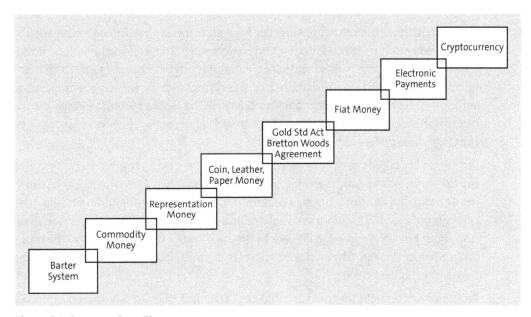

Figure 8.1 Currency Over Time

8.1.2 Introduction to Cryptocurrency

Every form of currency we've seen historically is centrally managed. The bartering system is the only exception to the central control. In commodity money, the commodity to use for transactions was decided by the king or the head of the community. In fiat money, the currency is heavily controlled and managed by the government or the central/reserve bank. This means that the value and the stability offered by the currency can change if the government collapses or other issues impact the country. Moreover, the fiat currency isn't backed by anything, so the government can print as much as needed. This could increase inflation and devalue the money. This ends up in a catch-22 economic problem for any country. To hold the value of anything, the item, including currency, must be in a limited or at least somewhat controlled supply. Gold is a prime example of a supply and value relationship.

Another common thread of weakness in the various forms of currency is the trust in one party or person. The accuracy of transactions is based on the third party such as a bank or other lending institution. The transactions can be forged or altered without detection. There is no guarantee of a successful transaction without a reliable third party. For example, if you're buying a house, you need to have a broker or use escrow to make sure the buyer and seller deliver what they have promised. Without escrow, someone can easily change their mind, and it would cost lots of money and time to solve the dispute. Things will go smoothly if the escrow is trustworthy; otherwise, even after having a third party, things could end up in court. On top of this, we end up paying fees for retaining the third party or to the bank for transactions. Even for the smallest transaction, banks charge fees.

Counterfeiting or creating a fake currency is as old as money itself. People have tried to duplicate metal coins, leather money, and paper money. The latest technological developments have made very fine printing easy, and, although it's not very easy, a fake currency can be printed and distributed in the system. Again, we rely on the central authority to curb counterfeit money circulation. The whole system is based on the trust of a third party and central control. If that fails or someone makes a wrong decision, the whole system can collapse.

Privacy and reliability are other issues with fiat money. Every transaction is recorded with all the details about the individual or the business involved. When a bank, a business, or a credit card company gets hacked, all the information about involved parties gets compromised. With more and more online transactions, this is a very serious concern about privacy. People can go back in time and change the ledger entry. The transactions are reversible. Theoretically, there is no reliability that once recorded, the transaction is guaranteed.

People always thought technology could help us. In fact, people have started exploring ideas because the computer has made its way into our lives. In 1970, Steven Wiesner proposed the use of quantum money that can be hard to counterfeit. However, given the difficulty of implementing quantum physics, the idea wasn't proven very practical. (You'll learn more about this in Chapter 10.) In 1982, David Chaum made the first (unsuccessful) attempt at the digital currency called eCash [Chaum, 1979]. You'll learn more about David Chaum in Section 8.1.3.

Since the mid-1990s, the internet fueled the growth of online businesses, and most transactions have been taking place online. Money transactions are already happening online, but there is a catch. Underneath the online transactions, people are still using fiat money, which is dependent on central control and third-party trust.

The bottom line is that people were looking for a viable, digital alternative. We needed a currency that wasn't centrally controlled, not dependent on third-party trust, not reversible, and untrackable. The answer was cryptocurrency!

It's almost undebatable that Bitcoin is the first popular cryptocurrency. Although there is no formal or official account, it appears that the work of developing Bitcoin started in 2007. In mid-2008, the *https://bitcoin.org* site was registered. In late 2008, a paper titled "Bitcoin: A Peer-to-Peer Electronic Cash System" [Nakamoto, 2008] was published under the name of Satoshi Nakamoto.

> **Satoshi Nakamoto**
>
> The inventor of Bitcoin has decided to keep his identity very confidential. Not much is known about Satoshi Nakamoto. In fact, it's not clear if the person is a male or female or is even a single person or a team of developers. For the sake of simplicity, we'll refer to Satoshi Nakamoto as "him"—male, singular.
>
> Nakamoto was reachable and worked with other developers until mid-2010. One report says that in 2010, a dark web, Silk Road, announced that they would do business only using Bitcoin—primarily because it's not traceable or reversible. Nakamoto responded using a post saying that he didn't expect Bitcoin to be famous this way. He has disappeared since that post.
>
> According to one estimate, Nakamoto owns between 750,000 and 1.1 million Bitcoins. At the time of writing, that's more than $75 billion US dollars
>
> Satoshi Nakamoto is honored by giving his name to the Bitcoin measurement. One unit of Satoshi is 0.0000001 Bitcoins.

In early 2009, Satoshi Nakamoto released the first version (version 0.1) of the Bitcoin software on SourceForge. He introduced the first ever block of the network. The first or the initial block of the blockchain is known as the genesis block—you'll learn more about blockchain in the next section. After the initial release, Nakamoto himself made all the modifications to the code, but, later in 2010, he gave all control of the source code to Gavin Anderson and some other key contributors. Anderson founded the Bitcoin Foundation in 2012 and served as a chief scientist. The Bitcoin Foundation was very similar to the Linux Foundation.

One of the main reasons Bitcoins are so sought after is that, like gold, they are also in limited supply. Satoshi has designed the network such that a maximum of 21 million Bitcoins can be available in circulation. Approximately, more than 19 million (of 21 million) Bitcoins are already in circulation—that is more than 90% of the total Bitcoins! The process of verifying the transaction and adding a new block to the blockchain is known as *mining* the Bitcoin. Mining involves solving complex mathematical problems. Once the Bitcoin is mined, it's added to the circulation. People earn Bitcoins by mining. Approximately every 10 minutes, a new Bitcoin is mined, and the Bitcoins are earned. About every 210,000 blocks or about every four years, the number of Bitcoins mined per block is reduced in half. When Bitcoins were first introduced, 50 Bitcoins were awarded for adding a block. Right now, it's 6.25 Bitcoins. By the end of 2024, it's expected to go down to 3.125 Bitcoins. It's expected that by 2136, the reward for mining will go down to 1 Satoshi (0.00000001 Bitcoin) and will be 0 by 2140 [Lewis, 2018]. Once all the 21 million Bitcoins are mined, miners won't be able to earn by mining anymore. The only way to earn Bitcoin after that is by the transaction charges.

If you're confused with all these new terms—mining, blockchain, transaction fees, and so on—don't worry, you'll learn about all of these in the next section.

8 Cryptography in Cryptocurrency

8.1.3 Primer on Blockchain

Blockchain is a type of database that stores data (transactions) in blocks. Once a transaction is confirmed, a block is connected to a previous block. A chain consists of a number of blocks connected to each other in chronological order. It's like an electronic ledger that is simultaneously updated and verified by the involved parties. Once a transaction is recorded in the blockchain, it's very difficult or almost impossible to change it. The transactions are cryptographically verified. Figure 8.2 shows the blockchain in its basic form.

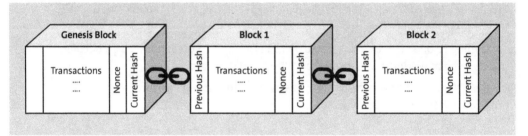

Figure 8.2 Basic Blockchain

Who Is David Chaum?

David Chaum published the first research paper outlining the concept of blockchain and digital cash in 1979, titled "Established, Maintained, and Trusted by Mutually Suspicious Groups." Chaum is credited for DigiCash and eCash. In 1982, he founded the International Association for Cryptography Research (IACR). IACR is considered a very reputable organization in the cryptography community. Chaum is known as the father of digital cash and the godfather of cryptography. If you're interested in or working on cryptography, you'll be amazed to know his achievements. Find out more at *www.chaum.com*.

Most people think that blockchain is a new technology, but David Chaum first proposed the concept in 1979. Since then, particularly since the early 1990s, blockchain has always been a topic of interest for many researchers. However, blockchain gained momentum only after it was used with Bitcoin. Some of the primary characteristics of blockchain are listed here:

- **Decentralization**
  The main characteristic and advantage of blockchain is everyone has equal control and transparency. Unlike conventional databases, where only one entity, such as a bank or escrow company, has control over the ledger or the transactions, every involved party can manage and see the transactions. Each participating computer is

known as a *node*. Decentralization also eliminates the need for third-party trust because each node can see the entire ledger or the history of the transactions.

- **Immutability**
  Immutability simply means that once the transaction is recorded, it can't be changed or altered. This makes the ledger irreversible and tamper-proof. If any participant makes a mistake, then the new, corrected transaction must be entered. The ledger will show both the corrected and the previous transactions in the history. This will stay in the ledger forever.

- **Consensus**
  One of the primary reasons for blockchain to not have central control or third-party trust is the characteristic of consensus. The blockchain doesn't confirm a transaction until it's verified. This means that the transaction must be approved by a majority of the participants before it can be included in the ledger. Because the transactions are transparent to everyone, and the majority has given consent, there is no need to have a third party to establish trust between the participants.

- **Anonymity**
  Although transactions are transparent in blockchain, the participants are identified by their public key. This actually hides the real identity and provides anonymity.

- **Security and integrity**
  Blockchain uses cryptography for the security and integrity of the data. If a hacker manages to steal the data, there is no real compromise of information as all the transactions are recorded using cryptographic keys.

As shown in Table 8.1, there are primarily three main generations or versions of the blockchain. The first generation, version 1.0, is the basic blockchain used with cryptocurrency. This version uses distributed ledger technology (DLT). The primary purpose is to make financial transactions transparent and facilitate the development of Bitcoin. The next version of blockchain, version 2.0, uses smart contracts. *Smart contracts* can be compared with predefined scripts that run when certain conditions are met, for example, sending a notification to the recipient when a package or shipment is delivered. In this case, as soon as the package or shipment delivered condition or flag is set, the script runs and notifies (sends an email or text) the recipient. This increases efficiency and reduces the cost. Ethereum runs on this version of the blockchain. The third version, blockchain 3.0, is capable of running decentralized applications (DApps). Conventional applications (or a shortcut of an executable) are run on the top of the blockchain. The app itself can be developed in any language but must have the ability to communicate with the underlying distributed network. This allows apps to use the decentralized blockchain network. BitTorrent and Tor are examples of DApps.

8　Cryptography in Cryptocurrency

First Generation Blockchain (Blockchain Version 1.0)	■ Provides transparency to transactions. ■ Used with Bitcoin.
Second Generation Blockchain (Blockchain Version 2.0)	■ Smart contracts are introduced in this version. ■ Reduces cost and increases efficiency. ■ Used with Ethereum.
Third Generation Blockchain (Blockchain Version 3.0)	■ DApps work on top of this version. ■ This is the future. The more traditional apps work with blockchain, the more it will improve.

Table 8.1 Development of Blockchain over Three Generations

The use of blockchain isn't limited to cryptocurrency as it's used in various industries and in many ways. Blockchain networks, as shown in Figure 8.3, can be classified into four main types of blockchain networks: private blockchain network, public blockchain network, hybrid blockchain network, and consortium blockchain network.

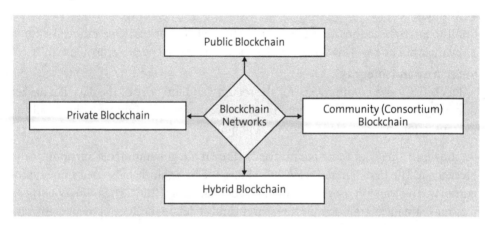

Figure 8.3 Types of Blockchain Networks

The types of blockchain networks are modeled after the public cloud types. Let's take a closer look at each:

- **Public blockchain network**
  Almost anyone can join and use the public blockchain network. This type of blockchain primarily runs as an open-source network, and developers can make changes as it evolves. This type of network offers high transparency and complete decentralization. However, the downside is that there is almost no privacy as everything that everyone does is completely transparent. Over time, the network can grow too big and can lead to a decrease in performance and an increase in cost.

- **Private blockchain network**
  Of course, the opposite of a completely public network is a private blockchain network. This type of blockchain is created and operated by a company or organization.

The transparency is within the organization only, which can be controlled based on the organization's needs and type of business. The major incentive to have a private blockchain is to make sure it's used by a single entity and the business has full control over it. This type of network does provide good performance and scalability, but they aren't truly transparent.

- **Hybrid blockchain network**
As the name implies, the hybrid blockchain network is made up of the hybrid configuration of public and private blockchains. An organization can set up a blockchain network in a way that part of the network is only accessible to their employees while the other half is open to everyone else for contribution. The concept is very similar to a company's website—a part is on the internet, and another part is on the intranet and accessible to employees only.

- **Consortium blockchain network**
More than one business or organization forms a consortium blockchain network to collaborate. In many implementations, this is also implemented as a hybrid consortium network. In this type, each organization manages a portion of the network for internal use, and the remaining network is part of the community work.

Computers or digital systems communicate with each other over the network. Data needs proper formatting and rules, called protocols, to travel over the network. We've already learned several protocols, such as Transmission Control Protocol (TCP), File Transfer Protocol (FTP), Transport Layer Security (TLS), Internet Protocol Security (IPsec), and so on. Blockchain technology also uses various protocols. Following are some of the popular protocols used by blockchain technology:

- **Hyperledger**
Hyperledger is an open-source protocol that offers various tools and libraries. It can be used with most blockchain projects, and enterprises use it because of its modularity and scalability features. The Hyperledger protocol is used to build a private blockchain network because it's a permissioned protocol, and only authorized users can access the blockchain. Hyperledger protocol development is backed by the Linux Foundation. The smart contract engine of the Hyperledger can be easily integrated and used.

- **Multichain**
Multichain is another blockchain protocol used for a private blockchain network. It offers very granular access control that makes it very attractive for private use or internal purposes for any organization. The primary advantage of the Multichain protocol is application programming interfaces (APIs). It provides APIs that make the development and integration of blockchain solutions very easy.

- **Ethereum**
Ethereum is a decentralized, open-source protocol. Anyone can contribute to its development, and anyone with a certain amount of Ether (ETH) can join the node

(Ethereum's cryptocurrency is Ether). According to Ethereum's website, Ethereum underwent an upgrade in 2022 and switched from proof-of-work to proof-of-stake. The website claims that this change reduced the energy consumption required to secure Ethereum by 99%. We'll discuss proof-of-work and proof-of-stake in more detail in Section 8.2.1.

Ethereum also offers an enterprise version for businesses. The use of smart contracts and DApps with Ethereum makes it very flexible and easy to integrate and use.

- **Corda**
  Corda is another open-sourced blockchain protocol. It stands apart from others because of its interoperability with other blockchain networks. Corda works well with private blockchain networks as it requires strong identity proof from each node before the node can join the network. It's primarily used in the banking and finance industry.

- **Quorum**
  Although Quorum is an open-sourced protocol derived from Ethereum, J. P. Morgan Chase has contributed a lot to its development. Quorum is primarily used in the banking industry. The primary difference between Ethereum and Quorum is the use of smart contracts. Smart contracts in Ethereum don't provide privacy as they duplicate the ledger. Quorum takes care of the privacy issue with smart contracts.

8.2 The "Crypto" in Cryptocurrency

If you're wondering what "crypto" is doing in a currency and why we need it, this section is for you. In this section, you'll learn about the use of cryptography in digital currency and how it makes transactions secure.

The fundamentals of cryptography—encryption, hashing, digital signature, and so on—that we discussed previously and the blockchain fundamentals covered in this chapter will be applied to cryptocurrency. Although cryptocurrency is a general term, there are many different types of cryptocurrencies being used. In this chapter, we'll use Bitcoin as the base to explain the technology.

In this section, you'll learn how transactions are recorded, verified, and protected using cryptography on the blockchain network. Next, you'll learn how to implement these cryptography algorithms, and, lastly, you'll learn about crypto wallets.

8.2.1 Cryptographic Transactions

To understand the use of cryptography in cryptocurrency, let's first understand how Bitcoin transactions are entered, verified, and validated. Assume that Person A is sending 2 BTC (BTC is a short version or a stock market ticker for Bitcoin) to Person B. Both Person A and Person B have a crypto wallet to store their cryptocurrency (for more information on crypto wallets, see Section 8.2.3). Person A has a cryptographic key pair, known as a public cryptography key pair, a private key, and a public key. The private key is kept secret with Person A, and the public key is to be shared with everyone. Person A sends 2 BTC to Person B. Person A uses a digital signature to sign the transaction. The digital signature is created using the Elliptic Curve Digital Signature (ECDSA) algorithm. You'll learn more about the ECDSA algorithm in Section 8.2.2. The transaction is then broadcast to all nodes on the network.

On the blockchain network, all the connected computers are known as nodes. A node is a computer that can validate a transaction and create or add the block to the blockchain. The block can't be added until the transaction is verified.

As soon as Person A broadcasts the transaction, in our example, sending 2 BTC to Person B, all nodes start working on verifying the transaction. The verification process can be summed up like this:

1. A new transaction is broadcasted to all nodes.
2. Nodes collect the unverified transactions to the block.
3. Nodes work on proof-of-work to solve the target hash value problem.
4. Once the target hash value problem is solved, a node broadcasts to all nodes.
5. The block is accepted by nodes only if all transactions are valid and the hash value is less than the target value.
6. If the transactions are accepted, the nodes include the hash of the new block in the proof-of-work calculation of the next block. This indicates that the block is accepted.
7. Usually, nodes assume that the longest chain is the correct one and keep working on extending it. If two nodes broadcast a new block at the same time, the node will work on the one that they received first (not all nodes will receive both at the same time). The other will be saved as a branch. The problem will be resolved when the block's proof-of-work is broadcast.

Let's explore each step further. First, as soon as the new transaction is broadcasted, it will be added to the pool of unverified transactions.

The node will see the amount, timestamp, and the public key or public address of the sender and the receiver. The identity linked to these addresses isn't revealed. Everyone on the network will see that the transaction took place from one public address to another for an exchange of 2 BTC, but no one will know who these people are, unless the corresponding private address is exposed or somehow identity is revealed by the owner of the public key. It's recommended that the new key pair should be used with every transaction. If the same key pair is used for many transactions, then the hacker can study the pattern and gather some clues about the identity.

Going back to our process steps, the unverified transactions are added to the new, unconfirmed block. At this stage, miners at various nodes will start working on validating the block. The node will use the hash value of the previous block, the transaction details, and a random number to compute the hash value of the new block. The random number is known as a *nonce*, a short form of "number once." The calculated hash value must be less than the target value for the block to be accepted. If the hash value of the new block is more than the target hash value, then the block is rejected. If the block is rejected, the miner can start working on hashing again by adjusting the nonce value. The miners can keep repeating this process until one of them solves the problem or achieves the hash value for the block that is lower than the target value. This process of calculating the block's hash value is called *proof-of-work*. The target value usually has some leading zeros, and the goal of proof-of-work is to ensure that the calculated hash value has fewer leading zeros than the target. On average, it takes about 10 minutes to solve this cryptographic puzzle. You would think that with technological advancement, the time to solve the problem should be reduced drastically since 2009. However, the difficulty of solving this problem is changing and adjusting every 2016th block, which happens in about two weeks. Lately, the difficulty has increased almost exponentially.

> **Why 2,016 Blocks?**
>
> There are 20,160 minutes in two weeks: $24 \times 60 \times 14 = 20,160$. If the average time to solve the problem is 10 minutes, it would take about two weeks to solve 2,016 blocks. Given this, the difficulty is adjusted based on this formula:
>
> *New difficulty level = Old difficulty level × 20,160 minutes / Time taken to solve last 2,016 blocks*
>
> If the average time to solve the last 2,016 blocks was more than 10 minutes, then the new difficulty level will be slightly easier to adjust the time. On the other hand, if the average time to solve the last 2,016 blocks was less than 10 minutes, then the new difficulty level will be slightly harder to adjust the time.

The detailed structure of the block and the proof-of-work process is depicted in Figure 8.4.

> **Proof-of-Work versus Proof-of-Stake**
>
> Conceptually, proof-of-work and proof-of-stake do the same thing: both confirm or validate the transaction. However, the way they arrive at this decision is very different. Proof-of-work is based on computation power, while proof-of-stake is based on the amount of coin in the wallet. Bitcoin uses proof-of-work, whereas Ethereum uses proof-of-stake. Proof-of-work uses lots of energy and is hard to scale. In proof-of-work, participants are known as miners, while in proof-of-stake, they are called validators. The validators must have coins in the wallet, but miners just need a fast enough computer to crunch out the hash value.

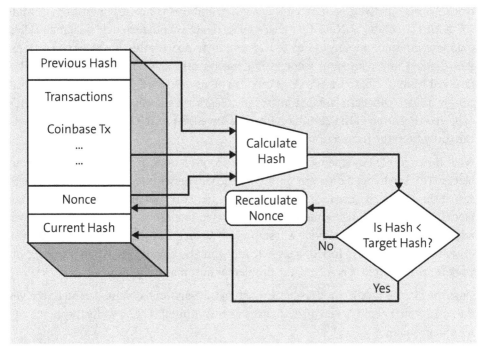

Figure 8.4 Proof-of-Work Process

The block's hash or digest is calculated using the SHA-256 hashing algorithm. We studied this algorithm's workings and implementation in detail in Chapter 4, Section 4.1.3. To recap, the SHA-256 algorithm produces a fixed 256-bit long hash value regardless of the message size. It's impossible to track the original message from the hash value. Even the smallest change in the original message will change the output 256-bit hash value.

If this is the first block, then the value of the previous hash is zero. The first block of the blockchain is also known as the *genesis block*. Satoshi Nakamoto created the first genesis block on January 3, 2009.

Once the hash is calculated, if the value is lower than the target value, the node will broadcast to all nodes indicating this. The nodes will check the transactions, and, if valid, they will accept the new block. The acceptance is usually indicated by using the hash value of the newly verified block in the proof-of-work calculation of the next block. The block being worked on is known as a *candidate block*.

The process of calculating proof-of-work and creating a new block is called mining, and people who do this are called miners. The miners have an incentive to earn Bitcoins for validating or mining each block. Miners get 3.25 Bitcoins for mining a block. This value was 50 Bitcoins in 2009. The incentive value is divided in half after every 210,000 blocks or roughly after four years. So, how do miners get these Bitcoins? This happens with a *coinbase* transaction. The coinbase transaction is the first transaction for every block. This is a network-generated transaction, and this is how the network adds Bitcoins into circulation. (Note that there is a private cryptocurrency exchange called Coinbase, but the company Coinbase has nothing to do with the coinbase transactions. Please avoid any confusion when using the terminology or names.) The Bitcoins received from coinbase transactions can also be used toward transaction fees. In this case, a miner can earn small transaction fees for validating the transaction. It's expected that after all the 21 million Bitcoins are added to the circulation, miners will earn only from the transaction fees.

What if two miners independently solve the proof-of-work problem for two different blocks? This can happen on a large public blockchain network. When two miners broadcast at the same time, because of the size of the network, not all nodes will receive the blocks at the same time. Some nodes will receive one block, while many nodes will receive the other block. The node will start working on the block it received first and the other block will create a fork or a branch and won't be worked on. When the proof-of-work for the next block is solved, the problem is automatically resolved.

Once the block is verified and sealed, the actual transaction happens. Person B receives the 2 BTC from Person A. Person B can use Person A's public key to verify the digital signature of Person A.

Calculating hash values or working on proof-of-work is a very intensive process. It requires lots of CPU and RAM power to calculate the hash. Graphical processing units (GPUs) can do the job more effectively than CPUs. Now, with all the latest technology, miners use application-specific integrated circuits (ASICs). The miners' goal is to keep the cost of mining lower than the value of the Bitcoins they earn.

The blockchain provides some key benefits over the conventional ledger or bookkeeping system:

- **Privacy**
  The first benefit is privacy. In the blockchain, although everyone on the network can see the transactions, no one really knows the identity. On the other hand, in our conventional system, the transactions aren't shared, and all the details are kept with the bank. However, if the bank is hacked, then all of your data can be stolen, or if a malicious bank employee decides to post all your data on Facebook, then your identity is compromised.

- **Consensus**
  The second advantage is the approval. For the transaction and the block to be accepted, at least 51% of nodes must provide consensus. This provides a great deal of security because it's very difficult for any hacker or even hacking gangs to hack 51% of the nodes—in a big public blockchain network, 51% is a lot of computers to manipulate. This feature helps even if a hacker attempts to start a new chain with malicious transactions. The remaining nodes won't accept this unknown chain of transactions.

- **Tamper-proof**

  The third advantage is being tamper-proof. Once the transaction is recorded, it's almost impossible to modify or change it. This is because as soon as the transaction is changed, the hash value of the block changes. The hash values of all the subsequent blocks will also change because the hash value of the previous block is included in calculating the hash of the block. Updating the hash value of every block is a power-intensive and time-consuming task, and it won't go undetected.

8.2.2 Cryptography Algorithms Used in Cryptocurrency

There are four cryptographic concepts used in cryptocurrency:

- Hashing algorithm (SHA-256)
- Public key cryptography (elliptic curve cryptography [ECC])
- Digital signature (ECDSA)
- Merkle Root

We'll take a closer look at each in the following sections.

SHA-256 Algorithm

The first cryptography concept used within cryptocurrency is *hashing*. As we discussed in the previous section, the hashing algorithm is used within each block of the blockchain to provide integrity to the transactions and blocks. We've already studied how to implement various hashing algorithms. We've covered several applications of SHA-256 algorithms, including the one with the blockchain, in the previous section. Refer to Chapter 4 to learn more about the SHA-256 algorithm.

Elliptic Curve Cryptography

The next important cryptography concepts used within cryptocurrency are public key cryptography and digital signatures. The cryptocurrency ecosystem heavily relies on public key cryptography. It uses a public key as an address to receive currency as well as for digital signatures to verify any transaction. In cryptocurrencies, the public key is hashed and encoded to create an address for the key holder. This address is then associated with the wallet and used to receive the currency. The private key is used to send the currency from the wallet or to manage the wallet and to sign the transactions.

Public key cryptography requires two keys—a public key and a private key. We've studied several algorithms to generate the key pair: RSA public key cryptography, Diffie–Hellman key exchange, Elgamal algorithm, and elliptic curve cryptography (ECC). Cryptocurrency uses ECC because of smaller key sizes and efficient computation. Even with a small key size, ECC provides more efficient operation and robustness than other algorithms. For example, to achieve the same level of security as a symmetric algorithm with a 128-bit key size, the RSA algorithm needs a 3,072-bit long key compared to a 256-bit long key for ECC.

The elliptic curves can be defined over real numbers, rational numbers, complex numbers, and integer modulo p. Cryptography works over the mathematics of a finite set of numbers, and it uses the elliptic curve over the integer modulo p. Let's consider some important characteristics of elliptic curves. They offer symmetry over the x-axis or horizontal symmetry. This means that any point on one side of the x-axis can be reflected over the other side of the axis. Another characteristic is that any nonvertical line can intersect the curve at three points. These characteristics sound simple but are very important.

In Chapter 3, Section 3.2.4, we studied the implementation steps of ECC in great detail. We'll skip the remaining basics of ECC and go straight to how ECC is implemented with cryptocurrency.

The elliptical curves are based on the equation

$y^2 = x^3 + ax + b$

where a and b belong to field K and $4a^2 + 27b^2 \neq 0$.

For cryptocurrency, the elliptic curve used is defined by the Standards of Efficient Cryptography Group (SECG). The curve used is known as SECP256K1, and it satisfies the following equation:

$y^2 = x^3 + 7 \bmod p$

This is the equation used to plot the elliptic curve to derive the public and private keys used with cryptocurrency. However, this equation is derived from the original elliptic curve equation by substituting $a = 0$ and $b = 7$, as follows:

$y^2 = x^3 + ax + b \bmod p$ where $a = 0$ and $b = 7$
$y^2 = x^3 + 7 \bmod p$ where p is a large prime number

Figure 8.5 shows the curve with these properties. Note that the curve isn't plotted by computing points on the equation. This is a drawing to explain the concept.

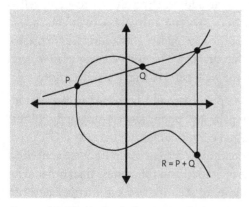

Figure 8.5 Elliptic Curve Cryptography

The rest of the process, such as selecting a number for the private key and calculating a public key, are the same as in other elliptic curve plots as discussed in Chapter 3, Section 3.2.4.

The recommended elliptic curve domain parameters over finite field F_p have been given nicknames and have predefined meanings [SEC, 2010]. In SECP256K1, the SEC denotes Standards for Efficient Cryptography, the name of the group. The P represents the prime field. The 256 represents the key size in bits. K identifies if it's a Koblitz curve or random parameters (r is used to denote random parameters), and the last digit indicates the sequence number. K1, in this case, means it's the first curve in this sequence.

> **Why Is the NIST Curve Not Used in Bitcoin?**
>
> There are two popular types of elliptic curves in use: SECP256K1 and SECP256r1. The K curves represent the Koblitz curves, while the r curves represent the random curves. NIST recommends using the SECP256r1 curve. This curve is supposed to be more widely tested and used, but surprisingly, Satoshi did not use the curve backed by NIST. Even though r curves are generated using pseudorandom bits, they are slightly more secure than the Koblitz curves.
>
> It's speculated that Satoshi did not want to use NIST curves because there might be a backdoor or way for the government to control them. However, we don't really know why Satoshi made the decision to use the Koblitz curve over the NIST curve.

Elliptic Curve Digital Signature Algorithm

The Digital Signature Algorithm (DSA; see Chapter 4, Section 4.3) works on public key cryptography. The owner of the key pair uses the private key to create a digital signature, and the receiver of the message or the transaction uses the sender's public key to verify the signature. The DSA can use any of the public key algorithms, such as RSA, Diffie–Hellman–Merkle, Elgamal, or ECC, to create the digital signature. In the case of cryptocurrency, the DSA uses ECC to generate the key pair, hence the ECDSA is the underlying DSA used with cryptocurrency.

Merkle Root

The last cryptographic concept used with cryptocurrency is the *Merkle root*. The basic idea behind using the Merkle root is to save disk space and make data transfer between the nodes faster. We've studied two more widely used applications of hashing—digital signature and MAC. The Merkle root is the third application of hashing. In the Merkle root, the hash values of the various data are combined and hashed again. The process of hashing and combining continues until all the hash values combine into one hash value. This single hash value is known as the root and the process of incrementally combing the hash values is known as the *Merkle tree*.

This is admittedly confusing. Let's dive deeper and understand the concept. The Merkle root, conceptually, is shown in Figure 8.6.

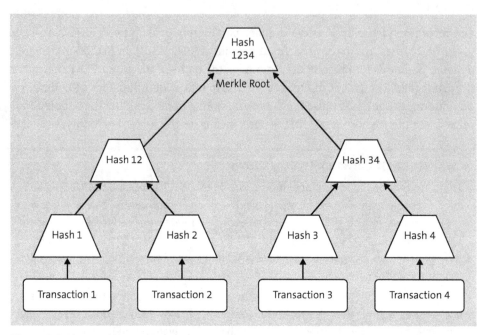

Figure 8.6 The Merkle Tree Concept

Each block has more than a couple of thousand transactions. The transactions are hashed and combined into the leaf nodes. The leaf node is again hashed and combined into the next level. The process continues until there is only one hash value. What if there are odd numbers of transactions? In that case, the hash of the last transaction is duplicated. The hash of the leaf nodes and nodes are all duplicated to get the pair.

The Merkle root is included in the block header. This makes nodes-to-node communication or transfer of data very effective. Plus, Merkle trees also make it easy to verify the hash value of any transaction in any block. For example, if one transaction is altered, the change in the hash value of that transaction is reflected in all hash values in that branch and will also change the hash value of the root. Just from the block header, which has the Merkle root, you can tell if any transaction has been altered. Another advantage is that the Merkle tree makes it easy to pinpoint which transaction has been altered. In addition, if only one transaction needs to be revalidated, then only that branch of the tree needs to be rehashed, and everything else stays unaffected.

8.2.3 Cryptocurrency Wallets

Since the beginning of this chapter, we've been discussing decentralization and third-party trust and their disadvantages. However, they offer one important service—keeping our money. In conventional or fiat currency, the money is stored at a central or third-party location—a bank, a brokerage firm, or something similar. In cryptocurrency—because of the complete decentralization, this isn't possible. We have to keep our own currency. But how and where are we going to keep the cryptocurrency? This is where cryptocurrency or blockchain wallets come into play. These wallets store the cryptocurrency owned by an individual or an organization (basically, for anyone who signs up for the wallet services).

How do these wallets work? The answer is, yet again, cryptography. But don't worry, we won't have to learn more mathematics or geometry.

These wallets work on the principle of public key cryptography. Once a user opens an account with one of the wallet services, the user is required to generate a key pair—a public key and a private key. The public key is hashed, encoded, and used as the wallet's address. Anyone can send cryptocurrency using this address. The private key, on the other hand, is confidential to the user. The private key is used to send the currency out of the wallet or to manage the wallet. This is very similar to PayPal, Zelle, or Venmo accounts. The user ID, phone number, or email address can be compared with the public key while the password for these accounts is a private key. Different crypto wallets can (and do) use different formats or styles of using public addresses. Most wallets can store various types of cryptocurrencies.

The crypto wallets don't actually save or keep the cryptocurrencies. This is just like your bank account. If you see in your bank statement that your account has $100, this doesn't actually mean that your bank has a box in the building, and it has $100 bill kept for you. Similarly, crypto wallets have only the transaction recorded by the blockchain. Crypto wallets also keep the private key and public key. The private key is used to authorize the transaction using the digital signature.

Most cryptocurrencies and crypto wallets use ECC to generate the key pair. As we saw in the previous section, Bitcoins use a SECP256K1 type of curve to generate the key pair. Other cryptocurrencies and wallets can use different types of curves or even different algorithms altogether to generate the key pair. The confidentiality, privacy, and integrity of the wallet is based on the underlying cryptography technique used.

The type of crypto wallet used is based on either the usage/application or implementation. There are two types of crypto wallets based on the usage or application:

- **Hot wallet**
  Hot wallets are connected to the internet and are always live. They are used when you need to transfer funds frequently and need quick access to the blockchain network. These wallets have lots of flexibility and ease of access, but they are also riskier. Hackers are always scanning hot wallets to see if they can exploit any unpatched vulnerabilities. By nature, hot wallets are online wallets and more susceptible to attacks. Web-based wallets, mobile app wallets, desktop app wallets, and most software-based wallets are examples of hot wallets.

- **Cold wallet**
  The cold wallet is an offline wallet, which means it's not active or constantly connected to the internet. The major reason for using a cold wallet is security. Compared to hot wallets, cold wallets are less likely to be hacked and attacked. If you used the internet in the early 1990s, you can compare cold wallets with dial-up internet. Desktop apps, hardware-based wallets, and paper wallets are some examples of cold wallets.

> **Hot versus Cold Wallets**
>
> IT professionals may be familiar with the terms hot backup and cold backup. In *hot backup*, the data is backed up to another database in real time, almost instantaneously. The advantage is that if one database, instance, or data center goes down, the other one can take over without any delay. On the other hand, *cold backup* is an offline backup. The data is copied over to the cold backup periodically—something like every night at 2 am.
>
> If you aren't working in the IT industry, you can compare hot and cold wallets with refrigerators used in the house. Many households keep two refrigerators, one in the kitchen and the other in a garage or the basement. The one in the kitchen is like a hot wallet; family members have easy access to a snack or drink even in the middle of the night. The one in the basement is more like a cold wallet, requiring extra effort to get to it.

It's important to recap that the crypto wallets keep the private and public keys. If someone can hack into the wallet, the hacker basically has the key to the kingdom!

There are three types of wallets based on the implementation:

- **Software wallets**
  Online wallets, mobile wallets, and other app-based wallets are implemented using software. Most software wallets are hot wallets, but some, such as desktop apps, could be cold wallets.
- **Hardware wallets**
  Hardware wallets are still implemented using software, but the difference is that they are implemented on removable hardware. Hardware wallets are implemented on USB drives and other detachable devices. The user can unplug the hardware when the wallet isn't in use. This is more like a cold wallet.
- **Paper wallets**
  Remember, crypto wallets don't really have funds but have the key pair to perform the transactions. In paper wallets, the keys are printed on paper, usually in the form of a scan code. The advantage of paper wallets is that it can't be hacked, or the phishing attempt won't work on the paper wallet, unless, of course, you can save the file on your computer or cloud storage. The downside is, if you lose the paper or the

paper gets damaged rendering the scan code unreadable, then you have no access to your keys and the funds.

> **What If You Lose Your Private Key?**
>
> The big question is—what happens if you lose or forget your private key? The answer is simple: you lose all your cryptocurrency tied to that key. The key pair is irrecoverable. There is no way to recreate the same key. There is no such thing as a "Forgot your password?" link.
>
> According to one estimate, more than 15% of Bitcoins (15% of $21 billion!) are out of circulation because people have lost their keys. These Bitcoins will never be added back to the circulation.

With more people using crypto wallets and advancements in technology, crypto wallet companies are developing a wallet name service. The idea is to replace long, encoded, difficult-to-remember wallet addresses with human-friendly, easy-to-remember, customizable wallet addresses. The concept is very similar to the Domain Name Service (DNS), where an IP address is replaced with human friendly website addresses.

8.3 Outlook: Cautiously Optimistic

After reading up to this point, you're probably thinking that cryptocurrency is a walk in the park. But that isn't the case. Cryptocurrency has its own challenges on all fronts—technical, economic, legal, and social. Let's review them:

- **Technical challenges**

  One of the most important concerns with cryptocurrency is security. Although cryptography makes the cryptocurrency very secure and tamper-proof, the infrastructure isn't ready. Crypto wallets, exchanges, and other stakeholders in the transactions are still not very secure and are vulnerable to cyber security attacks. The weakest link in the system weakens the entire ecosystem.

  Another technical challenge with cryptocurrency is scalability. As more users are added every day, the number of transactions will also grow, and blockchain technology could have difficulty handling a large number of transactions efficiently. Improving scalability is something cryptocurrency will have to work on very soon.

  Mining and proof-of-work consume lots of energy and pose environmental challenges for cryptocurrency. The use of customized ASIC chips and newer technology is helping somewhat with energy consumption.

- **Social challenges**

  Cryptocurrency has become popular among investors and tech enthusiasts, but the majority of the population of the world is still not aware of it. Many parts of the world still don't have internet and expecting them to use cryptocurrency for everyday life is going to be challenging. In developing countries, many people are very skeptical about the future of cryptocurrency. It's hard for people to digest the idea of managing and using money without a bank.

- **Economic challenges**

  People are nervous because of the volatility in the price. The price or value of cryptocurrency is a matter of speculation. For people to use it in daily life, the value has to be stabilized. Right now, it's more like a stock market investment. People can't live with this hype in everyday life.

  Businesses aren't ready or set up to accept cryptocurrency as a primary form of payment, which hampers its use. People are also unsure how and if financial institutions, the government, the Federal Reserve, and other central agencies will integrate cryptocurrency into the mainstream economy. Although cryptocurrency is based on the philosophy of decentralization, some agencies or companies will step in.

  In 2022 alone, Binance and Coinbase, two cryptocurrency trading companies, executed more trade than all other companies combined. This is almost centralization.

- **Legal challenges**

  Legal or regulatory challenges comprise the biggest challenge of all. Because cryptocurrency is decentralized, it's very difficult for the government to regulate and impose proper income tax and other laws. Creating any international laws or treaties around cryptocurrency is also challenging because most countries have very different views on it.

> **Bitcoin Pizza Day**
>
> On May 22, 2010, Laszlo Hanyecz asked on the Bitcoin chat forum if someone could buy him two large pizzas in exchange for 10,000 BTC. A man from England took him up on the offer and ordered two pizzas from Papa Johns. The cost of 10,000 BTC was estimated at $41 USD at the time, which is worth millions today! This is the first official commercial transaction using cryptocurrency, and May 22 is known as Bitcoin Pizza Day.

These challenges will likely be handled over the next few years. If improvement in CPU and RAM in our computers is any indication, then the technical challenges will be solved easily. The regulatory framework and the government's support in terms of regulations and laws will help with legal and regulatory challenges. The regulatory framework will also help with economic stability in cryptocurrency. Education and awareness are the only things required to meet the social challenges.

One possible technical challenge for blockchain and cryptocurrency is the looming threat of quantum computers. It's believed that quantum computers can break asymmetric cryptography algorithms, including the ECC algorithm and SECP256K1 curve public key cryptography. In 1994, Peter Shor proposed a quantum algorithm for prime factoring that could calculate the private key given the public key in asymmetric cryptography, if implemented using quantum computers [Shor, 1994]. In addition, in 1996, Lov Grover proposed a quantum algorithm that can perform an unstructured data search with quadruple speed using quantum computers [Grover, 1996]. As discussed earlier, cryptocurrency uses the SECP256K1 curve to generate the key pair, and Shor's algorithm can be implemented using quantum computers to find the private key from the public key—when quantum computers become available. Grover's algorithm, when implemented using quantum computers, can weaken the symmetric key algorithms. In cryptocurrency, the hashing algorithm (SHA-256) will be weakened but not broken.

ECC is considered more secure because for a given security level (e.g., security level = 128 means the symmetric key size is 128-bit), ECC needs a much smaller key size than RSA. Plus, using the classical way, it takes much longer to solve the discrete log problem (used in ECC) than a factoring problem (used in RSA). This gives ECC an edge over RSA. However, this isn't true with Shor's algorithm. It takes almost the same number of steps to solve either problem.

Researchers are working on using lattice-based public key cryptography algorithms that are quantum resistant. (You'll learn more about post-quantum cryptography in Chapter 10.) Only time will tell how cryptocurrency and blockchain handle these threats, but if history is any indication, we can safely say that security practitioners will prevail over the attackers!

As of now, blockchain technology is primarily used in the financial sector only (and by hackers to collect ransomeware money). Even when it's used in other industries, it's used to manage finances. Before cryptocurrency can stabilize and gain momentum, underlying blockchain technology must be used in society and widely accepted.

Many research organizations, such as the Ethereum Foundation, Web3 Foundation, Protocol Labs, and so on, are working on blockchain research. Researchers are hopeful that blockchain will be used in many more industries by the end of this decade.

One interesting proposed application of blockchain is in academia to store student records/transcripts. The blockchain technology will make any change/tampering almost impossible. Plus, it will give full privacy to students as the transcripts will be recorded with their public key and not their name. Another use case is voting. Votes can be recorded using blockchain. This will give full transparency and tampering will be almost impossible.

Blockchain and cryptocurrency have many challenges and it's unlikely that cryptocurrency will replace fiat money anytime soon. However, cryptocurrency has a lot of potential because of privacy, decentralization, and security, and we never know when it'll take off. After all, didn't we have the same kind of discussion in the early 1990s about using the internet for e-commerce and banking?

8.4 Summary

Although money can't buy everything, it's central to people's lives. We use and think of money almost all the time, but we hardly ever do anything to learn about it. Even in schools, kids aren't taught about finances. Cryptocurrency has changed that mindset and forced us, even adults, to look at money from a different angle. Cryptocurrency works totally differently from our fiat money and that leaves us no choice but to learn about it.

In this chapter, we reviewed the very brief history of money and the development of cryptocurrency. We studied the fundamentals of blockchain. The core of the chapter was focused on how cryptography is used with cryptocurrency. We learned two new cryptographic techniques in this chapter—ECC using SECP256K1 and the Merkle tree. Both of these techniques are vital for cryptocurrency. We concluded the chapter by examining the challenges faced by cryptocurrency and what the future holds for cryptocurrency and blockchain technology, including some potential innovative usages of the technology.

> **Key Takeaways from This Chapter**
> - Reviewed the fundamentals of cryptocurrency and blockchain
> - Learned how cryptography is used and implemented with cryptocurrency and blockchain
> - Explored various types of crypto wallets and how they are used

In the next chapter, you'll learn about the use of cryptography with artificial intelligence (AI). We'll explore answers to how cryptography can protect AI and how AI can help in automating cryptography.

Chapter 9
Cryptography and Artificial Intelligence

Artificial intelligence (AI) is here to stay. AI can perform many tasks just like humans, and it can execute them efficiently and accurately. These properties of AI can be used with information security—especially with cryptography. The synergy between AI and cryptography takes data security to the next level. In this chapter, we'll learn how AI and cryptography can be used to each other's advantage and why it's a win-win situation.

Since the launch of ChatGPT in 2022, AI has been the talk of the town, and everyone is trying to figure out how to use AI to their advantage. AI has already made its way into information security. Some examples are Intrusion Detection Systems (IDS), log analysis and incident response, and malware and virus detection. The ultimate goal of information security professionals is to use AI for data protection. The relationship between AI and cryptography is mutual. Using AI with cryptography for safeguarding data will give data protection a more efficient and secure alternative. Cryptography can also help protect AI, which not only improves information security but also benefits all AI applications.

Historically, cryptography is used for confidentiality, and, even today, when we talk about cryptography, we discuss how cryptography protects data while in transit and how it has helped us secure e-commerce. But the reality is very different. For more than 35 years, researchers have spent more time exploring and using other cryptography services, such as integrity, authenticity, and nonrepudiation. With all of these experiences, protecting AI with cryptography is a no-brainer for the industry. On the flip side, the application of AI in cryptography isn't widespread yet, but researchers have started exploring and implementing AI into various algorithms. It's just a matter of time before AI-powered cryptography algorithms are ready to be implemented.

> **Chapter Highlights**
> - Overview and fundamentals of AI/machine learning (ML)
> - Learn to use AI with cryptography
> - Analyze risks posed to AI and learn how cryptography can help protect AI

9 Cryptography and Artificial Intelligence

This chapter is divided into four sections, which are almost independent of each other, and you can choose to read in any order you prefer. The chapter will start with a primer on AI/ML in Section 9.1. We assume that you have some introductory-level familiarity with AI and ML. However, the primer will give you a brief introduction to the subject and will get your feet wet. You can skip this section if you understand how AI/ML works. Section 9.2 zooms into the use of AI in cryptography. To understand how AI can improve cryptography algorithms, you must understand how cryptography algorithms work. If you're not familiar, we recommend you read Chapter 2 and refresh the concept. Section 9.3 is focused on the risks posed to AI and how cryptography can help secure AI. The chapter concludes with an assessment of best practices and ethics involved with the use of AI in Section 9.4.

9.1 Primer on AI

Although most people started talking about AI only after the launch of ChatGPT, the concept of AI isn't new. In fact, the first chatbot, Eliza [Ireland, 2012] was introduced in the early 1960s—60 years ago! The idea of machines doing work like humans is believed to have been first conceived by Alonzo Church and Alan Turing in the mid-1930s. In 1950, Alan Turing proposed the concept of AI using the Turing test [Turing, 1950]. The Turing test was very simple. Turing proposed that if we have a machine and a human send signals from one side of the wall, and a person on the other side of the wall can't differentiate which signal came from a person and which came from a machine, then the AI is working as intended. Let's understand the Turing test with a present-day example. When you're chatting with a customer service representative on your computer, and you can't tell if the responses are coming from humans or responses are AI generated, then the AI is working as intended.

In the simplest terms, AI can be defined as intelligence displayed by machines rather than intelligence displayed by humans. The term *AI* was coined at Dartmouth College in 1956, and the first AI lab was established at MIT in 1959. Eliza was developed at the MIT lab. The history of AI is long and interesting!

If the concept of AI has been around for almost 90 years and even the first chatbot was launched 60 years ago, then why did AI take off only after the launch of ChatGPT in 2022? What was holding AI back? To answer this question, we must first understand how AI works.

> **High-Level AI Explanation**
> This explanation of AI is at a very high level and intended to convey the concept only. You should use other dedicated resources for a complete understanding of how AI works.

Machines learn in three simple steps, just as humans do: predict, feedback, and learn. We first predict and try something, and then based on the feedback, we learn from it. Let's consider an example of a child learning when to wear shoes. Say the child goes outside on a hot day. He takes his first step, realizes the ground is too hot, and decides he must wear shoes. Next time, he goes out with shoes on. Let's understand this analogy step-by-step:

- **Prediction step**

  The child first incorrectly predicts that he can go out of the house to play without wearing shoes.

- **Feedback step**

  As soon as he steps outside, he realizes that he made a mistake. The hot surface provides feedback that he made a mistake and must wear shoes.

- **Learning step**

  The child learns from the feedback and next time goes out with shoes.

The child learned from his experiences. Machines learn the same way. Compare this to an algorithm used to predict airfare prices for a travel destination:

- **Prediction step**

  Machines run the algorithm based on the information available and provide an output. The first time, the prices will be off.

- **Feedback step**

  However, the algorithm learns that the output was incorrect and adjusts the algorithm to predict better airfares in the next attempt.

- **Learning step**

  In the next iteration, the prices might be closer than the first time. After several iterations, the algorithm will predict very accurate airfares.

The human and the machine go through the same steps. However, the child learns from experiences while the machine or the algorithm needs data to predict accurate airfares. Although the concept of AI has been around since the 1950s, we didn't have much data to train the algorithms. The real data was created after the internet and heavy use of social media began. Until 2005, there wasn't enough data to train AI models. Thanks to Facebook, Google search, and all other social media platforms, there is lots of open-source information available. This Open Source Intelligence (OSINT) information is available to train. The lack of training data is the first reason why AI didn't gain any traction until very recently.

The second reason was technology. Until the late 2000s, computers weren't fast enough to execute complex algorithms quickly. In the past 15 years, hardware capabilities and cloud computing have increased exponentially, and that has helped run complex AI algorithms.

Together, the availability of training data and increased hardware capabilities have driven the rapid growth of AI-powered tools in the last several years.

Based on the traits or functionalities exhibited, AI can be divided into the following six domains:

- **Reasoning/problem solving**
  Reasoning or problem-solving is the ability of AI to make small but sensible decisions with a methodical approach. The AI algorithm developed a step-by-step approach for reasoning and problem-solving.

- **Knowledge representation**
  Knowledge representation allows AI to conclude or deduct information based on knowledge or previously trained data. This trait allows AI to ask related questions. For example, when you say "Alexa, remind me to check into my flight," knowledge representation makes Alexa ask you a counter question, "What time should I remind you?"

- **Natural language programming**
  Natural language programming (NLP) uses human communication or language. ChatGPT and Alexa are prime examples of NLP that write and speak to communicate with humans. Modern NLP techniques are so good that they not only communicate with us but are also able to distinguish one person's voice from others.

  When we use social media sites, search engines, or any other site like that, we often see a relevant advertisement on the side. Have you ever wondered how the social media site knows what we've searched? They all use NLP. The NLP is used to extract the relevance from the searched data and find the appropriate advertisement. Similarly, when you use an e-commerce site to purchase something, the site also suggests other things "that go" with your purchase or sometimes they display a list of suggested items next to "People who bought this item also bought similar items." This is happening because they put NLP to work.

- **Perception**
  Humans have senses—touch, vision, smell, listening, and so on. Machines use sensors to generate these senses and then make decisions. For example, self-driving cars, like Tesla, can detect if an object is on the road and immediately react to stop the car.

- **General and social intelligence**
  Social intelligence allows humans to act with emotions—anger, humor, and so on. General intelligence is overall intelligence that basically combines all other intelligence, including social intelligence and decisions. Machines haven't achieved social and general intelligence, but it's included as a future-looking domain.

- **Machine learning and deep learning**
  Machine learning (ML) and deep learning (DL) algorithms can be used to execute any

of the other five domains. Figure 9.1 shows all six domains. However, ML and DL are shown at the bottom to emphasize that they are used across all domains. DL is also known as artificial neural network (ANN). Very loosely speaking, DL is executing ML algorithms in several layers. DL is considered a subset of ML. ML solves problems by mathematics, while DL adds neural network to the mix to solve complex problems.

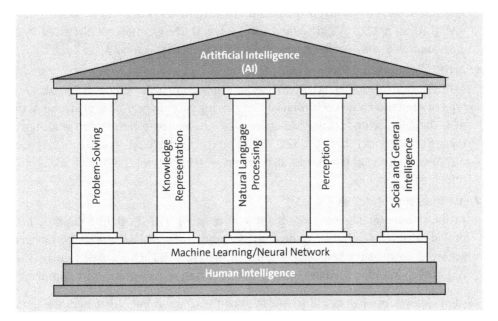

Figure 9.1 AI Domains

Robotics Is an AI Domain?

Robots are a very popular topic when people talk about AI. Boston Dynamics, a spin-off from the MIT Lab, is a very well-known robot manufacturing company, and they have launched some very successful, even commercially successful, robots. Boston Dynamics has played a big role in making robots popular and getting people's attention.

Most people confuse robotics with the AI domains, but they shouldn't for three reasons. First, robotics is a branch of physics that designs, builds, and operates robots. It has nothing to do with AI. They aren't dependent on each other, nor do they have any relationship. Second, robots can and do exist without AI. Third, a robot, in its basic form, is a mechanical device or just hardware.

To debunk the myth, robotics is *not* an AI domain. However, it's possible to use any or all of the AI domain functionalities with a robot to make it look like an "intelligent" robot. This is no different from using the AI domains with any other hardware, like Alexa, a self-driving car, or even a laptop.

AI functionality can be displayed using one domain or multiple domains. The more domains that are used, the more intelligent is the system. Based on the level of overall intelligence, AI can be divided into the following three stages (or types):

- **Artificial narrow intelligence (ANI)**

  When a system uses only one or maybe two domains to execute the task, it's known as artificial narrow intelligence (ANI). Informally, it's also known as *weak AI* or *narrow AI*. Alexa or ChatGPT are examples of weak AI. The example we discussed with an e-commerce site, search engine, and social media sites is weak AI.

- **Artificial general intelligence (AGI)**

  Artificial general intelligence (AGI) uses multiple domains to execute the task. This type of AI is also known as *strong AI*. Self-driving cars are examples of strong AI or AGI. The self-driving cars use NLP, reasoning, and perception to understand signs, make decisions, and "hear and see" things on the road. We can assume that a good AGI or very strong AI has almost all domains working except the general social and intelligence domains.

- **Artificial super intelligence (ASI)**

  Artificial super intelligence (ASI), is also known as *super AI*. It exhibits all properties of all domains, including a high degree of general and social intelligence. We haven't achieved ASI yet, but when we do, the machine will act and behave like a human. The AI-powered machine will have emotions, senses, and other behaviors like humans.

We don't know if or when AI will act and behave like a human, but researchers and technologists are working toward it. Chances are that a very high degree of super AI can perform even better than humans. Will ASI-powered machines have negative qualities of humans, like jealousy, anger, cheating, and so on? If that happens, it will be hard to predict what will happen to the human race in the long run—will ASI control us, or will we manage it? Or will there be war, like a science fiction movie, between men and machines for power? Only time will tell.

Lastly, we must understand or realize that most AI models are trained, developed, and used with some sort of underlying ML or DL algorithms. Following are the three types of ML algorithms:

- **Supervised learning**

  In supervised learning, the algorithm is fed with a labeled data set and desired output. In this case, the algorithm has a well-defined output goal. This is known as a supervised model because the model is, in a sense, "supervised" by the given data set and the expected outputs.

  In a real-life situation, we can compare this to a classroom setting where a teacher assigns a problem. The students know what is expected of them, and the problem and the output are predefined. The students learn by finishing the problem to a predefined result.

- **Unsupervised learning**

    This is exactly the opposite of supervised learning. In this case, the unlabeled data set is fed, and no formal output is defined. The algorithm is on the hook for defining it, and that could be the result or the goal.

    This can be compared with a research project where nothing is well defined or predefined. The researcher starts working on sorting the information and then arriving at a conclusion. There is no expectation as to what that conclusion might be.

- **Reinforcement learning**

    In this learning, the algorithm interacts dynamically with inputs and generates the output. Based on how close or far the expected output is, the algorithm is rewarded or punished. The self-driving car is one example of reinforcement learning being used. Based on the input from the surroundings, the algorithm drives the vehicle.

    Training a dog is an example of reinforcement learning. We give a treat to the dog every time it acts as expected.

This section provided a very high-level overview of AI and ML. We've omitted most technical details and kept this very brief. There are many books written on these topics, and even graduate courses are offered on this subject, so there is no way we can do justice in one section. The idea is to give you an introduction to the topic before we really start using the concept in the next section.

9.2 AI for Cryptography

AI is used everywhere, so there is no reason why cryptography shouldn't take advantage. The cybersecurity industry heavily relies on scripts, primarily written in Python, for automation and optimization. However, the automation offered by the scripts is limited. AI and ML algorithms, on the other hand, can do a lot more than scripts. This is where AI plays an important role.

In this section, you'll learn the role AI plays in cryptography and the AI algorithms that can be implemented with cryptography.

9.2.1 Role of AI in Cryptography

To determine how and where AI can play a role in cryptography, we first need to go back to the basics of cryptography [Rivest, 1993]. If you recall from Chapter 1, there are two main types of symmetric ciphers—stream ciphers and block ciphers. To recap, the stream cipher works by XORing the plain text with a key, bit by bit. The key is generated using a combination of a seed value and a random sequence. The random sequence is generated using linear feedback shift register (LFSR) and some Boolean functions. We've been using this technique or some variation of it for many years. The decryption process works the same way and uses the same keystream to decrypt bit by bit. AI can help us generate a random keystream.

The block cipher encrypts a message block by block. The size of the block is based on the cipher type used and varies by algorithm. The block is processed or encrypted using a substitution-permutation network (SP-network). In the SP-network, the message block goes through several rounds of the process. In each round, the message block goes through confusion and then diffusion. Confusion is performed by processing each block through the substitution boxes (S-boxes). Diffusion, on the other hand, is spreading the plain text over the cipher text. The purpose, if you recall from Chapter 2, of confusion is to hide the relationship between the plain text and the encryption key, while the purpose of diffusion is to conceal the relationship between the plain text and the cipher text. After the process of confusion and diffusion, the key is added for the round. The process is repeated for each round. The goal is to achieve stream and block cipher functions using AI. AI can perform the following tasks:

- **Algorithm optimization**
  If AI is used to implement the encryption/decryption algorithm, the AI algorithm works to optimize the cipher's speed and security. The goal of the optimization is to find the most appropriate solution, with minimal resources and in the shortest time. The AI algorithm can do that if used with a stream cipher or block cipher.

- **Performance optimization**
  Encryption and decryption processes are resource intensive. They consume lots of power, memory, and CPU time. If you recall from Chapter 7, resource consumption is one of the major reasons why we need lightweight cryptography. AI algorithms can partially solve the problem. AI algorithms can make sure the resources are optimized, the encryption and decryption processes don't consume lots of resources, and the resources are freed up as soon as possible.

- **Random number generator**
  AI algorithms can be very useful with stream cipher for generating random number streams. This random number stream can be used as a keystream with the stream cipher. The random sequence generated by the AI algorithm is very hard to predict and duplicate.

- **Key management**
  AI algorithms can help generate and manage the symmetric encryption key. They can also perform predictive analysis based on historical data and usage patterns.

Asymmetric cryptography is based on a difficult, one-way mathematical problem. We discussed asymmetric algorithms in detail in Chapter 3 and won't repeat them here. In an asymmetric algorithm, like RSA, AI algorithms can be used to find the safe prime numbers p and q.

There are many use cases for AI algorithms in cryptography [Batina, 2022]. However, our goal is to use AI algorithms to optimize encryption and decryption. The use of algebraic function–based construction of symmetric algorithms has limitations, and AI can help select and optimize the algorithms.

9.2.2 AI Algorithms for Cryptography

In the introduction of this section, we laid the foundation for where we can use the AI algorithm to enhance the security of the symmetric cipher. We identified three areas:

- Boolean functions
- S-boxes
- PRNG

In this section, we'll discuss various AI algorithms that can be used to achieve these properties of the symmetric ciphers [Mariot, 2023].

Genetic Algorithm

A genetic algorithm is an optimization technique that is somewhat modeled after Darwin's evolutionary theory. In AI, genetic algorithms are used to solve complex problems using optimization. The goal is to find an optimal population or candidate solution to a complex problem over several generations or, simply speaking, iterations. Genetic algorithms are primarily used to represent Boolean functions, such as LFSR and other Boolean operations. The key terms used in the genetic algorithm are explained in Table 9.1.

Key Terms	Meaning
Population	Candidate solution to the problem
Chromosome	A single candidate solution
Fitness function	How well it solves the problem
Crossover	Combination of two chromosomes to produce a chromosome that holds the characteristics of both original chromosomes
Mutation	Introduces random changes into a chromosome to maintain genetic diversity

Table 9.1 Key Terms of Genetic Algorithm

Figure 9.2 shows the following steps used in executing the genetic algorithm [Mariot, 2022]. To understand the algorithm, we can compare each of the following steps with the theory of human evolution:

9 Cryptography and Artificial Intelligence

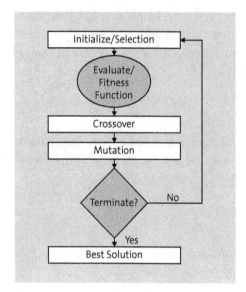

Figure 9.2 Generic Algorithm Process Steps

1. The first step is to generate the initial population or a set of candidate solutions. This could be based on some predefined criteria. If we compare this to humans, this is the initial set of people; we can take the initial set as the current generation.
2. Next, evaluate the chromosomes. The evaluation is done using the fitness function. Once evaluated, select the chromosome based on the fitness score. There are three selection methods:
   - Roulette wheel selection: This method of selecting the chromosome is based on a roulette wheel in a casino. Assume that the chromosomes are placed on the wheel. The wheel displays the fitness score of each chromosome. The higher the fitness score, the better the chances of being selected.
   - Tournament selection: This selection is modeled after a sports tournament. The chromosome with the best fitness score is selected from the subset of the chromosome.
   - Rank selection: This is the simplest method of selection. This is based on the ranking; the higher the fitness score, the earlier the selection.

   This step is similar to a physical fitness test for humans. Based on the score, we know the health of a person.
3. Once the chromosomes are evaluated and selected, the next step is to pair two chromosomes to generate a crossover. Mating can be achieved in three ways:
   - Single point crossover: The parent chromosomes are swapped at one particular point.
   - Two-point crossover: The swapping of the parent chromosomes happens at two points.

- Uniform crossover: Swapping is continuous.

   You could compare crossover to conception in humans.

4. In this step, the algorithm checks for the probability of the offspring (child). If the probability is low, apply the mutation, and introduce the new genes. Here are two primary methods for mutation:
   - Bit flip mutation: Flip a bit in a chromosome to achieve the mutation.
   - Swap mutation: Swap the genes within the chromosomes to achieve the mutation.

   The process of cell mutation is the same in humans. However, cell mutation has almost no effect on everyday life.

5. Use new offspring and replace the old chromosomes as needed. Hire young people and retire people over 65!

6. Lastly, terminate if the optimal solution is found; otherwise, continue the cycle again. Here's where the comparison with human life breaks down because, unfortunately, we can't start over!

Genetic Programming

Genetic programming is a twin brother of the genetic algorithm, and they both work very similarly. The difference is the way the individual is represented. In the genetic algorithm, an individual is represented by fixed-length bitstrings, whereas in genetic programming, the individual is represented as a tree structure.

> **Note**
> Genetic programming and our next topic, cellular automata, are new and upcoming areas of research that don't have many practical implementations as of the time of writing. For the purposes of this book, we'll provide insight into these topics at a high level.

In genetic programming, the Boolean truth table can also be encoded by the tree. The leaves at the end of the tree represent the input variables, and the nodes are the Boolean operators. The root node is the output.

Both genetic algorithms and genetic programming can be implemented using software. The genetic programming algorithm goes through the same evolution steps as the genetic algorithm—initialization, evaluation, selection, crossover, mutation, replacement, termination, and so on.

Although genetic algorithms have been researched in great detail over the years, other algorithms, like the simulated annealing algorithm and the swarm intelligence optimization algorithm, have also been explored. The genetic programming algorithm is perhaps the most promising for use with a symmetric cipher.

9 Cryptography and Artificial Intelligence

Cellular Automata

Cellular automata are discrete models consisting of a grid of cells. If the cells in the grid are in one row, then the grid is one-dimensional. If the cells in the grid are in an array, then it's a two-dimensional grid. Each cell in the grid is in a finite state, and the state is updated by evaluating a local rule. The neighborhood of a cell defines which cell affects its next state. The rules define the next step (generation) of the cell. This is based on the current state as well as the state of the neighbor.

The working of cellular automata is very simple. All cells are initialized with a predefined or initial state. The rules are applied to every cell. All cells in the grid are updated simultaneously when the rules are applied. The process continues iteratively until the desired result is achieved. These process steps remain the same regardless of the applications of the cellular automata, except the rule. Based on the rule, the functionality changes.

For example, to implement the substitution box (S-box) for symmetric encryption using cellular automata, Wolfram's rule 30 or rule 90 can be used. Rule 30 is one of the basic rules introduced by Stephen Wolfram in 1983. This rule updates the cell based on the status of the neighbor's cell. The process works to generate the nonlinear transformation that essentially does the same thing as LFSR in our traditional implementation. Once the rule is applied, all cells in the grid are updated simultaneously. The process should continue until the desired result is applied. This process can be subdivided into three steps: initialize the grid, update the grid based on the rule (this is similar to iterations or rounds in the traditional implementation), and, after the grid evolves, map the final states.

Cellular automata can be implemented using software. Let's go over the Python implementation. (Note that this is not a complete implementation, but just a sample.)

The first step is to implement Wolfram's cellular automation rule 30, as follows:

```
def rule30(left, center, right):
    return left ^ (center | right)
```

Next, iterate through the steps, using the initial state as starting point and setting the state to result of the rule 30 calculation, as shown in Listing 9.1. This evolves the cellular automaton for a given number of steps or iterations.

```
def evolve_ca(initial_state, steps):

    state = initial_state[:]
    for _ in range(steps):
        new_state = [0] * len(state)
        for i in range(1, len(state) - 1):
```

```
        new_state[i] = rule30(state[i-1], state[i], state[i+1])
    state = new_state[:]
return state
```

Listing 9.1 Evolve Function

Use the evolve function with ca_grid output and steps as the input to get the final state, as shown in Listing 9.3.

```
def s_box_ca(input_bits, steps=5):
    # Initialize the CA grid (1D array)
    i = 0
    ca_grid = []
    for bit in input_bits:
        ca_grid[i] = int(bit)
        i = i + 1
```

Listing 9.2 Final State

Evolve the CA and then map the final state to the S-box output, as shown in Listing 9.3.

```
final_state = evolve_ca(ca_grid, steps)
mapped_sbox = map(str, final_state)
final_output = ''.join(mapped_sbox)
return final_output
```

Listing 9.3 Map to S-Box Output

You can refer to Python CA library at *https://pypi.org/project/cellular-automaton/* for more information.

This is a very high-level example to show the feasibility of implementing S-box using cellular automata; for complete details, you can refer to Picek, 2017, where the authors used genetic programming to design the S-box using the cellular automata rules.

Cellular automata have multiple applications in cryptography—as pseudorandom number generators (PRNGs), keystreams in stream ciphers, S-boxes in block ciphers, and hashing.

Cellular automata provides security by generating very complex dynamic behavior and implementing it efficiently. The complexity depends on the local rule.

9.3 Cryptography for AI

In the previous section, we learned that AI algorithms can help cryptography by generating pseudorandom numbers and implementing S-box and Boolean functions. Using AI is a big step forward for cryptography in improving the security and efficiency of symmetric ciphers and key generation in asymmetric ciphers. However, what if the AI model is manipulated or attacked? Our use of AI with cryptography assumes that AI is very secure. In practical applications of AI, we found that there are some risks associated with AI. In this section, we'll survey these various risks and discuss how cryptography can help mitigate them.

9.3.1 Security Risks of AI

Everything we do in information security is for data (security), and anything we gain from it is because of data. Our lives revolve around data, and data is the lifeblood of AI. Although the biggest risk associated with the AI model is data privacy, the AI model is impacted by any manipulation of data. Because people are using more AI-based tools and applications, attackers have realized that the best way to access our data is through AI tools, and they have begun launching attacks against AI models. This is why it's very important for cybersecurity professionals to understand attacks against AI and how to secure AI models against various attacks.

The attacks on AI can be classified based on the following three criteria:

- **The attacker's influence**
  The attacker's influence decides at what stage the attack can take place—at the training stage or testing stage.

- **The attacker's goal**
  The attacker's goal decides what is the target or what the attacker intends to achieve from the attack—the targeted attack or the untargeted attack.

- **The attacker's knowledge**
  The attacker's knowledge of the system can have a crucial impact on the severity of the attack. The more the attacker knows about the system, the more damage the attacker can cause to it—white-box attacks versus black-box attacks.

In this section, we'll review various attacks and risks against AI and their effect on the output of the AI models. Table 9.2 summarizes the various attack types on the AI models.

Attacks Based on Attacker's Goal	Attacks Based on Attacker's Knowledge	Attacks Based on Attacker's Influence
Targeted attack	White-box attack	Training-time attack
Untargeted attack	Black-box attack	Testing-time attack

Table 9.2 Summary of Attacks on the AI Models

Let's walk through them:

- **Adversarial attack**
  An adversarial attack is launched by applying malicious input to the AI model to

manipulate its output so that it doesn't generate the intended output. This poses a big risk when using the model because the user may not know the output of the model being used isn't the correct one. Adversarial attacks can be classified based on the attacker's goal, knowledge, and influence [Hernandez-Castro, 2022].

Based on the attacker's goal, there are two types of adversarial attacks:

- *Targeted adversarial attack*: In a targeted adversarial attack, the AI model's input is manipulated to achieve the specific, but incorrect output. This type of attack is carried out with a specific purpose in mind. For example, if an organization is using a face recognition method for authentication, the attacker can manipulate the input (the employee's image) so that the attacker can get access to the system.
- *Untargeted adversarial attacks*: In an untargeted adversarial attack, the AI model's input is manipulated so the AI model produces any incorrect output. For example, if the attacker distorts the output of the face recognition system that is used for surveillance, the distorted output won't help identify anyone. The goal is to disrupt the operation, and there's no specific purpose in mind.

Based on the attacker's knowledge, adversarial attacks can be classified as the following:

- *White-box adversarial attacks*: In a white-box adversarial attack, the attacker has full (or enough) knowledge of the system being attacked. This gives the attacker an edge to plan and carry out much more severe attacks.
- *Black-box adversarial attacks*: In a black-box adversarial attack, the attacker has no knowledge of the AI model to be attacked or the system. This limits the amount of damage the attacker can cause to the system.

Based on the attacker's influence, adversarial attacks can be classified as the following:

- *Training-time adversarial attacks*: In training-time adversarial attacks, the attacker tries to manipulate the training data. You'll learn more about this type of attack later in this section.
- *Testing-time adversarial attacks*: In testing-time adversarial attacks, the attacker manipulates the testing data. The model is trained with valid data. This type of attack is used to explore the system and learn more about the AI model.

An adversarial attack is one of the most common types of attack.

- **Data poisoning attacks**

  In a data poisoning attack, the training data of the model is manipulated to compromise the performance or even bias the AI model's output. The data poisoning attack is a training-time attack. The data is manipulated at the training stage to influence the output. If an attacker can manipulate the training data, then the attacker basically controls the model's output.

The biasing of the AI model is a much deeper problem than just the data poisoning attack. Even legitimate developers can bias the training data to influence the outcome. For example, in stock price predicting models, companies can tweak the training to control the sell or buy trend of the stock. In addition, data can be poisoned to be biased against or in favor of a particular product in the market, to be used in profiling people, and so on. The list can be very long, but the point is, this attack has a much deeper impact than it appears.

The data poisoning attack isn't easy to launch. For this attack to work properly, the attacker must have access to raw training data, which means the attacker must have access to a database. In addition, the attacker must have access to the AI model and the training methodology. In other words, this attack has to be a white-box type attack to be successful.

A malicious insider or intentional biasing is usually the root cause of data poisoning attacks.

- **Backdoor attacks**

Like software, AI models are also vulnerable to backdoor attacks. These attacks usually result from known loopholes in the model. The backdoor is typically created for debugging or testing purposes. The attack is triggered when an attacker inserts or injects a malicious payload into the input. The attack can be crafted so that regardless of the input, when it's inserted from the backdoor, the model behaves in a certain way.

The backdoor attack can create a significant impact on the behavior of the AI model if crafted properly.

- **Model stealing and reverse engineering**

The attacker can steal the AI model itself. This attack can be to understand the architecture of the model or just to reuse the model to influence the user base in one particular direction. Another possible motive for this attack is to get ahold of the training data. The stolen model can be reverse engineered to study further. Model stealing and reverse engineering can be categorized as two different attacks, but we've combined two attacks here because one attack, model stealing, leads to another attack, reverse engineering.

- **Evasion attacks**

Simply put, this type of attack tricks the AI model into believing something similar but true. A common example of this attack is using a picture of a person to unlock a phone instead of facial recognition. If the attacker can trick the AI model with such small changes, eventually, he can learn about the model's behavior and plan to launch a bigger attack.

- **Privacy attack**

One of the biggest concerns people have in using AI models is the privacy of their information. Privacy attacks can be divided into two subattacks:

- *Model inversion attack*: In a model inversion attack, the attacker tries to find sensitive information used in the training data by studying the output of the model. The attacker tries to correlate the input-output relationship.
- *Membership inference attack*: In this attack, an attacker could try to find out if the specific data set is used as part of the model's training data. This is done to gain insight into the model's workings. The attack is also used to relate training data to a specific person or entity—a direct attempt to steal or obtain personally identifiable information (PII).

These are some of the common cybersecurity-related attacks on AI models. In the next section, you'll learn about the mechanism for defending against these attacks.

9.3.2 Securing AI Models with Cryptography

In this section, you'll learn how to safeguard the AI/ML models using cryptography. Some of the basic defense mechanisms for the AI models can be classified as follows [Hernández-Castro, 2022]:

- **Guards**
  Guards works as "gatekeepers" to the AI/ML models. This defense method tries to detect the malicious input before the input gets to the AI model. The idea is to detect malicious input before it's processed and reduce the potential impact.

- **Defense by design**
  In the defense by design method, the model is kept updated with the latest technologies and software. This technique is similar to the hardening process and is also called "model hardening."

- **Certified defense**
  This technique guarantees the defense in a given condition. In the certified defense method, the model is tested against a given set of training data. The model is then certified as secure within that data set range. The certified defense method works only on the training data and not on the test data.

These defense techniques will improve the security of the AI/ML model. They are general good hygiene practices. The security of AI models using cryptography can be divided into four categories:

- **Training data security**
  Securing training data is one of the most important security tasks in protecting AI models. At the least, all the input training data must be sensitized and validated before use. This will help eliminate the malicious input right off the bat:
  - The training data must be kept confidential. Encrypting the data achieves data confidentiality. Encrypting data also meets the compliance requirements if the training uses sensitive or PII data.

- Cryptographic hash functions and keyed hash or hash-based message authentication code (HMAC) must be used to verify the data integrity. This is a very important step if the training data is downloaded/uploaded or transferred between the databases.
- Access to the training data must be restricted and granted only on a need-to-know basis. The digital signature should be used to have the nonrepudiation property on the training data. Although it's not a cryptographic property, multi-factor authentication (MFA) must be used to access the training data.
- The differential privacy (DP) technique can help protect the training data by adding "noise" to it or the model's output data. By adding the noise, the data is hard to correlate with any particular data points. This will improve privacy, and if a data breach occurs, the data won't give any meaningful reference to the hacker.
- Lastly, homomorphic encryption can be used to process the data without decrypting it. This will allow the model to be trained using data without actually decrypting it. Homomorphic encryption will be discussed in detail in Chapter 11.

- **Model security**
  AI/ML models are developed using languages such as Python, Java, and so on. In fact, Python is one of the popular languages used to develop AI models. This basically means that all the secure coding practices that we apply to other software development must be applied to AI models:
  - The AI model (underlying code file) must be encrypted to protect it from anyone accessing it. To add another layer of security, the models can be digitally signed and can also use the hash function to validate the integrity.
  - As discussed previously, all access control measures should be implemented to protect against hackers or malicious insiders.

- **Communication security**
  Data and the model files should be transferred using secure protocols, such as Transport Layer Security (TLS), virtual private networks (VPNs), or point-to-point connections, and encrypted during the inference phase. The TLS protocol and VPN were already discussed earlier in Chapter 6 if you need a refresher.

- **Other security measures**
  Other security measures, such as logging and monitoring the training data and the models, should be implemented. This can provide an indication of suspicious activities in the environment.

Using cryptography to protect AI is a step in the right direction. Cryptography techniques such as symmetric encryption, asymmetric encryption, hashing, digital signatures, and so on provide confidentiality and integrity to AI models and training data. The use of AI is growing every day, and protecting the training data and AI models is very important.

9.4 Best Practices and Ethical Use of AI

Whether man or a machine, gaining power is accompanied by the danger of going rogue. Ethics or moral principles, hopefully, can control rogue behavior. The definition of ethics, according to the Oxford online dictionary is, "moral principles that govern a person's behavior or the conducting of an activity." If AI is going to do our work and act like humans, then we must make sure that we apply this definition to AI (or the use of AI).

First of all, why is it a concern? AI tools can be very powerful and can do a lot more than any other machine or device we've used so far. The concern is that bad actors can use this power to their advantage. That is why it's our responsibility to use AI ethically.

So, how can AI be used unethically? The main unethical use of AI is *biased output*. The AI models are trained and generate output based on the training data. It's very tempting to behave unethically and train the AI model to produce the output to our advantage. For example, Amazon stopped using AI-based candidate scoring tools to aid in the hiring process because Amazon used their previous hiring data (résumés and applications received over the years) to train the tool. The data from the past 10 years had more men than women. This data trained the tool in a way that gave men an advantage. Based on the training data, tools concluded that the men are better for the job [Dastin, 2018]! Obviously, we don't have all the internal details about the model and the training data used, but the point is a malicious developer in the company could develop a model and train the tool to exclude a particular race or gender from the hiring process or score them low. This would be the most unethical use of the tool.

Let's take an example that can be dangerous—a self-driving car. What if a malicious developer intentionally tweaks the AI model to not stop when a person is detected on the road? Can you just imagine how many people will be killed before the model is corrected? Obviously, this is an extreme example, but the point is, if your company is developing or using AI models, the company should have a policy as to what the AI tool can and can't do. The policy must include detailed guidelines on the ethical use of AI and thorough testing requirements.

The second most concerning issue with AI after bias is *privacy*. The training data used can include PII and other personal information. If the tool makes an error or misbehaves, this personal information can leak or end up in the wrong hands.

The best way to use AI ethically is to establish and follow the best use practices for AI for your organization:

- Create and enforce the AI usage policy and AI ethical use policy.
- Don't enter any personal information with public AI tools like ChatGPT. Anything you ask ChatGPT is probably going to stay in Microsoft Azure forever and can be used for training in the future.

- Use your own judgment. Although AI-based tools are very smart, they are developed by people, and there is always a possibility that the tool will make a mistake. You must verify the validity of the information.
- Most creations generated by public AI tools are copyright-free; however, this isn't guaranteed. If you must use AI-generated work or incorporate AI-generated work in your work, it's better to confirm the copyright details.

These are some examples of AI best practices. Your ethical use policy and best practices must be tailored to your organization's needs.

9.5 Summary

The concept of AI isn't new, but the use of AI has been picked up over the past several years. This opened up the possibility of using AI models with cryptography primitives and cryptography in securing the AI models. This scenario created a true win-win situation. If people understand the risks associated with the use of AI and use AI responsibly, then the AI and crypto combination will make a difference in people's lives.

In this chapter, we discussed the fundamentals and basic concepts of AI. We examined various AI models that can be used to create PRNG, S-box, and Boolean functions. We surveyed the risks associated with AI and the risk mitigation techniques. Lastly, we discussed the downside of not using the AI responsibly and some best practices to use AI.

> **Key Takeaways from This Chapter**
> - Learned AI models to implement cryptography primitives
> - Surveyed risk associated with AI and how cryptography can help
> - Reviewed fundamentals of AI and the ethical use of AI

In the next chapter, we'll tackle another new area in cryptography: post-quantum cryptography. Is there any risk to cryptography when quantum computers become a reality? If so, are we ready for it? How are we going to secure our data and information? Let's find out in the next chapter.

Chapter 10
Post-Quantum Cryptography

Quantum computing is a rapidly developing area of computer technology. When quantum computing becomes a reality, quantum computers will have the processing power to break modern cryptography. The focus of this chapter is to discuss the risks of quantum computing on our security and privacy and to explore post-quantum cryptography algorithms that can protect our data.

Quantum computing is no longer considered a future technology. When quantum technology is available for commercial purposes or everyday use, our society will reap many benefits but also face some challenges. This chapter discusses the fundamentals of quantum computing, the potential risk it brings to cryptography, and ways to safeguard your data in the post-quantum era.

Section 10.1 starts with a brief history and introduction of quantum mechanics. Section 10.2 focuses on the core of the chapter: the potential risk of quantum computing on cryptography and ways to protect against it. One way to protect against post-quantum risk is to implement quantum cryptography. However, the technology isn't developed yet, and, at this stage, it's not very practical to use quantum cryptography. Alternatively, we can protect our data using the quantum-resistant or post-quantum cryptography algorithms. We'll dive deeper to understand and implement these algorithms. Section 10.4 discusses the strategies to migrate to post-quantum cryptography. The chapter concludes with a short survey of the future of cryptography in the post-quantum era.

10.1   Primer on Quantum Computing

Quantum computing, based on the quantum mechanics branch of physics, results in exponentially faster processing and computing compared to classic computing. Although much research and development around quantum mechanics has been done in the past 50 years, the concept of quantum mechanics is almost 125 years old.

Moore's Law states that computing speed and processing power double every two years, and we've seen them double in just 18 months. To increase the processing power, we can increase the CPU speed, bus speed, and memory speed. However, there is a limit.

Quantum computing is beyond that limit because it doesn't solve complex problems merely with speed but by using quantum mechanics.

In this section, we'll go over a brief history of quantum mechanics and learn the fundamentals of quantum computing. The section will also introduce various technologies available for quantum computing.

10.1.1 History of Quantum Mechanics

The word *quantum* is derived from the Latin word *quanta*, which means "how much or amount." The Oxford Dictionary defines quantum mechanics as "the mathematical description of the motion and interaction of atomic and subatomic particles developed from the old quantum theory."

When metal is heated, it starts to emit light at the red end of the visible spectrum. If we continue heating the metal, eventually it will start to change colors—to yellow, white, and blue. This is because as we heat the metal, it continues to emit light but at higher frequencies or at a shorter wavelength. This is known as thermal radiation. This was demonstrated with experiments by many physicists in the late 1800s. However, there was no theoretical proof for this behavior. Max Planck, a German theoretical physicist, was the first to propose a model that defines the amount of energy emitted at each frequency in December 1900. He also suggested that the change in the amount of energy from one frequency to another isn't continuous but changes in steps or jumps from one frequency to another. This is because atoms can absorb only a certain amount of energy. This amount of energy or quantized energy theory is considered the first work in quantum physics or quantum mechanics. This is also known as *Old Quantum Theory*. Planck was awarded a Nobel Prize for his work in 1918.

> **What Is a Quantum Leap?**
>
> The term *quantum leap* describes a sudden increase or a dramatic change. It's believed that the roots of the term are in physics, and the usage is echoing what Max Planck theorized—a sudden, unexplainable change in the amount of energy from one frequency to another.

In 1905, Albert Einstein extended Planck's proposed model of quantized energy to light. Einstein suggested that when light is radiated from a surface, it doesn't spread as a continuous wave in the extending space but rather spreads as a *quanta of energy*. The light that we see is traveling as a continuous wave; however, the radiation of light is spread in the stages or in quanta—fixed energy at a time. (Both Planck's model and Einstein's theory are overly simplified here for explanation purposes.)

Research on quantum mechanics was continued after the First World War into the 1920s and 1930s. Physicists described various counterintuitive properties of quantum mechanics, like wave-particle duality, superposition, and entanglement.

In *wave-particle duality*, light can travel as a wave and as particles. In *superposition*, atoms, electrons, or photons can be in two states at the same time and collapse to one state when they are observed or measured. In *entanglement*, two particles remain connected and work in reverse actions, even if they are physically apart. Yes, this isn't a typographical error. We did say that light can travel as waves and particles, electrons can be in two states at the same time, and two particles remain connected even if they are physically apart! And we haven't yet discussed Schrodinger's famous thought experiment. Erwin Schrodinger, who worked independently from Einstein and other scientists on quantum physics, theorized the superposition property with his thought experiment where a cat is dead and alive at the same time (more on this later).

If you're confused by these discoveries and experiments, don't worry. Quantum mechanics defies common sense. Even Einstein called the property of entanglement "spooky action at a distance." We'll shed some more light on these concepts in Section 10.1.2, but it's best to embrace the seemingly illogical nature of quantum mechanics.

Planck, Einstein, and Schrodinger are famous for their work on quantum physics, and their work is often quoted; however, they aren't the only scientists who contributed to the research and development of quantum physics. The history of quantum mechanics is very interesting but beyond the scope of this book.

> **Weirdness of Quantum Physics**
>
> Are we saying that in superposition, atoms can have multiple states at the same time? Is it true that electrons can be physically separated but remain connected in entanglement? If this is difficult for you to understand, you're not alone—in fact, you have lots of company. Most people who try to learn quantum mechanics are baffled by the weirdness of it all. You can take this at face value or are welcome to research more on the subject, but for the purpose of quantum computing and post-quantum cryptography, we need to know only a little bit about the fundamentals of quantum mechanics.

10.1.2 Quantum Computing 101

Our conventional computers work with binary bits 0 and 1, while quantum computers work on quantum bits. The quantum bits are called *qubits* (pronounced "cue bits"). Holevo's theorem states that the amount of information stored in quantum bits is the same as the amount of information stored in conventional binary bits [Holevo, 1973].

So, why is there such big hype around quantum computing if qubits store the same amount of information? What makes quantum computing faster at solving problems?

To answer these questions, we'll have to understand the properties of quantum mechanics used with quantum computing. Quantum computing uses wave-particle duality, superposition, and entanglement principles of quantum mechanics. In this section, we'll briefly discuss each of these properties.

10   Post-Quantum Cryptography

> **Is Quantum Mechanics Really Useful in Everyday Life?**
>
> Up to this point, we've confirmed one thing—quantum mechanics is bizarre and defies common sense. But, do we have any evidence that it will be useful? The fact is, we already heavily use the principles of quantum mechanics in everyday life. Some common examples of quantum mechanics are lasers, MRI machines, fluorescent lights, and GPS.

Wave-Particle Duality

In the early 1800s, Thomas Young performed a double-slit experiment to demonstrate the behavior of light. At that time, he probably had no idea about his contribution to the fundamental property of quantum mechanics. The double-slit experiment is used in modern physics to demonstrate the wave-particle duality of photons. As shown in Figure 10.1, we place a continuous source of light, a flashlight, or a bulb, on one side of the screen with a double slit. The continuous source generates the waves. The light will pass through both slits and create a dark and bright stripe pattern on the wall behind the screen. This dark-bright stripe pattern on the wall aligned with our expectations. All good so far!

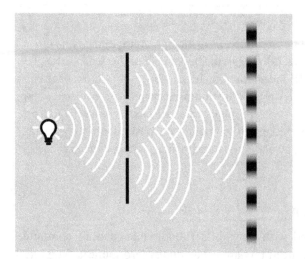

Figure 10.1 Wave-Particle Duality with a Continuous Light Source

Now look at Figure 10.2. In this experiment, we changed our source to send one photon every minute instead of a continuous source of light. We continue firing photons every minute at the double-slit screen for a few hours. We expect that the photons will either pass through slit 1 or slit 2 and create two bright spots on the wall: one behind each slit. We expected a different pattern by photons than the pattern created by the continuous source of light in Figure 10.1. However, that isn't the case. The pattern we see in Figure 10.2 is very similar to what we saw in Figure 10.1. This means that the photons passed from both slits at the same time! When the continuous source of light was used, light traveled as waves and passed through both slits. In the second experiment, the photon gun was used instead of a continuous source of light. However, this time, we also got the same pattern on the wall. This tells us that light travels as a wave and particle. This is the principle of wave-particle duality.

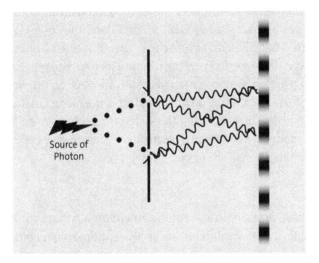

Figure 10.2 Wave-Particle Duality with a Photon Gun as a Source

Superposition

How can conventional physics explain one photon passing through two slits at the same time? There is no real explanation for this behavior of photons. Quantum mechanics steps in where conventional wisdom defies. Quantum mechanics explains this behavior of photons—passing through the same slit at the same time—with a phenomenon called superposition. Superposition is the ability of electrons, photons, and atoms to be in two states at the same time. For example, electrons can be spinning up and down at the same time, or photons can be polarized vertically and horizontally at the same time. The superposition continues until the state of the electrons or photons is observed or measured. When observed, the superposition collapses to one state.

Schrodinger's thought experiment comes to our rescue to understand superposition. Let's try to understand the thought experiment with a simplified explanation. In Schrodinger's thought experiment, he describes the cat as dead and alive at the same time. Suppose we put a cat and some poisonous food in an enclosed cage. We lock the cage for a couple of hours. We can't see inside the cage. During these couple of hours, the cat could be dead because she ate poisonous food, alive because she didn't eat the food, or she ate the food and, as a result, is very sick. We have no idea about the state of the cat. The cat could be in any of these (alive, dead, or sick) states. That is what happens to particles during the superposition state. However, as soon as we open the cage, we know the state of the cat, whether alive, dead, or sick. When we measure or observe the particles, they collapse into one state—just like the state of the cat. The superposition phenomenon of quantum mechanics is used in quantum computing, as we'll see later in this section.

Entanglement

Entanglement is another quantum mechanics property that is used in quantum computing. Once two quantum systems interact with each other, they enter the entanglement state. As defined earlier, in entanglement, two particles are connected even if they are physically apart. Moreover, these particles, while in entanglement, work in the reverse direction. For example, if two electrons are entangled, and one electron is spinning up, we know that the other must be spinning down. In addition, in entanglement, we can find out the state of one particle by measuring the state of the other—even if they are physically apart. These entangled or correlated electrons or protons give exponential power to quantum computers, as we'll see next.

Qubits

Before we use the phenomenon of superposition and entanglement for quantum computing, we must know about qubits. Qubits are different than conventional binary bits. Although conventional binary bits have two states 0 and 1, they can hold only one state at a time, either 0 or 1. On the other hand, qubits can hold both states 0 and 1 at the same time. To see the power of exponential growth with qubits, assume we have a computer with 3 qubits. A conventional binary bit computer with 3 bits can hold up to 8 states ($2^3 = 8$) while the 3 qubits quantum computer can hold 24 states simultaneously.

As shown in Figure 10.3, qubits are three-dimensional. Electrons, photons, and atoms are used as qubits with modern technologies. In conventional computers, we use bits to identify between 1 and 0, while in qubits we use the properties of these atomic particles to define 1 or 0. The electrons can spin up, 1, or spin down, 0. The photons use their polarity—a vertical or left-angle polarized photon denotes 1, and a horizontal or right-angle polarized photon denotes 0. Atoms are ionized, and ionized atoms acquire a positive or negative charge, which decides the digital state 1 or 0.

Modern technology uses these properties to develop quantum computers using qubits. One of the primary challenges in building qubits and quantum computers is the *coherence time*, which is the time during which quantum particles maintain their quantum property of superposition. In other words, when the quantum particles are in the superposition state, they are coherent. When we measure or observe the state, the quantum particle de-coheres or exits the superposition state. This is a critical property of qubits because to meet the expectation of quantum computing, we need a longer coherence time. Longer coherence time allows qubits to calculate or solve huge problems in a much shorter time compared to traditional binary computers. Let's return to our example of a quantum computer with 3 qubits. In this case, during the coherence time, qubits will be calculating or solving the problems with 24 bits! As soon as we measure the state, the coherence state collapses. That is why the longer the coherence time, the faster the calculation. As you'll see later, maintaining coherence time is challenging.

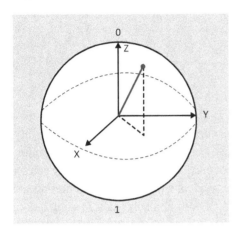

Figure 10.3 Qubit

Quantum Computers Aren't Supercomputers

Don't confuse quantum computers with supercomputers. Supercomputers are conventional computers that run using transistor-based silicon chips. Supercomputers are fast and have lots of computing power because of faster CPU speed, fast data transfer, and very low memory access times. However, supercomputers run using conventional binary bits and solve the problem sequentially. On the other hand, quantum computers use qubits and quantum properties and don't rely on hardware to increase the speed. The power of quantum computers comes from their ability to solve problems using superposition and entanglement.

10.1.3 Quantum Computing Technologies

Quantum computing is a rapidly developing field. The technology is still in its infancy, and researchers and developers are still exploring to find the best approach. At this stage, several technologies exist to develop qubits and quantum computers. Qubits are made up of either atoms, electrons, or photons.

In this section, we'll briefly discuss some of the commonly used technologies. This isn't an exhaustive list, and technology is continuously evolving so we encourage you to do more research on the subject. The commonly used quantum technologies discussed in this section are superconducting, trapped ion, photonic, quantum dots, and neutral atoms. Table 10.1 summarizes the major qubit technologies with types of qubits.

Qubit Type	Photon	Electron	Atom
Qubit Technology	Photonic	• Superconducting • Quantum dots • Topological	• Trapped ion • Neutral ion

Table 10.1 Types of Qubits and Technologies

Let's take a closer look at these technologies:

- **Superconducting**

  Superconductor technology uses electrons as a qubit. Superconductors have zero resistance. That means in a closed loop, the current can flow infinitely. In reality, there is always some tiny resistance and hence the loss of current, but superconductors offer "almost" or very close to zero loss of current. The superconducting state is achieved at a very low temperature. Several types of superconductor materials are used to implement the qubits. This was the first commercially realized qubit. Major players like Google and IBM are exploring and using superconductor qubits. The major downside is the need for cryogenic cooling and short coherent times.

- **Trapped ion**

  Trapped ion uses atoms as qubits. Atoms become ions when they gain or lose an electron and get an electrical charge. The trapped ion can process the information using an electromagnetic field. One issue that researchers face with qubits is to keep them in a coherent state. Trapped ion qubits can't stay in a coherent state for a long time.

- **Photonic**

  As the name indicates, photonic uses photons as qubits. Photons can also remain in a coherent state for a long time. Photons (i.e., light particles) are used to process quantum information, and, as a result, photonic qubits are preferred for many quantum applications.

- **Quantum dots**

  Like superconductors, quantum dots use electrons as qubits. Quantum dot qubits are created using a semiconductor material like silicon. The advantage of this type of qubit is the industry's familiarity with silicon usage in computers. This also requires cryogenic cooling. This type of qubit isn't very popular so far.

- **Neutral atoms**

  Needless to say, a neutral atom uses atoms as a qubit. Atoms are suspended in a vacuum by arrays of tightly focused laser beams to form the neutral atoms qubits. These qubits have a longer coherent time and that makes them an attractive choice for many applications.

Other qubit types are defect based, topological, and nuclear magnetic resonance. These technologies are only theoretical (i.e., haven't been developed yet or are in the very early stages), so we won't discuss them in more detail.

10.2 Quantum Computing and Cryptography

Quantum computing can solve many problems at a much faster pace than traditional computers. However, quantum computing also has another side—it can put our secrecy and privacy at risk. In the first part of this section, you'll learn about Shor's and Grover's algorithms and the risk of quantum computers on cryptography. Then, we'll discuss, at a high level, quantum cryptography and various quantum-resistant algorithms.

10.2.1 The Risk of Quantum Computing

As we've seen in the previous section, keeping quantum particles in a coherent state for a longer period and building qubits/quantum computers are very difficult tasks. Even the slightest noise or interference can collapse the superposition state. This may result in an erroneous observation or reading. On top of these two obvious hurdles, quantum computing also faces other challenges, such as technology being very expensive and a lack of programs to execute the desired tasks on quantum computers. Quantum computers aren't ready for prime time yet, but we need to prepare now—and we'll discuss why.

Cryptography currently works on the difficulty of solving the mathematical problem. One of the most difficult and time-consuming problems was factoring a large number to find its prime integers. Even with the most advanced and powerful computers available today, factoring a large number takes a very long time. However, that was an asset for cryptography. As you learned in Chapter 1, with Kerchoff's principle, the cryptography algorithm is known to everyone. The encryption key is the only secret in cryptography. Public key cryptography algorithms, like RSA, rely on mathematical difficulty and use the two prime numbers to compute the key. Traditional computers can take up to, at least, hundreds of months to factor a long (100 digits or more) number. That is why our data and information were secured with the current encryption methods.

Encryption was considered one of the best data protection mechanisms. Even if all other layers fail, data is still protected if it's encrypted. However, quantum computing puts a dent in our security mechanism. The first quantum computer programmable algorithm was presented by Peter Shor in 1994. Shor developed an algorithm to factor a large number to find the prime integers [Shor, 1994]. Shor's quantum algorithm uses exponential speed to solve the discrete and elliptic curve logarithmic problems. The algorithm was no doubt a reason for cryptographers to panic, but it was too far ahead of its time. In the late 1990s, technology wasn't ready to build any usable quantum computers. Even after researchers started building quantum computers, the error rate was high, and, even today, we don't have a quantum computer that successfully implements Shor's algorithm. This isn't to say this won't happen. In fact, at the pace of current technological advancement, it's just a matter of time, and cryptographers must prepare for the worst-case scenario.

Shor's algorithm leverages the superposition property of quantum computing to calculate all the possible solutions using all bits. Shor uses a randomized algorithm, known as the Las Vegas algorithm. The Las Vegas algorithm always terminates with a correct result or by indicating a failure. Shor states,

> *"This paper gives Las Vegas algorithms for finding discrete logarithms and factoring integers on a quantum computer that take a number of steps which is polynomial in the input size, for example, the number of digits of the integer to be factored." [Shor, 1994]*

Shor's algorithm, although a problem for cryptographers, was a great breakthrough for quantum computing, and a reason for cryptanalysts to be hopeful!

Soon after Shor, Lov Grover in 1996 developed a quantum algorithm to search for an item on the list at a very high speed. Grover's algorithm [Grover, 1996] can search unstructured data. Unlike Shor's algorithm, which provides exponential speed over conventional computers, Grover's algorithm provides quadratic speed over conventional computers. For example, if we search the database of x items with classical computers, it will take an average of x/2 number of searches, and, on a bad day, the search will have to go through all x items before the match is found. However, Grover's algorithm can perform this search, thanks to the superposition property, in a square root of x searches. This really helps speed up the search in a large database. Although Grover's algorithm was intended to help perform the search of a large database, it also helps cryptanalysts perform brute-force attacks or weaken some of the symmetric cryptography algorithms. Grover gave another reason for cryptanalysts to celebrate.

We won't go into the mathematical details of Shor's algorithm and Grover's algorithms, but it's enough to know that these algorithms created a panic for cryptography researchers and practitioners. These algorithms, when implemented, will pose a risk to the security and privacy of our data and information.

Table 10.2 shows the cryptography algorithms, the impact of quantum computing on these algorithms, and the potential solution.

10.2 Quantum Computing and Cryptography

Cryptography Algorithm	Type of Algorithm	Impact of Post-Quantum Computing	Potential Solution
RSA	Pre-quantum computing	Broken	Use National Institute of Standards and Technology (NIST) recommended algorithms
Diffie–Hellman		Broken	
ECC		Broken	
AES		$\sqrt{2^n}$ weakened	Migrate to longer key 128-bit → 256 bit → 512-bit
SHAn		$\sqrt{2^n}$ weakened	
McEliece	Post-quantum computing	Not broken yet	N/A
NTRU		Not broken yet	N/A
Lattice-based		Not broken yet	N/A

Table 10.2 Effect of Quantum Computing on Various Cryptographic Algorithms

If quantum technology isn't fully prepared for prime time, and quantum computers may not be available for the general population for several years, then why is it considered a risk right now?

Since Shor published his algorithm, cryptanalysts have been very optimistic that one day they will decipher the data. Over the past decade, there has been a lot of traffic and data on the internet. Hackers have already started harvesting this data. Most of the information in the data will be outdated by the time hackers are able to use quantum computers and Shor's algorithm to decipher the harvested data. However, there is data with a long shelf life. For example, consider hackers who are collecting data traffic from a bank or a credit card company right now to extract the data later. When quantum computers are available, people's bank account numbers or credit card numbers will no longer be applicable. However, if the bank or credit card data includes people's Social Security numbers (SSNs), hackers can use the SSN even 10 years down the road since those numbers aren't going to change.

What Is Long-Shelf-Life Data?

Data that doesn't change and is useful and valuable over a long period of time—typically more than 10 years—is considered data with a "long shelf life." Some examples of long-shelf-life data include SSNs, date of birth, mother's maiden name, birthplace, intellectual property, trade secrets, and so on.

This is the why security professionals should start thinking now about protecting data with post-quantum computing. However, if quantum computing is putting our security and privacy at risk, then how can we protect against it? Can quantum mechanics come to our rescue? The answer is yes. There are two ways to protect our data and information from quantum computing, which we'll cover in the next sections:

- Quantum cryptography
- Post-quantum cryptography

10.2.2 Quantum Cryptography

The concept of quantum cryptography isn't new. The superposition and entanglement properties of photons are at the heart of quantum cryptography. To recap, in superposition, particles can be in two states simultaneously, and in entanglement, particles are correlated with each other and exhibit opposite behavior as long as they are correlated. Photons can be polarized in two ways—rectilinear or diagonal. (Quick recap of high school physics—polarization, simply put, is the direction of the electromagnetic waves.) Moreover, in rectilinear polarization, photons can be either vertically polarized or horizontally polarized, while in diagonal polarization, photons are either left-angle polarized or right-angle polarized. This property of photons means that photons can be in four states, and, when in superposition, they can be in all four states. For example, if we assign 1 to vertically polarized and left-angle polarized photons and 0 to horizontal and right-angle polarized photons, then we can transmit our binary message with photons. Figure 10.4 shows the polarization schemes.

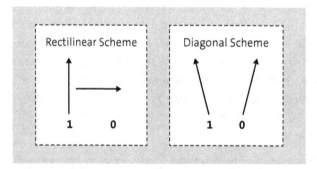

Figure 10.4 Photon Polarization Schemes

In 1984, Charles Bennett and Gilles Brassard proposed one of the first quantum cryptography algorithms [Bennett, 1984], named BB84, using the photon polarization property. The working of the BB84 algorithm is shown in Figure 10.5 with an example and explained (in very simplified terms) next:

1. The sender transmits 1 and 0 using photons. The sender uses either a rectilinear or diagonal scheme for each bit. For example, the sender transmits 1 0 1 0 using vertical, right-angle, left-angle, and horizontal polarization.

2. The receiver has no idea of the sequence transmitted. The receiver applies the best guess and measures the photons using his polarization. He may get some right and some wrong. Only bits for which the sender's and receiver's basis are matched are kept; the others are discarded,

3. At this point, the sender and receiver communicate by phone, email, or text messages. The sender reveals the polarization scheme used. In our example, the sender tells the receiver that in the current communication, a rectilinear, diagonal, diagonal, and rectilinear scheme is used. Now even if an eavesdropper is to tap this conversation, this message won't help as the sender isn't disclosing whether the first rectilinear scheme is vertical (1) or horizontal (0). The eavesdropper will have to guess between the two polarizations for every photon.

4. Once the understanding between the sender and the receiver is established, that string of bits can be used as the secret key.

5. The key is used to encrypt the message. The communication happens on the open channel.

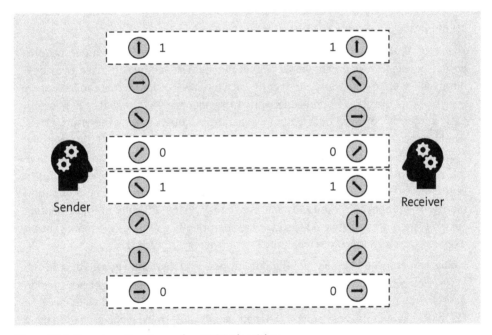

Figure 10.5 Quantum Cryptography: BB84 Algorithm

If an eavesdropper intercepts the communication and guesses a wrong polarization filter, the sender and receiver will know that someone is trying to listen to the conversation. On the other hand, if the eavesdropper happens to guess and use the right polarization filter, the sender and receiver won't know about the eavesdropper. The probability of an eavesdropper guessing the correct filter is very small and close to impossible.

On the other hand, if an attacker carries out a man-in-the-middle attack (i.e., intercepts and takes control of the communication), then it won't be detected. In most cases, the attacker receives the sender's message and relays the modified message as if the attacker is a sender—the attacker basically impersonates the receiver and sender. It won't be detected by the principles of quantum mechanics, but there are other methods available to easily detect this type of attack.

Another quantum cryptography algorithm was developed by Artur Ekert in 1991. Ekert used the properties of entanglement. In his algorithm, the sender sends the entangled photons to the receiver, and then both decide on the scheme. Because the photons are entangled, the receiver's photon will be polarized in the opposite direction. If this isn't the case, they know that the message is intercepted. Ekert's algorithm is based on Bell's inequality theorem, while the BB84 algorithm is based on Heisenberg's uncertainty principle.

Quantum cryptography is also known as *quantum key distribution (QKD)* because the primary goal is to decide on the secret key.

> ### Stephen Wiesner versus Muhammad bin Tughlaq
>
> The seeds of quantum cryptography go back to the 1960s, when Stephen Wiesner proposed the idea of using photons to protect money. The idea was to have two serial numbers—one a regular serial number for each bill, like F1234567K, that can be seen by everyone, and the other an embedded quantum sequence, i.e., a sequence of photons. Each photon is polarized differently to generate a sequence. Only the issuing bank would know the correct sequence for each bill. Counterfeiters can't find out the sequence and can't create fake currency. Wiesner's idea was rejected for a long time but was eventually published in 1983. The concept of using a sequence of photons to send a message, a secret key, in quantum cryptography is based on Wiesner's quantum money. The problem, in the 1960s and even today, is the difficulty of implementing such a system. Not only would it cost more than the face value of the money, but it would also be very difficult to maintain. The technology isn't ready.
>
> Indian emperor Muhammad bin Tughlaq introduced brass and copper coins in the 1300s. However, just like we're not ready to print quantum money right now, Indian mints weren't capable of producing metal coins in the 1300s. The idea failed because his mints couldn't produce properly shaped metal coins, so everyone started making irregularly shaped coins at home. It was a huge failure for him, and historians labeled him "The Mad Sultan." Today, we do use metal coins. The idea was good, like Wiesner's quantum money, but way ahead of his time. Maybe one day we'll be using quantum money!

Quantum cryptography won't be mainstream or commercially successful cryptography anytime soon; it has its own share of problems. Quantum cryptography works on the physical properties of photons. It is, like qubits, still in its infancy and very sensitive.

Even a slight interference can affect the values. It's expensive and difficult to implement quantum cryptography.

10.2.3 Quantum-Resistant Cryptography Algorithms

Quantum-resistant cryptography, also known as post-quantum cryptography, is another approach to secure data against quantum algorithms. Post-quantum cryptography, like conventional cryptography or pre-quantum cryptography, is based on the difficulty of solving a math problem. As we discussed earlier, pre-quantum cryptography (like RSA) works on the difficulty of factoring a large number into two prime integers. This math problem is difficult to solve by conventional computers and can be solved in a relatively short period of time by quantum computing. Math used in post-quantum cryptography takes much longer to solve than conventional cryptography algorithms, even using quantum algorithms. One advantage of post-quantum cryptography over quantum cryptography is that it works even on conventional computers. This allows us to start implementing and testing these algorithms now instead of waiting until we have access to quantum computers. Table 10.3 summarizes the difference between quantum cryptography and post-quantum cryptography.

Quantum Cryptography	Post-Quantum Cryptography
Also known as quantum key distribution (QKD)	Also known as quantum resistant cryptography
Implemented on quantum computers	Runs on conventional computers but also protects against quantum attacks
Can't be cracked without detection	Can be cracked without detection
Photons used to transfer data	Conventional bits transfer
Based on the principle of physics	Based on the difficulty of math problems
Very expensive, sensitive and difficult to implement	Easy, inexpensive, and stable implementation

Table 10.3 Quantum versus Post-Quantum Cryptography

In this section, you'll learn about five different types of post-quantum cryptography [Bernstein, 2008] algorithms: code-based, lattice-based, multivariate, hash-based, and isogeny-based cryptography.

Code-Based Cryptography

Code in code-based cryptography doesn't mean the software or the secret code; it means error-correcting code. In code-based cryptography, an error is introduced intentionally to create a stronger cipher. Error-correcting codes detect errors in the message and correct them. One way to introduce an error is by adding some redundancy. Redundancy is removed from the message during the decryption process. One common error correcting code mechanism is a Goppa code. A *Goppa code* is a linear error correcting code used in code-based cryptography.

Let's understand this with some examples. Suppose the sender sends a message, "I need a swimsuit." Anyone who reads the message will understand this. But what happens if we change the message to "I need a swimsuit, jacket, and boots." Now let's add some error-correcting mechanisms, which is common sense in this case. A person can swim without a jacket or boots, but not without a swimsuit. So, the error-correcting mechanism will discard the other two, and the message will be delivered as "I need a swimsuit."

Now let's take an example in binary. We want to send a four-digit message 1010. Now add some redundancy or error to make it complicated for any eavesdroppers. We'll send a message like 0111 1000 1011 1000. The error-correcting mechanism at the receiving end will have to pick the digit in each group with the most repetitions and discard other digits. The message will be received as 1010.

These are very basic examples of what it means to add errors intentionally to the message to confuse unauthorized readers if they happen to tap the message. The implementation of code-based cryptography uses math-based error correcting codes.

The concept of code-based cryptography is very old. Robert McEliece proposed a code-based algorithm in 1978, and it's still unbroken. McEliece's algorithm is a public key encryption (PKE) that generates a private code using a random binary irreducible Goppa code and the public key using a generator matrix. The message is encrypted using a public key and some random errors or error codes. The private key, which includes the error-correcting mechanism, is used by the receiver to decrypt the message. Only the use of the private key can remove the errors from the cipher text.

McEliece's algorithm is considered a quantum-resistant cryptography algorithm because it relies on the hardness of the linear code. It's considered an NP-complete problem (where NP-complete is a complexity class).

The biggest advantage of the McEliece algorithm is its low complexity and fast encryption and decryption. The drawback of the McEliece cryptography is its large key size—from 100 KB to several megabytes. However, technology has come a long way since 1978, and increased processing speed and memory size have made this algorithm a good candidate for post-quantum cryptography.

Lattice-Based Cryptography

Lattice-based cryptography is gaining momentum because of its potential ability to withstand quantum computer attacks. Like all other public key cryptosystems, lattice-based cryptosystems also rely on the difficulty of solving a math problem. The concept of lattice-based cryptography was first introduced by Miklos Ajtai in 1996.

First, let's discuss what lattice is and the hard problems to solve in a lattice-based system. A *lattice* is a set of points or coordinates on an infinite plane. Points or dots on the vectors are integers. A vector is an object with magnitude and direction. The coordinates of the vectors allow us to manipulate the direction or the magnitude of a vector. The basis is a collection of vectors at any point that forms the lattice.

One difficult problem to solve in lattice-based math is to find the lattice point that is closest to, but not equal to, the origin. This problem is referred to as the *shortest vector problem (SVP)*. The SVP gets difficult to solve as the dimensions increase—meaning it can be solved relatively easily for a two-dimensional lattice, but the difficulty increases as you increase the dimensions. Another difficulty, similar to the SVP, is the *closest vector problem (CVP)*. The CVP is to find a lattice point that is closest to a particular point. The collection of these problems is believed to be very difficult for classical as well as quantum computers to solve.

How do these lattice math problems work with cryptography? To generate the public and private key, as shown in Figure 10.6, set up a lattice with two different bases that generate the same lattice. One basis has vectors parallel to each other, while the other basis has vectors perpendicular to each other. The basis with vectors perpendicular to each other is considered a good basis, while the basis with vectors parallel to each other is a bad basis. The CVP is a lot more difficult to solve with a bad basis because it's very difficult to arrive at the closest point on the lattice with information about a bad basis. It's relatively easier to find the closest point on the lattice if you know a good basis.

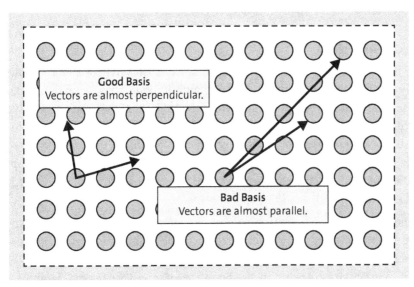

Figure 10.6 Lattice-Based Cryptography Concept

What does this mean for cryptography? The good basis is our private key, and the bad basis is our public key. If an attacker or malicious user tries to decipher a message using a bad basis, it will be practically impossible to solve—meaning to arrive at the closest point (in the case of cryptography, it's the closest point to the message point). Because it's easier to arrive at the closest point with a good basis, it's kept secret—via a secret key—and the owner of the private key can easily decrypt the message. This concept was originally proposed by Oded Goldreich, Shafi Goldwasser, and Shai Halevi in 1997, and was known as the GGH cryptosystem [Goldreich, 1997]. They also developed another algorithm called the GGH signature scheme. However, shortly thereafter in 1999, a cryptanalyst, Phong Nguyen found a way to solve the CVP with a bad basis and cracked the GGH cryptosystem. The GGH signature scheme was also cracked later [Nguyen, 1999].

Although this algorithm was broken, it provides a good foundation for the understanding of the lattice problems and sets the stage for another lattice-based algorithm—learning with errors (LWE). By adding errors, we make it harder to solve the problem. Without errors, they are simply linear equations and can be solved with faster computers. The errors are part of the secret but added in computing the public key. This makes it very difficult to solve the equation for the public key—even for quantum computers—unless the errors are known.

NTRU is another example of a lattice-based cryptography algorithm developed by Hoffstein, Pipher, and Silverman [Hoffstein, 2008]. NTRU uses a lattice with a special structure. The security of NTRU is believed to be somewhat stronger than LWE.

The lattice-based cryptography algorithms are robust (with LWE) and efficient. They also work with fully homomorphic encryption (FHE) schemes (see Chapter 11). One disadvantage of the lattice-based algorithm is its large key size.

Multivariate Cryptography

As the name indicates, multivariate cryptography is based on solving equations with multivariable unknowns. It uses multivariate quadratic polynomial equations. The security of multivariate cryptography comes from the NP-hard of the problem to solve nonlinear equations. The public key is generated using a set of multivariate equations. The solution to these equations is our private key. The probability of using multivariate cryptography as a post-quantum cryptography has fueled the research and development in the field of multivariate public key systems over the past several years, which is expected to continue.

Although there aren't many known uses of multivariate cryptography, the multivariate signature scheme offers a definite advantage because of the short signature. Slow encryption and decryption processes, significantly above average consumption of memory, and large key size are primary disadvantages of multivariate cryptography.

Hash-Based Cryptography

The primary purpose of cryptography was the confidentiality of the data, and that's how it has been used for centuries. However, modern cryptography has expanded the use of cryptography in authenticity, integrity, and nonrepudiation by using hash functions. The building block of hash-based cryptography is the hash functions and not the difficulty of solving math problems. The collision-resistant hash functions are at the heart of hash-based security. Quantum computing can't crack the hash functions that are based on finding the collision. Refer to Chapter 4 for a detailed discussion of hash functions.

Hash functions take the message as input and generate a fixed size value, known as a digest or a hash value. This value is irreversible. Some collision-resistant hash functions are listed here:

- **Lamport-Diffie One-Time Signature (LD-OTS)**
  LD-OTS uses a pair of public and private keys. The message is divided into bits. Each bit of the message is signed with a part of the signature. Combining all the signed parts of the message generates the full signature. The drawback is the large key size and inefficiency.
- **Winternitz One-Time Signature (WOTS)**
  WOTS, like LD-OTS, also divides message into small parts. However, instead of signing each bit of the message, WOTS creates small groups and signs groups. This makes the scheme a bit more efficient and manageable. It's an improvement over LD-OTS, but it's not easy to use because there could be many small groups to sign and manage.
- **Merkle signature scheme (MSS)**
  MSS works on the concept of a Merkle tree (see Chapter 8). MSS, like LD-OTS and WTOS, also generates multiple key pairs. One pair is to be used with each message. However, in MSS, all the public keys are hashed and rolled up to the root. The root public key can be used to verify and decrypt the message.

The hash-based cryptosystems are fast, efficient, and can work with any hash function. The downside is the large size of the signature compared to some other post-quantum cryptography signature algorithms.

Isogeny-Based Cryptography

Isogeny-based cryptography [De Feo, 2011] is a relatively newer approach in cryptography; it gained some traction only in the early 2000s. It's based on the concept of elliptical curve cryptography (ECC). Much of the attention was given to ECC and its variants until cryptographers realized that quantum computers can crack ECC. This gave opportunity to isogeny-based cryptography as it's believed to be quantum resistant. These algorithms use the properties of isogenies between the elliptical curves and maps between elliptical curves.

The supersingular isogeny Diffie-Hellman (SIDH) algorithm [De Feo, 2011] is based on the specific type of elliptic curve—supersingular elliptic curve. Solving the problem of isogenies between the supersingular elliptic curve is, at least theoretically, quantum resistant. The advantage of the SIDH algorithm is the short key length. The SIDH protocol is a key exchange protocol and proposed as a drop-in replacement of the current Diffie-Hellman key exchange protocol.

10.3 Preparing for Post-Quantum Cryptography

In this chapter, you've learned about quantum computing and the risk it's posing to our security and privacy. We also studied various post-quantum cryptography algorithms that can protect our data and information from quantum computing attacks. So, you have all the information about the risk and options to prevent the risk, but you're faced with some questions:

- How do you know which option is the best for your organization?
- How do you plan the implementation?
- Is it too early to start?
- Will the implementation work on conventional computers?

We'll address all of these questions in this section, which is divided into two parts. In the first part, you'll learn about what the government has done so far or is doing about post-quantum cryptography. You'll also briefly learn about the post-quantum cryptography algorithms that are selected for standardization as well as the algorithms selected as alternates. In the second part, you'll learn the step-by-step process of implementing post-quantum cryptography in any organization.

10.3.1 NIST: Initiative on Post-Quantum Cryptography

Given the continuous research and development in the field of quantum computing, the arrival of quantum computers is just a matter of time. As discussed in the previous section, this poses a severe threat to the current cryptography algorithms. To protect against the threat of quantum computing, the National Institute of Standards and Technology (NIST) announced the intention to standardize the post-quantum cryptography algorithms in 2016.

NIST started a new competition process and announced a call for papers in December 2016. NIST focused on the four criteria to evaluate the algorithms—security, cost, performance, and algorithm design and ease of implementation. *Security* of the algorithm is obviously very important as we're going to secure our data using these algorithms. The selection criteria also require that the algorithms can withstand common attacks such as side-channel attacks, multi-key attacks, misuse, and so on. The standardized algorithm should work securely with other commonly used protocols. *Cost* is almost always a factor in developing new feature or technology, and that doesn't change for post-quantum cryptography. In this case, cost not only includes the cost of writing software to implement the algorithm but also the cost of the key, including rotating, destroying, and managing the key. *Performance* is the third evaluation criteria. Apart from efficiency of encrypting and decrypting the messages, performance considers the size of the key and the memory utilization as well. *Algorithm design and ease of implementation* are the last criteria. The algorithm should be flexible enough to accommodate changes and should be able to be implemented on any platform.

NIST received phenomenal response to their call for papers and received 82 submissions—this is a big jump compared to the 42 submissions received in 1998 for the Advanced Encryption Standard (AES) competition and 64 for the SHA-3 competition in 2008. In December 2017, NIST announced the list of 69 algorithms out of 82 that made the cut in the first round. Out of the 69 algorithms, 49 were for PKE, and 20 were for digital signature algorithms. In January 2019, NIST selected 26 algorithms to go to the next step, the second round of the standardization process. At this time, NIST also invited comments on the algorithms from the submitters and community members. In July 2020, third-round finalist as well as alternate candidates were announced. Exactly two years later, in July 2022, NIST announced the first set of algorithms to be standardized along with alternate algorithms that will be going to the fourth round for evaluation. A brief description of these algorithms is discussed next:

- **CRYSTAL-Kyber**
  Kyber is a module learning with errors (MLWE), which is a type of lattice-based key encapsulation algorithm. Lattice-based algorithms are theoretically proven to be quantum resistant, and Kyber is also considered quantum resistant. Kyber has fast key generation, encapsulation, and decapsulation. Kyber interstates all the advantages of lattice-based cryptography discussed previously plus adds a layer of security because of the MLWE. Strong Kyber is selected for standardization based on strong security (difficult to crack with conventional and quantum computers) and excellent performance. Kyber is the only PKE algorithm selected for the standardization.

- **CRYSTAL-Dilithium**
  Dilithium is selected for signature standardization. Dilithium uses lattice-based cryptography. It's efficient and is, theoretically speaking, quantum resistant. Additionally, Dilithium has very easy implementation.

- **FALCON**
  FALCON—Fast Fourier Lattice-Based Compact Signature over NTRU—is another lattice-based signature algorithm selected for the signature scheme. FALCON is secured. The primary reason for the selection is Falcon's small bandwidth, which is required for certain applications.

- **SPHINCS+**
  SPHINCS+ is the only non-lattice-based algorithm selected for standardization. It's a stateless hash-based signature algorithm. SPHINCS+ has a short public key size but a long signature.

The following algorithms were selected for the next round for further evaluation:

- **BIKE**

  Bit Flipping Key Encapsulation (BIKE) is based on binary linear quasi-cyclic moderate-density parity check. BIKE has the best performance in non-lattice-based algorithms. The key size is comparable to lattice-based algorithms, but slightly larger. BIKE remains in the competition as NIST is looking to standardize a lattice-based algorithm.

- **HQC**

  Hamming Quasi Cyclic (HQC) is, like BIKE, based on binary linear quasi-cyclic moderate-density parity check. One downside of HQC is the large key size. HQC offers good security assurance but doesn't offer the best performance.

- **SIKE**

  Supersingular Isogeny-based Key Encapsulation (SIKE) uses isogeny-based difficult math problems, which are different from other math problems being evaluated. Isogeny-based cryptography is a relatively new field. SIKE offers a small key size but is slow in performance.

- **Classic McEliece**

  Classic McEliece is a code-based cryptography algorithm. McEliece is a public key encapsulation algorithm, which offers very strong security but lacks performance. It has a very large public key and a slow key generation process.

NIST released a draft of the Post-Quantum Cryptography (PQC) standards in August 2023 [FIPS 203, 204, 205, 2023] and was open for comments until November 2023. The process is wrapped now, and it's expected that NIST will release the PQC standard in 2024.

10.3.2 Preparing and Implementing Post-Quantum Cryptography

Preparing for and implementing post-quantum cryptography is a huge, time-intensive endeavor. In a big enterprise, changing an encryption key could take a long time, so changing the cryptography strategy obviously will take a long time—potentially, multiple years. Organizations should start preparing for the migration to post-quantum cryptography right now to protect the data with a long shelf life. Figure 10.7 summarizes the process and steps to migrate to post-quantum cryptography.

Let's walk through these steps in more detail:

1. **Research**

   It's very important to research the various algorithms available for post-quantum cryptography and study the one you're most interested in or that fits the needs of your organization. It's also important to research the standards related to cryptography and keep in touch with the standard development organizations. The research will allow you to check out the available options, make you aware of any protocol changes, and get you ready for the migration.

10.3 Preparing for Post-Quantum Cryptography

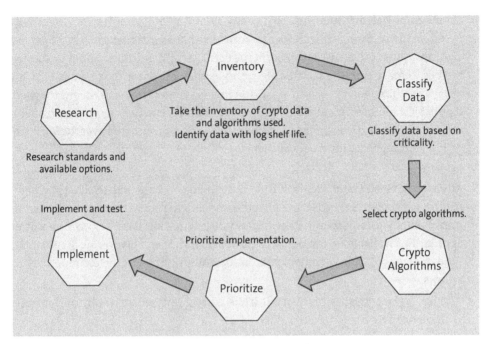

Figure 10.7 Steps to Migrate to Post-Quantum Cryptography

2. **Inventory**
   This is the key step in the process because it's where the actual work starts. In this step, the organization should spend time collecting crypto-related data. Find out what is already encrypted now and how. It also helps to inventory all data and see if all the sensitive or important data is encrypted. This is the time to tweak the process or expand the scope. High-value assets (HVA) and long shelf-life data should be identified in this stage. This phase is also used to inventory the cryptography algorithm in use by the organization.

3. **Classify data**
   The next step is the classification of the data found during the inventory. It's a crucial step, and it helps later in prioritizing the implementation. Organizations should classify the data based on the criticality. Usually, three or four types of data classifications are used. Military and government organizations use four categories to classify the data, while industry or private organizations use either three or four data classes depending on the business need. The data classification or criticality also identifies how the data should be protected. This is very important to understand as every security control takes a toll on performance and costs money to implement and manage. More sensitive data, such as personal identifying information (PII), intellectual property, customer data, and other important data, should be classified as sensitive or critical data. This data and its backup should always be encrypted. As the criticality of the data goes down, the protection should also be implemented accordingly. This step in the process helps identify the data to be encrypted.

4. **Cryptography algorithms**

    Based on the research earlier, we've identified various post-quantum cryptography algorithms that can be used to encrypt the data. These algorithms should be studied further to make sure the selected algorithm is the best fit for the organization. It's always recommended to use an algorithm that is standardized and widely accepted, if possible. Moreover, it's also advisable to select more than one algorithm when you can. That will help if one method is compromised or gets outdated over the long run, giving your organization a backup plan.

5. **Prioritize**

    After we have all the ducks in a row—identified the data and selected the algorithm—it's time to prioritize and plan the implementation. The priority of what to implement first is purely the organization's internal decision—no cookie cutter approach is available or can be used. However, asking and answering some of the fundamental questions can help you decide and prioritize. Some examples are as follows:

    - What is the criticality of the data? If the data is very sensitive, obviously it gets first priority.
    - Is the system externally facing—can data be modified by users from outside the firewall? If this is the case, even if the data criticality is lower, data should be encrypted first.
    - What is the life of data? The priority differs when the data life is short versus long.

    Once management and technical teams agree on the priority of the organization, this stage primarily becomes a design phase. Business and security requirements are analyzed, understood, and embedded into the design. This stage can also be used to identify and perform threat modeling. This stage is also an opportunity for the organization to update the crypto-related policy and standards. Based on the new algorithm selected and the revised policy, this stage can also offer a chance for the organization to update the processes. Organizations can also use this phase to educate developers by providing sample crypto implementation codes and examples of how to use various crypto libraries.

6. **Implement and test**

    This is the most critical stage in the entire process. Research, planning, and design are of no value if the implementation isn't carried out properly. It's a good idea to implement the prototype or sample code and perform various tests before starting to develop the production software. On top of regular testing, organizations should schedule additional testing to find out how quantum-safe new algorithms and software are. This is crucial because many errors could be introduced during the implementation. Once the prototype is proven satisfactory, developers can start implementing the production code.

The secure software development and operations lifecycle (SDOL) must be followed throughout the lifecycle (refer to Chapter 1).

10.4 Future of Cryptography in Post-Quantum Computing

Based on what we know so far, quantum computing will become a reality at one point, and we must prepare for it. Quantum computing isn't a question of *if*, but *when*.

> **Quantum Supremacy**
>
> In 2019, Google introduced Sycamore—a 53-qubit quantum computer. According to Google, Sycamore can perform calculations in 200 seconds that conventional computers can take up to 10,000 years to perform. This was the first breakthrough by quantum computing and is considered *quantum supremacy*.

Because we don't have a crystal ball to predict the future, we can't say when exactly it will happen; however, we can say with certainty that we don't need to lose sleep over it. It's not happening very soon because quantum technology has its own share of problems:

- **Error rate**
  First and foremost is the error rate. One major difference between conventional computers and quantum computers is the way they arrive at the solution. Conventional computers provide the absolute solution or one answer, while quantum computing works on the probability distribution. Meaning it provides the probability of a solution. The solution with the highest probability is, most likely, the right answer, but that may not be always the case. This introduces errors, and reducing these errors is one of the challenges designers are facing right now.

- **Noise sensitivity**
  Qubits are also affected or very sensitive to noise. Even a slight noise could change the state of the qubits. This could give the wrong calculations, or the superposition could collapse into a premature state.

- **Error correction**
  It's necessary to run error–correcting algorithms to derive the solution using quantum computers. In fact, algorithms such as Shor's algorithm are difficult to run without using an error-correcting algorithm.

- **Scalability**
  Another challenge is scaling the qubits. Qubits are maintained at very low temperatures, and it's extremely difficult to scale this technology. Quantum computers will also need new software because the software we're running on conventional computers won't work on quantum computers.

> **Interesting Quotes About Quantum Mechanics**
>
> *I think I can safely say that nobody understands the quantum mechanics.*
>   —Richard Feynman
>
> *Spooky action at a distance.*
>   — Albert Einstein
>
> *I don't like quantum mechanics and I'm sorry I ever had anything to do with it.*
>   —Erwin Schrodinger
>
> *If quantum mechanics hasn't profoundly shocked you, you haven't understood it.*
>   —Niels Bohr
>
> *If you aren't confused about quantum mechanics, you haven't understood it.*
>   —Richard Feynman
>
> *If you think you don't understand quantum mechanics, just get married.*
>   —Sandip Dholakia, author

Despite all the challenges faced, researchers and engineers are working on making quantum computers a reality. As mentioned earlier, it's just a matter of time. Cryptographers were always aware of the fact that every math problem, no matter how difficult it is, will be solved one day in a reasonable time. Cryptographers have always relied on the slow progress of technology, and the strategy worked well until now, but quantum computing is catching up to cryptographers. Although we don't know what the future of computers and quantum computers will hold, we know for sure that the future of cryptography is strong. Cryptographers should work toward and achieve the following to make sure that cryptography protects data and information even in the post-quantum computing era:

- **Crypto-agility**

  Cryptographers should make sure that the crypto policies, standards, and overall approach have crypto-agility. Agility provides the flexibility to adapt to the new methodology and the ability to seamlessly transition to new algorithms. This is crucial, and we must work toward it now to make it happen.

- **Vigilance**

  Keep eyes and ears open to the new developments in the field and in related technologies. Technology is changing fast, and it's very important that we're vigilant about these changes and make sure to identify which way we want to move and which technology we want to explore and adopt.

- **Vision**

  Vision is the ability to look ahead. Set short-term and long-term goals about post-quantum cryptography. The short-term goals should be around conducting research, implementing agility, and planning, while the long-term goals could be selecting and implementing the post-quantum algorithm. These plans should also include various scenarios of how the change to post-quantum algorithms will impact organizations and how to plan for the impact.

- **Education**

  Things could go wrong even if we plan everything and have enough agility, but we don't implement it well. It's critical to educate developers to make sure they understand the new algorithm and implement the software properly. Most errors are introduced during the software implementation. Rigorous testing must also be part of the strategy.

- **Hybrid implementation**

  One way to go about implementing the post-quantum algorithm is to use a hybrid strategy. In other words, keep using the conventional or pre-quantum cryptography algorithm as it is and add a wrapper of the post-quantum cryptography algorithm. That way, if something goes wrong with the new post-quantum algorithm, the original pre-quantum or conventional algorithm can protect the data.

> **Google's Experiment of Hybrid Implementation**
>
> In 2016, Google announced the use of hybrid implementation—a combination of pre-quantum and post-quantum cryptography—in Google Chrome. Google used post-quantum cryptography algorithm, New Hope, with the existing encryption elliptic curve key exchange algorithm. This was to make sure if New Hope is cracked or if anything goes wrong with this new, quantum-resistant algorithm, data is protected with the existing algorithm. Google removed this hybrid implementation after a short period.

- **Other considerations**

  NIST advises against using the nonstandard post-cryptography algorithms. One reason for the advisory is the patent issues. If the cryptography algorithm we used is patented, then we would end up paying for the usage.

> **Research to Reality**
>
> Google, IBM, AWS, and some others around the world have started building quantum computers. Explore quantum computing with some of the major players in the field:
>
> - Google has a state-of-the-art QuantumAI lab in Santa Barbara, California: *https://quantumai.google*
> - Amazon Braket is a quantum service offered by AWS: *https://aws.amazon.com/braket/*
> - Azure Quantum is offered by Microsoft: *https://quantum.microsoft.com*
> - IBM offers a quantum utility: *www.ibm.com/quantum*

10.5 Summary

In this chapter, you learned that quantum mechanics defies common sense, and it's not hard for only us but also for great scientists like Einstein, who said it was "spooky." This chapter gave you a brief overview of quantum computing and post-quantum cryptography.

The chapter introduced quantum mechanics and quantum computing technology. We learned about the risks faced by cryptography because of quantum computing and how that puts our data and information at risk as well. An overview of quantum cryptography and various post-quantum cryptography algorithms gave a glimpse into how we'll be able to protect our data and information in the post-quantum computing era. After learning about various post-quantum cryptography algorithms, you should be able to dive deeper into the subject and select the most suitable algorithm for your organization.

This chapter also explained what NIST is doing regarding post-quantum cryptography and how you can prepare your organization for the migration to quantum-resistant cryptography. The chapter wrapped up by outlining the difficulties with the implementation of quantum computing and the future of quantum cryptography.

> **Key Takeaways from This Chapter**
> - Learned about quantum mechanics properties—duality, superposition, entanglement—that are important to quantum computing
> - Explored quantum cryptography and quantum-resistant cryptography algorithms
> - Walked through the process to migrate to post-quantum cryptography

If reading about post-quantum cryptography piques your interest, then wait until you read about another type of encryption. The next chapter opens the door to yet another new technology in cryptography—homomorphic encryption! You'll learn about what homomorphic encryption is, how it's used, and a lot more.

Chapter 11
Homomorphic Encryption

Encryption provides confidentiality as long as the data is encrypted. However, the data must be decrypted before it can be used or processed. This not only limits the usage of encryption but also impacts security because the decrypted data is in plain text and can be vulnerable. This is where homomorphic encryption comes to our rescue. In this chapter, we'll review the homomorphic encryption technique and some practical examples.

We have secure storage and secure communication. Yet data is still vulnerable because it lacks security while in process. There was a dire need for data security during computing, and homomorphic encryption is a direct result of that need. You learned in Chapter 7 that cloud computing didn't pick up in the beginning because cloud users weren't sure if their data would be safe in the cloud. Even now, users secure data with client-side encryption (CSE), but if the data needs to be processed in the cloud, the data has to be decrypted. This creates a twofold problem: (1) the data is vulnerable while decrypted, and (2) the key has to be transferred securely to the cloud.

Homomorphic encryption is a cryptography primitive that allows mathematical operations without decrypting the data. Homomorphic encryption solves these problems and extends cryptography's confidentiality properties from at-rest and in-transit to in-process! This enables us to use homomorphic encryption on a wide range of applications such as cloud computing, private information retrieval, two-party computation, zero-knowledge proof (ZKP), and so on.

The use of homomorphic encryption isn't as common as you might expect. This is because security is a balancing act between speed and performance—and, unfortunately, they rarely move in the same direction.

> **Chapter Highlights**
> - Review of homomorphic encryption: concept, history, and types
> - Learn about practical applications and implementation libraries of homomorphic encryption
> - Understand the major roadblocks and future of homomorphic encryption

This chapter is divided into three sections, which are almost independent of each other, and you can choose to read in any order you prefer. Section 11.1 starts with a primer on homomorphic encryption. We assumed that you have very little or no familiarity with homomorphic encryption. You'll learn about the history, concept, evolution and types of homomorphic encryption. You can skip this section if you're familiar with the fundamental concepts of homomorphic encryption. To understand how homomorphic cryptography primitives work, you must understand how symmetric and asymmetric cryptography algorithms work (we recommend you review Chapter 2 and Chapter 3 as a refresher). Section 11.2 will discuss the possible practical applications of homomorphic encryption. Finally, Section 11.3 will discuss the advantages, challenges, and future of homomorphic encryption.

11.1 Primer on Homomorphic Encryption

Homomorphic encryption is a cryptography primitive that performs mathematical operations on the cipher text of the data and encrypts the results. The decryption of the result produces the same answer as the mathematical operation performed on the plain text data. The concept of homomorphic encryption can be viewed as an extension of asymmetric public key cryptography.

In the following sections, we'll review the history behind homomorphic encryption, how it works, and the different types.

11.1.1 History of Homomorphic Encryption

The concept of computing over encrypted data isn't new. Ron Rivest proposed computing encrypted data in 1978 in his paper "On Data Banks and Privacy Homomorphisms" [Rivest, 1978]. However, it took almost three decades to figure out how to implement fully homomorphic encryption. In 2009, Craig Gentry proposed an implementation of fully homomorphic encryption [Gentry, 2009]. The homomorphic encryption schemes developed during this 30-year period are known as pre-fully homomorphic encryption (Pre-FHE). During this time, many schemes were proposed, and they were all considered partial homomorphic encryption. The development of FHE since 2009 can be divided into various phases or generations.

Gentry's proposal is considered the first generation of the FHE algorithm. These first-generation schemes or algorithms essentially allowed both operations—addition and multiplication—over the cipher text without any limits. The addition is performed using an exclusive OR operation, that is, mod 2 addition, and the multiplication is performed using a logical AND, that is, mod 2 multiplication.

However, there was a catch. The addition of the cipher text is fast but the multiplication operation over the cipher text is very slow. This was happening because some randomization or noise is added to the plain text during the encryption process. The reason for adding the noise is to make the resulting cipher text difficult for the cryptanalyst to crack. Without the noise, the encrypted cipher text can be cracked. The noise increased the security, but the security came at a cost. When a homomorphic scheme performs an operation on the cipher text, the operation is, of course, also performed on the noise. During the multiplication operation, the noise was also multiplied. The noise must be kept below the threshold level. If the noise exceeds the threshold, the homomorphic primitive loses its correctness.

We've encountered a problem! We cannot afford to lose the primitive's correctness. If that happens, the whole operation becomes meaningless. This is where Gentry came up with the brilliant idea of *bootstrapping*. The process of bootstrapping uses the homomorphic decryption of the noisy cipher text. Specifically, it uses the encrypted secret key to homomorphically decrypt the cipher text. In other words, the encrypted secret key is used to decrypt the noisy cipher text, but this happens within the homomorphic circuit. Once decrypted, the evaluation key or the public key is used to re-encrypt the cipher text. This is also called *refreshed* cipher text. The refreshed cipher text has the noise level reset to a minimum level. The process works great; the function can perform unlimited multiplications. However, the process is slow.

The second generation of homomorphic schemes was an extension of Gentry's proposed construction. They are based on learning with errors (LWE) problems. The homomorphic schemes of this generation used the *leveled approach*. This means, in simple English, these primitives can perform any operation until they reach a certain level. At this point, bootstrapping can be applied to move to the next level or to continue to perform more operations. The work of the second-generation homomorphic schemes happened between 2010 and 2012. The second generation of fully homomorphic encryption was led by Brakerski, Gentry, and Vaikuntanathan, commonly known as the BGV scheme [Brakerski, 2011].

The third generation of fully homomorphic encryption originally proposed by Gentry, Sahai, and Waters [Gentry, 2013], known as GSW, wasn't much of an improvement over the second generation. However, a new version of GSW, called *Fastest Homomorphic Encryption in the West (FHEW)* [Ducas, 2015] was introduced, and it was a significant improvement over the second generation. It reduced the bootstrapping time for one Boolean gate to nearly 1 second—the bootstrapping time for Gentry's original construction was close to 30 minutes for a single multiplication operation [Paillier, 2020]! The implementation is realized by the FHEW, an open-source library by Leo Ducas and Daniele Micciancio on GitHub. Because FHEW uses the Fastest Fourier Transform in the West (FFTW) library, it was named Fastest Homomorphic Encryption in the West (FHEW). The FHEW uses a method that is equivalent to the method of Faster Fully Homomorphic Encryption (FFHE), proposed by Chilloti and others [Chillotti, 2016].

Chilloti's paper, "Faster Fully Homomorphic Encryption: Bootstrapping in Less Than 0.1 Seconds," is actually an implementation of Fully Homomorphic Encryption over the Torus (TFHE). (Torus is an R/Z, a set of real numbers modulo 1.) This is considered the fourth generation of fully homomorphic encryption. TFHE affords to bootstrap every operation because the bootstrapping takes only 0.1 seconds. TFHE offers a security level of 128 bits, and the security is based on LWE and Ring Learning with Errors (RLWE).

In 2017, Cheon and others proposed in "Homomorphic Encryption for the Arithmetic of Approximate Numbers" [Cheon, 2017] a homomorphic encryption scheme commonly known as CKKS, based on the initials of contributors. The CKKS scheme proposed a method that rescales the encrypted messages after the multiplication operation. This process is very similar to the bootstrapping in the previous schemes.

11.1.2 Understanding Homomorphic Encryption

So now that you understand the concept, you need to know how it works and how to implement it. To understand the concept of homomorphic encryption, we'll start with some simple mathematics [Armknecht, 2015]. Mathematically speaking, we can define homomorphic encryption as follows:

$c = enc(m)$

where

m is the message in plain text and

c is the generated cipher text after encryption

enc is encryption

$c' = enc(f(c)) = enc(f(enc(m)))$

where

f is a mathematical operation to be performed on the cipher text c and

c' is the resultant cipher text

$f(m) = decr(c')$

This value $f(m)$ is the same if we had formed the operation $f(m)$ without encrypting. Figure 11.1 shows the conceptual process of homomorphic encryption.

Let's zoom into the process with a real-life scenario. The data owner generates a public key pair, a private key, and a public key. The public key is shared with everyone, including the cloud service provider (CSP). The owner encrypts the data using the public key and uploads the data to the cloud. The CSP can perform the computation on the encrypted data or the cipher text and encrypt the result again using the public key of the data owner. The computation was performed on the data in the cloud without the data owner sharing the secret key or decrypting the data. The data owner can download the computed data and decrypt it with the private key. The result would have been the same even if the data had been downloaded, decrypted, and then computed.

Figure 11.2 shows three different scenarios comparing the addition operation. Let's assume that P_1 is 3 and P_2 is 4. The first one is using homomorphic encryption. In this case, P_1 and P_2 are encrypted, and then the addition is performed on the encrypted values of P_1 and P_2. In this situation, the CSP has no idea what the values of P_1 and P_2 are.

When the result of the addition is decrypted, the answer is 7. In the second scenario, P_1 and P_2 are added in the plain text, and the answer is encrypted. When the result is decrypted, the answer is 7. This is how conventional encryption works today. The operation must be performed on the plain text. Lastly, if we don't use any encryption, the addition will still result in 7.

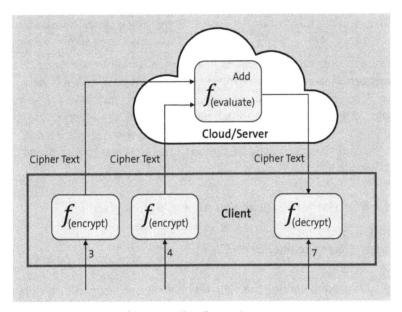

Figure 11.1 Homomorphic Encryption Concept

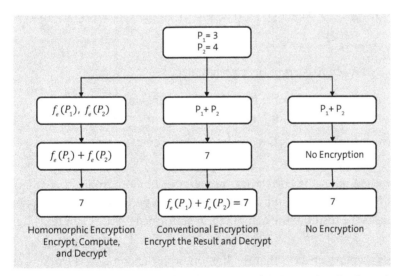

Figure 11.2 Comparing the Results of Homomorphic Encryption for Correctness

The public key used to encrypt the computation's output is known as the *evaluation key*. This key is the same as the public key of the data owner.

11 Homomorphic Encryption

> **Client-Side Encryption and Homomorphic Encryption**
>
> The process of encrypting the data and then uploading to the cloud, if you recall, is known as client-side encryption (CSE; see Chapter 7, Section 7.1.2). Without the use of homomorphic encryption, the data owner has to download and decrypt the data and then compute. Alternately, the data owner can share the private key to the CSP, but, in that case, the purpose of CSE is defeated. Homomorphic encryption solves these problems.
>
> For example, say you're a financial planner, and your client calls while on the road asking for the average performance of his investments over the past five years. Your data is in the cloud and encrypted. Without homomorphic encryption, you'll have to go to your office, download the data, and decrypt the data to compute the average. However, with homomorphic encryption, you can download the data on the phone and give the client the average numbers (assuming the app on the phone is able to decrypt and display the result in plain text).

Homomorphic encryption is formally defined as comprising four processes:

- **Key generation:** (S_k, P_k)
  Generating a private(secret)/public key pair, S_k, P_k.

- **Encrypt:** $(P_k, m) \rightarrow c$
  The public key, P_k, is used to encrypt the plain text message, m, generating a cipher text, c.

- **Evaluate:** $(EV_k, f(c)) \rightarrow c'$
  The evaluate key, EV_k, is used to encrypt the computed output $f(c)$ and generates the cipher text c'.

- **Decrypt:** (S_k, c')
  The secret key, S_k, is used to decrypt the cipher text, c'.

This is a very informal definition; a more formal definition is given by Shai Halevi in his tutorial on "Homomorphic Encryption" [Halevi, 2017].

Following are some of the key properties of homomorphic encryption:

- **Correctness**

  We already touched upon the correctness property of homomorphic encryption. Correctness is high when the result of the computation performed on the cipher text is the same as the result of the computation performed on the plain text after decryption.

  The correctness property ensures that the homomorphic encryption is working as intended. If the primitive can't achieve correctness, that means the encryption function isn't working according to the requirements. Although correctness doesn't guarantee the primitive's security, a lack of correctness could indicate potential vulnerability.

- **Robustness**
  Robustness is a security property of homomorphic encryption that indicates the primitive's ability to withstand adversarial attacks or abusive usage. This property confirms that the design of the homomorphic encryption algorithm is robust against various cryptographic attacks.

- **Strong homomorphism**
  A strong homomorphism property indicates a high degree of correctness without introducing noise. The lesser the noise, the stronger the homomorphism.

- **Circuit privacy**
  The circuit privacy property is another important security property of homomorphic encryption. The attacker shouldn't figure out anything about the circuit from the cipher text generated by the circuit. This is a key property because if the attacker figures out the computation performed from the cipher text, then they can eventually trace it back to the algorithm. This will create a huge security risk.

- **Compactness**
  Compactness is a key to success. The primitive must be able to perform the computation and generate cipher text in a reasonable time with a compact size of cipher text. The size and time must be within the control regardless of the complexity of the computation. The compactness property helps with the overall efficiency of the algorithm. This property is very important in practical applications.

- **Targeted malleability**
  The cryptography primitive is considered malleable if the attacker can generate cipher text from another cipher text without knowing or changing the original plain text. In cryptography, this property is considered a weakness or an undesirable property. In homomorphic encryption, somewhat counterintuitively, this is exactly what we want to achieve. However, malleability should be applicable only to the desired function and not to anything else. For example, if the homomorphic function is supposed to evaluate addition, then the primitive should be malleable to the addition function only. It should be able to derive another cipher text if a multiplication operation is performed. In other words, the malleability should be targeted to achieve specific functionality.

11.1.3 Types of Homomorphic Encryption

There are four types of homomorphic encryption [Armknecht, 2015] based on what the scheme can compute and how much the scheme can compute. Homomorphic encryption is designed to perform Boolean exclusive-OR addition or Boolean AND multiplication over the encrypted data. However, not all schemes can perform both operations. Additionally, there could be a limitation on how many operations can be performed.

Plus, not all primitives can perform both operations. Following is the summary of each type of homomorphic encryption function:

- **Partially homomorphic encryption**

  Partially homomorphic encryption schemes compute only one type of operation on the encrypted data—either addition or multiplication. The number of operations isn't limited.

  Following are some examples of partially homomorphic encryption:
  - RSA: Unlimited multiplications
  - Elgamal: Unlimited multiplications
  - Paillier: Unlimited additions

  All of these examples are part of the pre-FHE generation.

- **Somewhat homomorphic encryption**

  Somewhat homomorphic encryption allows both addition and multiplication operations but only in limited numbers. Multiplication usually limits the number of operations because the noise multiplies every time and grows exponentially, reaching the threshold faster. Once the threshold is reached, the decryption doesn't work as intended.

  Following are examples of somewhat homomorphic encryption schemes:
  - Van Dijk-Gentry-Halevi-Vaikuntanathan
  - Boneh-Goh-Nissim

- **Leveled fully homomorphic encryption**

  Leveled fully homomorphic encryption can compute the arbitrary number of operations of multiple types with a predefined depth or level. This is more like a limit on the number of multiplication operations that can be performed. The cipher text can't be decrypted once the level is reached.

  Following are examples of leveled homomorphic encryption schemes:
  - Brakerski-Gentry-Vaikuntanathan (BGV)
  - Brakerski-Fan-Vercauteren (BFV)

- **Fully homomorphic encryption**

  Fully homomorphic encryption allows an unlimited number of operations of addition and multiplication on encrypted data. A fully homomorphic state is achieved by using bootstrapping to control noise generated during the operations. These primitives are computationally intensive.

  Following are examples of fully homomorphic encryption schemes:
  - Gentry's scheme
  - Cheon-Kim-Kim-Song (CKKS)

Table 11.1 summarizes the types of homomorphic encryption.

Type of Homomorphic Encryption Scheme	Number of Operations	Type of Operations
Partially homomorphic encryption	One – addition or multiplication	Unlimited
Somewhat homomorphic encryption	Two – addition and multiplication	Limited
Leveled homomorphic encryption	Two – addition and multiplication	Predefined level or depth
Fully homomorphic encryption	Two – addition and multiplication	Unlimited

Table 11.1 Summary of Homomorphic Encryption Types

11.1.4 Homomorphic Encryption Using a Symmetric Key

Homomorphic encryption uses asymmetric cryptography because the cloud provider must have a key to perform the computation. The public key can be used as an evaluation key by the cloud providers to perform the evaluation or computation of the function over the homomorphic circuit. The public key is also used to encrypt the data before it's uploaded to the cloud. The problem is that asymmetric cryptography is relatively slow, and it would be impractical to encrypt large amounts of data with an asymmetric key. As we've seen in Chapter 6, the primary use of an asymmetric key pair is to exchange the symmetric key. An asymmetric key is not really used for the encryption of the data, but rather is used only to encrypt the symmetric key. The overall efficiency of homomorphic encryption improves if we encrypt and decrypt data at the client site using a symmetric key; however, this isn't easy to achieve.

Lots of research has been done in this area, and one interesting concept is *hybrid homomorphic encryption*, proposed by Dobrauning in 2021. Hybrid homomorphic encryption is very similar to envelope encryption, which we covered in Chapter 5. Symmetric encryption is used to encrypt the data and then the symmetric key is encrypted with the homomorphic public key and sent to the cloud along with encrypted data. The homomorphic public key or the evaluation key is then used to homomorphically decrypt the symmetric key and the data and then perform the calculation. The client can use the symmetric key again to decrypt the computed data on the client's end. The word *hybrid* is included in the name to indicate the use of both types of cryptography: symmetric and asymmetric.

Mathematically, we can summarize hybrid homomorphic encryption as follows:

1. **Key generation:** (S_k, P_k, K_s)
    First, generate a symmetric key, K_s. The symmetric key is used to encrypt and decrypt the data.

Second, generate a homomorphic key pair (private [secret]/public key pair, S_k, P_k), as follows:

$P_k = E_k$

Here, E_k is the evaluation key. The public key can be used as an evaluation key and to encrypt the symmetric key.

2. **Encrypt:** $Enc(K_s, m) \to c$ and $Enc(P_k, K_s) \to c_k$

   Encryption is a two-step process. First, the encryption function uses the symmetric key, K_s, to encrypt plain text data m and then uses the public key, P_k, to encrypt the symmetric key, K_s.

3. **Decompress and evaluate:** $Dcomp(E_k, Dec(S_k, c, c_k)) \to c'$ and $Eval(E_k, f, c') \to c'$

   In this step, the encrypted symmetric key and data are first decompressed using the evaluation key. The evaluation or computation function f is then run on the cipher text c' to generate the computed and encrypted cipher text c'.

   The cloud service provider performs the decompression and evaluation steps without actually being able to read the data in plain text.

4. **Decrypt:** $Dec(S_k, c')$

   Finally, the data owner or the client can decrypt the data using the symmetric key, K_s.

This is a very high-level overview of the proposed concept. For further reading, refer to Bakas, 2022.

11.2 Practical Applications of Homomorphic Encryption

Homomorphic encryption is often referred to as "the Swiss Army knife of cryptography" or "the Holy Grail of cryptography" because it has many advantages and can be used in a wide range of applications. Although homomorphic encryption is extremely promising in solving our two main concerns—security and privacy—it's still not very widely used. The technology is still evolving, but it has a lot of potential. In this section, we'll review some of the probable applications of homomorphic encryption based on what is available and feasible today. Note that, in this section, we won't discuss why we need homomorphic encryption because the reasons are the same as we've discussed previously.

11.2.1 Healthcare

One of the major data privacy concerns is in healthcare. It rarely happens that people and the government are on the same side when creating a law, but that is the case when developing privacy laws around healthcare data. People and the government have a valid reason to be nervous. People want to keep their health data private not only for social reasons but also for economic reasons. People don't want insurance companies to keep checking on this data and raise the premiums nor do they want CSPs to share the data with third-party companies. The government has different concerns. They don't want hackers to use this data for biometric information or blackmailing people.

Regardless of the reason, it's clear that this data must be protected. Right now, the Health Insurance Portability and Accountability Act (HIPAA) and other health laws around the world require that healthcare data be encrypted. However, the data isn't protected if the CSP or other support people have access to the data.

Homomorphic encryption can help in this situation. We'll discuss a few examples in the following sections.

Healthcare Application

Figure 11.3 shows a scenario where homomorphic encryption is very helpful. Let's say a patient goes to the doctor for a checkup. The doctor uses cloud-based storage and a software as a service (SaaS) application to analyze results. The doctor orders additional blood tests and other tests from labs and imaging centers. The doctor, labs, and imaging centers require uploading the patient's data securely to the cloud for storage and analysis.

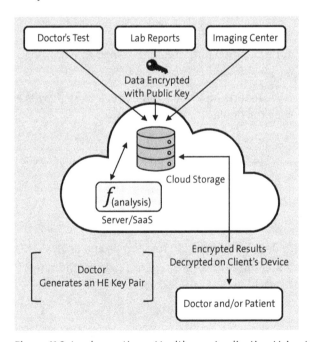

Figure 11.3 Implementing a Healthcare Application Using Homomorphic Encryption

The local system—doctor's office or hospital—generates a public cryptography key pair. The public key is shared with everyone—doctor's staff, labs, and imaging centers. Only the doctor and patient have access to the private or secret key. The doctor's staff, labs, and imaging centers encrypt the data using the public key and upload it to the cloud storage. Once all data is received in the cloud storage, the SaaS application on the cloud analyzes data. The analyzed data is then sent back to the patient portal with recommendations and/or predictions. The doctor and patient have a secret key to view the data on the portal.

In this example, homomorphic encryption plays an important role because it can retrieve the encrypted data from the storage and analyze it to generate recommendations/predictions without decrypting it.

Let's walk through the steps to implement the healthcare application using homomorphic encryption:

1. The doctor's office or the hospital generates a key pair:

    $KeyGen\_HE \rightarrow (K_s, K_p)$

2. The doctor/hospital publishes the public key, K_p, and shares it with all stakeholders, including labs, imaging centers, CSPs, and so on.

3. Each stakeholder uses the public key, K_p, and encrypts the data before uploading it to the cloud. The resulting cipher text C_n is uploaded to the cloud:

    - $C_1 = EncK_p (bloodwork)$
    - $C_2 = EncK_p (LabReport)$
    - $C_3 = EncK_p (xRay)$

4. The CSP uses the public key, K_p, for evaluation purposes. This makes $K_p = K_{eval}$.

5. The doctor's office and the patient keep the private key, K_s.

6. The CSP performs the analysis on the reports (uploaded in the previous step) by executing the function *f(analysis)* using the public key K_{eval}, as follows:

    $C_{analysis} = EncK_{eval}(f(analysis), C_1, C_2, C_3)$

7. The doctor and/or patient can decrypt the analyzed reports by using their private key, K_s, as:

    Doctor/Patient: Health Report = $DecK_s (C_{analysis})$

    Enc and *Dec* are the encryption and decryption functions, respectively.

In this scenario, the patient's health data is with the CSP for storage and analysis, but the CSP or any of the support personnel can't access the patient's data. Homomorphic encryption delivers security and privacy to very sensitive data.

Genome-Wide Association Study

Another popular and effective use case of homomorphic encryption is with genome-wide association studies (GWAS) [Blatt, 2020]. A genome study or DNA study helps researchers identify characteristics of a disease based on the DNA. This study is usually conducted for several people, and then the DNA pattern of each person is studied. However, this genome data must be kept very confidential. Homomorphic encryption can help us achieve the privacy of participants. The architecture of the GWAS is shown in Figure 11.4.

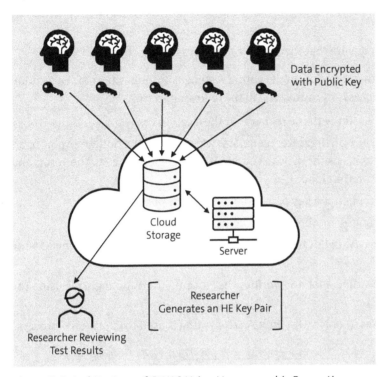

Figure 11.4 Architecture of GWAS Using Homomorphic Encryption

The test administrator or researchers generate the key pair. The administrator publishes the public key to the CSP or the application server. The researchers also use the public key to encrypt and upload the encrypted test results of all participants to the cloud storage. The analysis application in the cloud accesses the storage to analyze the results. The application server uses homomorphic encryption and the public key as the evaluation key to analyze/evaluate the results. The encrypted analyzed results are stored on the database.

The researchers (or the test administrators) or participants have the secret key to the data, and they can decrypt and access the results. The architecture completely protects the privacy of every participant.

11.2.2 Electronic Voting System

Since the COVID-19 pandemic, many countries and government offices are exploring the possibility of implementing remote or online voting systems. Implementing such a system is tricky because it needs to interact with several security domains to execute the voting task successfully:

11 Homomorphic Encryption

- First, the voter must register and authenticate. How does the system know that the ballot came from a particular voter?
- What if a voter denies after voting that he voted? How can the system prove if the voter has voted?
- How does a voter ensure that the ballot comes from the registrar?

Figure 11.5 shows the general flow of the voting system using homomorphic encryption [Yuan, 2023]. The process is summarized in the following steps:

1. All voters must register with the registrar or the election center.
2. The registrar (or election organizer) generates a key pair for the homomorphic operation. The registrar keeps the private key and issues the public key to the voters and the voting server on the cloud.
3. Every voter also generates a key pair and publishes the public key.
4. The registrar generates another key for digital signature.
5. The voter receives the ballot from the registrar. The ballot is digitally signed by the registrar.
6. The voter receives the ballot and verifies that it came from the registrar using the registrar's public key.
7. The voter votes Yes (1) or No (0). (Only Yes/No voting is used to explain the process.)

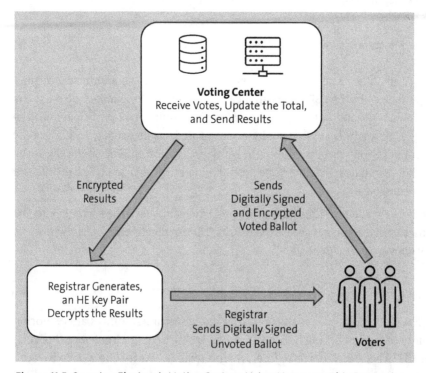

Figure 11.5 Securing Electronic Voting System Using Homomorphic Encryption

11.2 Practical Applications of Homomorphic Encryption

8. The voter encrypts the ballot with the homomorphic public key received from the registrar.
9. The voter also digitally signs the ballot and sends it to the voting center.
10. The voting center receives the ballot and verifies if it came from the registered voter. If verified, then the voting center homomorphically adds the vote to the count.
11. The voting center updates the database with the new count and stores it. The final tally is encrypted with the evaluation key.
12. When voting is over, the registrar decrypts the results (final tally) with the private key and announces the total count.

Let's walk through the simplified mathematical steps to implement the electronic voting system using homomorphic encryption:

1. The registrar creates a key pair that is used to digitally sign the ballot:

   $KeyGen\_Registrar \rightarrow (K_{sr}, K_{pr})$

2. The registrar publishes the public key, K_{pr}, to all voters so that they can validate that the ballot came from the right source.
3. The voter's key pair is used to digitally sign the ballot:

   $KeyGen\_Voter \rightarrow (K_{sv}, K_{pv})$

4. The registrar creates a key pair to encrypt the ballot for homomorphic encryption:

   $KeyGen\_HE \rightarrow (K_{she}, K_{phe})$

5. The homomorphic evaluation key is as follows:

   $K_{phe} = K_{eval}$

6. Each voter uses their public key, K_{phe}, to encrypt the ballot, and generates the respective cipher text C, as follows:
   - $C_1 = EncK_{phe}(Voter1)$
   - $C_2 = EncK_{phe}(Voter2)$
   - $C_3 = EncK_{phe}(Voter3)$

7. Voters encrypt the ballot and send it to the voting center, as follows:

   Voting center: $C_{votes} = EncK_{eval}(f(eval), C_1, C_2, C_3, .. C_n) = C_{votes}$

8. The voting center uses this key, K_{eval}, to encrypt the vote tally.
9. The voting center evaluates the function homomorphically, that is, performs the addition on the encrypted data.
10. At the end, the registrar decrypts the result:

    $Results = DecK_{she}(C_{votes})$

11. The registrar decrypts the data and announces the final results.

This is a simplified process for the purpose of explaining the mechanism.

11.2.3 Artificial Intelligence and Other Use Cases

In this section, you'll learn the application of homomorphic encryption with artificial intelligence (AI), private information retrieval (PIR), and outsourced computing and storage.

Artificial Intelligence

In Chapter 9, we've already briefly touched on the use of homomorphic encryption with AI. In this section, we'll discuss how exactly homomorphic encryption helps AI.

AI models work on training data. The more data we have to train the model, the more mature it will be. As we know, the concept of AI has existed for many decades, but its use took off only in the past 10 or 15 years. This is because we didn't have enough data to train the models. We not only need the training data, but we also need useful data that produces the result that is the intended output of the model.

Following are some of the major concerns with the training data:

- Training data must have proper distribution. Biased data can produce misleading results.
- Data must be sanitized and free of malware and other viruses to train the model properly.
- When AI models are trained and used on the cloud infrastructure, it raises some concerns about the training data:
    - How secure is the training data in the cloud?
    - Who has access to training data in the cloud?
    - How can we find out if the data in the cloud has been manipulated?
- Tampering and biasing the training data are the two major concerns in the cloud.

Homomorphic encryption solves this problem. The idea is simple: encrypt the training data before uploading it to the cloud. The training happens over the encrypted data. Because the training data is encrypted, adversaries can't bias, manipulate, or tamper with it.

Private Information Retrieval and Database Search

Private information retrieval (PIR) is another important use case of homomorphic encryption [Halevi, 2017]. This is very helpful when we want to search a large database and retrieve a record without the server's knowledge.

The user can encrypt the index of the record that he wants to retrieve. The server evaluates the query using the encrypted index and returns the encrypted results or record to the user. The user then decrypts the record and uses it in the plain text. The same logic can be extended to the SQL query to search larger databases.

We can use the same query concept with our healthcare examples in Section 11.2.1. If a health provider wants to retrieve a health record of a patient without the knowledge of the CSP or the server, the encrypted index concept can be used.

Another private search application is a password search. Many major CSPs compare user passwords against a large database of compromised passwords. When a user enters a user ID and a password, the information is hashed and encrypted. This information is then homomorphically encrypted. The large database of the compromised password is searched and compared. If the user password is found in the compromised password database, then the user is informed. This technique is actually used by Microsoft and Google right now.

Outsourced Storage and Computing

Throughout this section, we've discussed various scenarios of cloud computing and storage—so there isn't much left to discuss. Just to recap the concept, cloud customers want to protect their data in the cloud. The data is encrypted in transit and at rest. The only time data is exposed to the CSPs is while in process. While in process, support staff and others can access the data in plain text. This is critical, especially if the data contains any sensitive information. Homomorphic encryption is the way out of this situation for cloud customers by protecting the data in use.

Earlier in this chapter, we also discussed the disadvantages of CSE and how homomorphic encryption can help.

11.3 Advantages, Challenges, and the Future

Homomorphic encryption is a relatively new field, but one that already shows great potential for success. It's useful in many applications and offers many advantages, but these advantages come with some challenges too. In this section, we'll review the advantages, challenges, and future of homomorphic encryption.

11.3.1 Advantages

Following are the key advantages of homomorphic encryption:

- **Privacy**
  The first and foremost benefit of homomorphic encryption is its ability to preserve the privacy of sensitive data and information. It allows users to query databases and perform computations on the encrypted data. The data is never in plain text and is never shared with anyone except the rightful owner—the person with a decryption key.
- **Collaboration**
  Homomorphic encryption can use encrypted input from two different parties to

compute or evaluate the encryption data. The function is evaluated, and the encrypted result is shared. The party with the decryption key will be able to get the result.

- **Compliance**
  Homomorphic encryption meets the compliance and regulatory requirements of many industries, governments, and local laws.

- **Audit**
  This is a big win. The auditor can evaluate or perform the testing without accessing the data. In other words, the organization need not share any sensitive information or expose the data, and the auditor can verify the functionality.

- **Data monetization**
  The organization can sell the information about the data without revealing any information or data. For example, the organization can sell how many people are looking to buy a house in a particular zip code or what kind of restaurant is popular on Mother's Day without even looking at the actual data. No personal or sensitive information is shared. In fact, no information of the data owner is ever shared or exposed, but the organization can still monetize.

- **Outsourcing**
  We've discussed this several times, but we must mention it one last time because this is one of the major advantages of homomorphic encryption in the cloud world: homomorphic encryption ensures data privacy even when the computation of the data is performed by a third-party or CSP. All computations by third parties are performed on the encrypted data, and this guarantees the privacy of the data.

- **Security**
  All defensive controls, including all types of encryptions, provide data security. So, why do we have to mention this? Yes, this is true, but here it goes a step further. Because data is never in plain text or unencrypted when it's in use, there is no possibility of someone injecting viruses, manipulating data, or even stealing data. In other encryption methods, the attackers get a small window of opportunity, but that window is shut in homomorphic encryption.

11.3.2 Challenges

Following are some of the disadvantages or challenges of homomorphic encryption:

- **Speed and performance**
  Homomorphic encryption is slow. Computing a function on plain text data is much faster than computing on encrypted data. Additionally, the function's output or the evaluated result is encrypted again, adding an extra layer of processing. Gentry's initial proposal of bootstrapping took almost 30 minutes to compute one simple logical operation! Most applications of homomorphic encryption that we discussed use

asymmetric cryptography. Public key cryptography is slower compared to symmetric cryptography to start with, and, on top of that, we need an extra layer of encryption. This makes homomorphic encryption much slower.

- **Resource intensive**

  Homomorphic encryption consumes lots of resources in computing over the encrypted data. The extra consumption of memory and CPU could impact other operations on the server.

- **Expensive**

  It's expensive to maintain the high-end server with enough resources to support the homomorphic encryption. If it's used in the cloud, it has an additional cost for key management and storage.

- **Less explored**

  Homomorphic encryption is relatively newer, and lots of research is still happening. Researchers and academia are developing new primitives and use cases. It's hard to find people who are experienced with these primitives. Because the road is less traveled, users have a little more work to do than using something like Advanced Encryption Standard (AES) or RSA.

- **Implementation and design**

  Designing and implementing an algorithm is complex and can be difficult if you don't have experience. Complex mathematics and the use of newer libraries make it difficult.

- **Efficiency**

  Homomorphic encryption could be a bit challenging to implement efficiently if it's used to compute or evaluate on large, encrypted databases.

- **Security**

  Yes, this is also a disadvantage! The security of homomorphic primitives is as strong as the weakest link if the targeted malleability is designed poorly. Malleability is the ability of cryptography algorithms to create a cipher text from another cipher text without revealing the plain text. This is what we want homomorphic encryption to do. However, we want it to work on a specific, narrow functionality. If this is too broad or poorly designed, the security can be compromised.

- **Key management**

  The key management in homomorphic encryption is dependent on the use case. For example, in the example of electronic voting in Section 11.2.2, the registrar generates the key pair and is the owner. He also decrypts the results. In this situation, there is no critical key management needed as the secure key stays with one person/entity. On the flip side, in the example of healthcare records, the decryption key must be shared with the doctor and patient. In this situation, careful planning of the secret key or some level of key management is required.

  Additionally, after the key rotation, it's important to ensure that all the involved third parties who have evaluation/public keys update their keys.

11.3.3 Future of Homomorphic Encryption

The future of homomorphic encryption is very bright. The industry was very slow to warm up to implementing homomorphic encryption, primarily due to its performance, complexity, and cost, as we just discussed. There was very little development or progress between 1978 and 2008. However, things have changed since 2009.

A lot of research has been done in the past few years, and it has helped tremendously in improving the efficiency and performance of homomorphic encryption. Academia and industry are working jointly to standardize the implementation and usage of homomorphic encryption. The growth of open-source libraries is enough to see and prove the progress.

For use in your development of homomorphic encryption, Table 11.2 presents a list of some popular open-source libraries that can be used to implement applications using FHE.

Library	Description
cuFHE	CUDA-accelerated FHE library
FHEW	FHE library based on FHEW
FINAL	C++ FHE library based on the NTRU and LWE scheme
HEAAN	Scheme with native support for fixed-point approximate arithmetic
HEAANPython	Python binding for the HEANN library
HElib	BGV scheme with bootstrapping and the Approximate Number CKKS scheme
lattigo	Go library for lattice-based crypto that implements various schemes
libScarab	C library implementing an FHE scheme using large integers
libshe	Symmetric somewhat homomorphic encryption library based on the DGHV scheme
Microsoft SEAL	C++ FHE library implementing BFV and CKKS schemes
NFLlib	NTT-based fast lattice library specialized on the power of two polynomials
nodeseal	JavaScript/WebAssembly port of Microsoft SEAL
NuFHE	GPU-accelerated homomorphic encryption library, faster than cuFHE, that implements the TFHE algorithms

Table 11.2 FHE Libraries (Source: https://github.com/jonaschn/awesome-he)

Library	Description
OpenFHE	FHE library with all features from PALISADE, merged with selected capabilities of HElib and HEAAN (all major FHE schemes)
PALISADE	Lattice encryption library (superseded by OpenFHE)
Pyfhel	A Python wrapper for SEAL, HElib, and PALISADE
python-paillier	Partially homomorphic encryption based on the Paillier scheme
SEAL-python	Python binding for the Microsoft SEAL library
TenSEAL	Library for homomorphic encryption operations on tensors, built on Microsoft SEAL, with a Python API
TFHE	Faster fully homomorphic encryption
TFHE-rs	Rust implementation of the TFHE scheme for Boolean and integers FHE arithmetic by Zama

Table 11.2 FHE Libraries (Source: https://github.com/jonaschn/awesome-he) (Cont.)

The use of homomorphic encryption is truly a win-win situation as it's helping both AI and cryptography. The application of homomorphic encryption techniques in securing training data and models has taken the security of AI to the next level.

The usage and applications of homomorphic encryption are growing, and homomorphic encryption will likely emerge as the primary form of data protection mechanism in the cloud.

11.4 Summary

Homomorphic encryption has gained considerable momentum over the past 10 years. The need to secure data from third-party vendors, CSPs, and auditors while allowing them to process it simultaneously is rapidly increasing. Homomorphic encryption is one of the main solutions in sight right now. The increased use of cloud computing will fuel the need for homomorphic encryption and its development.

In this chapter, you learned the fundamentals and basic workings of homomorphic encryption and studied its four generations of development. We also explored some applications of homomorphic encryption that can be used in a real-world situation. We took a detailed dive into the healthcare and electronic voting applications. Lastly, we looked at the benefits of using homomorphic encryption as well as some of the roadblocks that explain why the progress of homomorphic encryption is slow. We concluded with a brief commentary on the future of homomorphic encryption.

11 Homomorphic Encryption

> **Key Takeaways from This Chapter**
> - Learned what homomorphic encryption is and how it can be used
> - Understood some practical applications of homomorphic encryption
> - Reviewed a list of open-source FHE libraries

In the next chapter, we'll survey various attack techniques on the cryptography algorithms. We'll also dissect the anatomy of ransomware.

Chapter 12
Cryptography Attacks and Ransomware

Cryptanalysis is a very old technique for attacking cryptography algorithms. The study of cryptanalysis is as important as the study of cryptography itself. To prepare for defense against attacks, we must understand how attacks work and how attackers think. In this chapter, we'll learn about cryptanalysis, various cryptanalytic attacks on cryptographic algorithms, and ransomware.

Money, revenge, and love are the greatest motivators for most crimes, and cybercrimes are no exception. Cryptography's history is more than 2,000 years old. Throughout the history of cryptography, people have tried to crack it. The reason people invented and used cryptography in the first place—a method to hide the message so that only the intended recipient can understand—is very simple. There are people who want to eavesdrop on others' communication. If everyone minds their own business, we don't have to worry about hiding the plain text message with a cipher text. But many people want to poke their nose into someone else's communication—and business. This could be for various reasons. Up until the Second World War, it was primarily for espionage. Enemies were spying on Caeser's messages in an earlier example, and Allied forces were trying to break Enigma—the German electromechanical device to generate ciphers—during World War II. And now, it's about hacking into your bank accounts and stealing anything from credit card numbers to Social Security numbers. The use of online transactions has exploded in the past 30 years, and hackers are taking advantage of it when people aren't careful.

Security architects recommend defense-in-depth or layered security approaches and not putting two eggs in one basket. It's necessary to use multiple controls to protect the data. The last control or the bottom layer comprises cryptography controls or data encryption. This way, if everything else fails and the hacks do end up getting their hands on the data, the data won't be of any use to them. Encrypted data is no good to hackers.

Hackers also realize what they are up against and that's why they do everything they can to attack cryptography controls. In this chapter, you'll learn about various cryptography attacks.

12 Cryptography Attacks and Ransomware

> **Chapter Highlights**
> - Learn about cryptanalysis and various cryptanalytic attacks
> - Survey other cryptographic attacks
> - Understand ransomware and ransomware as a service (RaaS)

This chapter is divided into two sections, which are almost independent of each other, and you can choose to read in any order you prefer. Section 12.1 starts with cryptanalysis and cryptography attacks. To understand these attacks, you must understand the basic cryptography concepts. We assume that you are familiar with these concepts. If not, we recommend that you review chapters in part one of this book. Section 12.2 covers ransomware and ransomware as a service (RaaS). Interestingly, ransomware isn't a cryptography attack, but it uses cryptography to execute the attack—one more application of cryptography (unfortunately)!

12.1 Primer on Cryptography Attacks

The study of cryptography attacks is done for two different reasons and by two different groups of people. One group is, obviously, hackers. Their purpose is very clear: to unethically gain from the information. The second group of people who study cryptography attacks are security architects. Security architects study various attacks and attack vectors to understand how they can make algorithms more robust. Both groups use similar techniques to crack the code, but for different end goals—hackers want to break it, while security architects want to test it.

Cryptography attacks can be divided into three categories: cryptanalytic attacks, man-in-the-middle attacks, and other attacks (those that don't fit the first two categories). We'll cover each of these categories next, and then review and compare them to round out the section.

12.1.1 Cryptanalytic Attacks

Cryptanalysis is a study of cryptography algorithms and related factors for decrypting cipher text without using the authorized encryption key. In other words, it's a technique for deciphering a message based on the knowledge of the primitive and the message text but without the knowledge of the encryption key. The person who performs the cryptanalysis is known as a cryptanalyst. *Cryptanalytic* attacks are the strategies to crack or weaken the cipher using cryptanalysis.

As you learned in Chapter 1, the first formal and documented cryptanalysis approach was presented by an Arab scholar, Al-Kindi, in the ninth century. He proposed frequency analysis, which finds a pattern in the cipher text and relates it to the possible and most frequent letters or words, assuming the cryptanalyst knows the language of communication. Although frequency analysis is a cryptanalysis technique, it's manual and was a threat at that time to the classical ciphers. Modern ciphers use heavy mathematical computations and high-end computing power, and such manual analysis stands no chance. For further explanation of the frequency analysis technique, refer to Chapter 1.

> **Deterministic Ciphers versus Probabilistic Ciphers**
>
> Let's clarify the concept of deterministic ciphers versus probabilistic ciphers:
>
> - **Deterministic cipher**
>   The deterministic cipher produces the same cipher text every time for a given encryption key and the plain text. In other words, the encryption algorithm is deterministic if the algorithm generates the same cipher text for the same plain text as long as the key remains the same.
> - **Probabilistic cipher**
>   The probabilistic cipher generates a different or random cipher every time for a given key and the plain text. In other words, even though the key and the plain text remain unchanged, the resulting cipher text changes.

In this section, we'll discuss the five most common cryptanalytic attacks: cipher text only attack (COA), known plain text attack (KPA), chosen plain text attack (CTA), and chosen cipher text attack (CCA).

All of these attacks use cryptanalysis techniques. Throughout the discussion of these attacks, it's assumed that the attacker, at the least, has access to the cipher text. The idea of encrypting the message was to be able to communicate on the public or unprotected network. Regardless of what algorithm or the key size used, the cipher text is transmitted on the public network.

Cipher Text Only Attack

In the cipher text only attack (COA) [Biryukov, 1998], the attacker has access to the following information:

- Encryption algorithm
- Cipher text

COA is one of the most difficult attacks to crack the code for because, as we mentioned earlier, the cipher text is always available to everyone. In other words, the attacker has very little to work with other than the cipher text and the algorithm—both of these are publicly known.

Stream ciphers have experienced many COAs. One well-known example is the Wired Equivalent Privacy (WEP) protocol, which was the first Wi-Fi protocol. WEP has experienced many COAs [Fluhrer, 2001]. Fluhrer and others demonstrated a cryptanalytic, in particular, COA, attack on WEP.

WEP used an RC4 cipher for encryption. The RC4 is a stream cipher that must generate a unique keystream. The uniqueness or randomness of the keystream is based on the initialization vector (IV) or the nonce. The WEP implementation didn't allocate enough bits for the nonce, which means the nonce was repeated after a certain value [Ferguson, 2010]. This has a direct effect on the key. The keystream was repeated and resulted in the same cipher text. This helped the attacker identify the key or the plain text with relatively minor effort. RC4 was already cracked and proven vulnerable a while ago, but it's important to note that the issue explained here is primarily attributed to the incorrect implementation of the cipher.

Although a COA is difficult to crack, most attackers are likely to attempt it first simply because cipher text is always available to work with. This is a COA, so technically no plain text is available to the attacker, but depending on the system, the attacker could get help. For example, if it's an email, the attacker can assume there will be a header and could look for some standard words like "from," "to," "subject," and so on.

Following are some of the important measures to defend against COA:

- The use of a robust and well-studied algorithm helps against the COA. For example, the community has researched and studied Advanced Encryption Standard (AES), and the chances of attackers cracking it with a COA are much less compared to a newer or proprietary algorithm.
- The larger the key size, the less likely it is to be compromised. The larger key size also allows more randomness in the key itself.
- One easy way to protect against this attack is to use a virtual private network (VPN)/Internet Protocol Security (IPsec) tunnel. The cipher text passes through this tunnel, and the attacker or eavesdropper won't be able to intercept it. See Chapter 6, Section 6.2.2, for more information.

Known Plain Text Attack

The attacker has access to the following information in known plain text attacks (KPAs):

- Encryption algorithm
- Cipher text
- Some known plain texts with a corresponding cipher text

In a KPA, the attacker knows some words or phrases in the plain text and the corresponding cipher text. This doesn't mean that the attacker has access to plain text also to start with. The attacker may launch the attack with COA but with some analysis, the attacker may find out some plain text by intercepting the message, eavesdropping, or simply analyzing the pattern. For example, if the message came in the morning, then the chances are the salutation is "Good morning," and when the message responded in the afternoon, it may be "Good afternoon." The attacker can also go a step further and launch a social engineering attack to somehow lure the victim to reply to a spam email and get some words of the response.

Following are some of the important measures to defend against the KPA:

- The use of a robust and well-studied algorithm helps against the KPA. To successfully defend against the KPA, the encryption algorithm must be probabilistic and generate a different or random cipher text every time, even if the key and the plain text are the same.
- The larger the key size, the less likely it is to be compromised. The larger key size also allows more randomness in the key itself.
- Obviously, using the VPN tunnel can help as the attacker can't get the corresponding cipher text.

Chosen Plain Text Attack

As the name indicates, in the chosen plain text attack (CPA), the attacker chooses the plain text message to be encrypted and matches it with generated cipher text to derive further information.

The attacker has access to the following information in CPAs:

- Encryption algorithm
- Cipher text
- Plain text of attacker's choice

As we mentioned earlier, it's safe to assume that the attackers of the ChatGPT era have basic tools to deal with modern cryptography and can always access the cipher text.

What exactly is the difference between KPA and CPA? In a KPA, the attacker works to find out the plain text based on the analysis and iterations. In a CPA, on the other hand, the attacker decides what plain text should be encrypted.

Does it mean that the attacker has access to the system if the attacker can pick the message to be encrypted? Probably not. If that is the case, there won't be any need to launch the attack.

In most cases, the attacker tricks a user into encrypting the message for him. For example, the attacker can send an email to the user if he knows that the emails are encrypted. Once the email reaches the server, the attacker can hack into the server to retrieve the encrypted message. Once the attacker has the encrypted message, he can compare his original email (plain text) with the encrypted message. (For all practical purposes, the attacker doesn't even care if the user has opened his email, as long as the email is encrypted, and he can hack into the server.)

The CPA could become a KPA if the attacker starts evaluating the pairs of plain texts with the corresponding cipher text.

The important measures to defend against the CPA are the same as the measures for the COA and KPA:

- The use of a robust and well-studied algorithm helps against the KPA.
- Using a probabilistic cipher reduces the likelihood of CPA. (Deterministic ciphers are more vulnerable to them.)
- The larger the key size, the less likely it is to be compromised. The larger key size also allows more randomness in the key itself.
- Obviously, using the VPN tunnel can help as the attacker can't get the corresponding cipher text.

Common sense says that if the attacker has access to snoop around the mail server or trick users into obtaining the cipher text for the chosen plain text, then the system is already compromised. The attacker will likely figure out the encryption key sooner or later. Because the attacker already has so much information, the CPA is more fatal than the KPA.

Chosen Cipher Text Attack

The chosen cipher text attack (CCA) is carried out on the receiving or decryption ends. The attacker chooses the cipher text and compares it against the generated plain text to find a pattern.

The attacker has access to the following information in a CCA:

- Encryption algorithm
- Cipher text
- Chosen cipher text

To carry out the CCA, the attacker must have access to the user's computer or the system where the decryption happens. This makes the attack a bit more difficult and less practical. For this attack to be successful, the attacker must inject the chosen cipher text into the encrypted message. To recap from Chapter 11, malleability is the property of the cipher to generate one cipher from another without knowing the plain text. We realized the importance of malleability with homomorphic encryption, but it's a weakness with the CCA.

In this scenario, the attacker uses malleability to intercept the cipher text and generate the new, chosen cipher text. For example, person A sends $100 from account number 123 to person B in his account number 789. The attacker has seen the cipher text of the bank transaction and knows what the account number is. The attacker can change the account number 789 to his account number 456 and generate a new cipher text, which will be sent to the bank. The transaction will go through, but the bank will deposit $100 to account number 456 and not 789. This is a malleability. The new cipher text was generated from the original cipher text without having any information about the plain text. The bottom line is that if the cipher is malleable, it's vulnerable to the CCA. It's comparatively easier on the asymmetric key system because the attacker can get ahold of the public key to re-encrypt the message.

The CCA is completely opposite of the CPA, but they aren't related. If the cipher is vulnerable to one, that doesn't necessarily mean it's vulnerable to the other. Obviously, using the VPN tunnel is the surefire way to protect against this attack as the attacker can't get his hands on the cipher text to manipulate and inject his chosen cipher text.

Linear versus Differential Cryptanalysis

We won't do justice to this section if we don't discuss linear and differential cryptographic attacks [Heys, 2002]:

- **Linear cryptanalysis attack**
  The symmetric encryption algorithm works with a probability of 50% (refer to Chapter 2 for more details). The bits are added using mod 2 addition. This is nothing but an XOR operation in the logical implementation. In simple English, without using mathematics, the attacker's goal is to deviate the cipher from the 50% probability to one end (closer to 0) or the other end of the probability (closer to 1). To do this, the attacker must have some known plain text and corresponding cipher text. Once the linear relationship is detected, the primitive can be exploited further to crack the key. In summary, the goal is to force the cipher to behave linearly rather than randomly and try to find the key.

- **Differential cryptanalysis**
  Differential cryptanalysis [Biryukov, 1998] uses the changes in the pattern to detect the relationship between the bits. In this attack, the attacker can choose some pairs of plain texts and see the difference in the resulting cipher text. The attacker continues the exercise until the resulting change is somewhat predictable. Once the pattern is understood, the attacker can further exploit the cipher to crack the code.

Security Considerations

To improve the security of the primitive against most cryptanalytic attacks, consider including the following properties in the cipher design or system implementation:

- **Indistinguishable encryption**
  If one can't tell the corresponding plain text from the cipher text (or vice versa), then the cipher is indistinguishable. If the attacker chooses a pair of plain texts to encrypt but can't tell which resulting cipher text corresponds to which plain text (meaning can't distinguish between the cipher text based on the chosen plain text), then the

cipher is indistinguishable. This is also applicable in decryption. The attacker can't launch a successful CPA or CCA if there is no way to find the corresponding plain text/cipher text pair. This property also safeguards against linear and differential cryptanalytic attacks.

- **Authenticated encryption**
  We've already discussed this earlier and won't repeat it here, but to quickly refresh, authenticated encryption provides both confidentiality and integrity. This prevents the attacker from injecting or using the cipher's malleability. The integrity feature of authenticated encryption helps against the CCA and all other attacks where the attacker needs to modify the message to launch the attack.

- **Probabilistic cipher**
  We've already discussed this earlier and won't repeat it here, but to recap, if the cipher generates random cipher text every time for a given plain text and the key, then it's difficult for the attacker to find the pattern or analyze the cipher.

We've already discussed other considerations that can make the cipher design robust. We'll provide a quick recap here. First, having a large key size/space and random nonce (IV) is critical to the design. Using a vetted algorithm also helps compared to inventing your own. And, lastly, using a VPN tunneling protocol, where possible, can protect from most cryptanalytic attacks.

12.1.2 Man-in-the-Middle Attacks

Although all attacks aren't classified as man-in-the-middle attacks, most of the cryptography attacks are the result of direct or indirect intercepting or eavesdropping. It's almost impossible to launch any attack without absolutely zero information. That means most of the attacks can be classified under the man-in-the-middle category. However, in this section, we discuss attacks that require direct intervention of the attacker (man) to launch an attack. The man-in-the-middle attacks can take many forms, but we'll briefly review some of the popular or most commonly talked about attacks in this section:

- **Man-in-the-middle**
  In this type of attack, an active hacker or attacker tries to intercept the message to retrieve the key, cipher text, or any other useful information. The attacker can establish sessions with the sender and/or receiver and, depending on the motive, manipulate the information. This is the umbrella category of the attacks. We'll see how the attacker can manipulate the information and the damage he can cause.

- **Meet-in-the-middle attack**
  This attack is carried out between the double encryption or two iterations (here, iteration is one full computation of a block cipher, including all rounds) of a block cipher. The attacker uses the known plain text and then encrypts it with all possible keys. The equivalent cipher text is then decrypted using all possible keys. The keys

are considered found for both encryption iterations when the plain text and cipher text match is found. This was used with Double DES (or 2DES). The goal was to double the strength of the Data Encryption Standard (DES) by repeating the encryption process twice, but because of the meet-in-middle attack, it didn't really double the strength. This was when the concept of Triple DES (or 3DES) took place (see Chapter 2, Section 2.2.2).

- **Passive eavesdropping**

  This attack takes place when the man in the middle acts as an audience and passively watches the traffic but doesn't actively launch the attack. This isn't an active attack. In fact, at this stage, it's not really an attack. If the data is encrypted, then all the attacker can see is some cipher text.

  Passive eavesdropping doesn't create any problems, but it could be the start of them. Most attacks start with information gathering. Unfortunately, the downside of technological advancement is the availability of cheap (even free) and easy-to-use tools for information gathering.

- **Replay attack**

  This is an active man-in-the-middle attack. In this attack, the attacker forces the repetition of the earlier action. Here is how it works: The attacker intercepts the message and collects the information or records the message. Later, the attacker sends the same message again. For example, the attacker can intercept a bank transaction or an e-commerce transaction. The attack saves the transaction. Later on, the attacker pretends to be a legitimate sender and initiates the transaction again. The receiver would think of this as a new transaction and perform the action—such as paying money.

- **Downgrade attack**

  The downgrade attack, as the name suggests, downgrades the security level or protocol and forces the communication on a weaker protocol. The downgrade attacks are primarily used with Secure Sockets Layer (SSL)/Transport Layer Security (TLS) protocols. The security of SSL/TLS protocols kept increasing from SSL 3.0 to TLS 1.3. However, the older versions are still active to allow backward compatibility. The attackers take advantage of this fact. The malicious actor on the network takes control and forces communication with the older version of the protocols. The older versions were upgraded because they had known vulnerabilities. The attacker exploits these vulnerabilities.

  Some well-known downgrade attacks are as follows:

  – *POODLE attack*: The Padding Oracle on Downgraded Legacy Encryption (POODLE) attack is targeted against block ciphers, particularly cipher block chaining (CBC) mode. The block cipher uses padding to meet the exact block size. If the padding is incorrect, an error message is returned. This message is used to detect more information about the plain text.

- *FREAK attack*: In the Factoring RSA Export Keys (FREAK) attack, the attacker convinces the client to use the weaker protocol and then gains insight into the network traffic. In the 1990s, the US government limited the RSA key size to 512 bits for the export-grade SSL. Later, the rule was changed, and the RSA key size was increased for the export-grade SSL. However, if the older SSL is still in use by the client, an attacker could take advantage of it and force the use of the weaker key. This allows an attacker to intercept the RSA connection. You'll be surprised to know that 20 years later, in 2015, more than 35% of browsers still used the export-grade cipher suites [DigiCert, 2015]!
- *Logjam attack*: This attack affects the key exchange protocol. The attack downgrades the version of the protocol to be used and forces the use of a weaker protocol. The attack is named after the analogy of logs jamming in the river.
- *BEAST attack*: The Browser Exploit Against SSL TLS (BEAST) attack allows the decryption of client-server sessions. The TLS protocol uses a symmetric block cipher in CBC mode for symmetric encryption or session key. The TLS protocol generates the IV for the block cipher. The BEAST attack exploits a vulnerability in the way the IV is generated. Although the attack is difficult to launch in the wild, if successful, the attack allows an attacker to eavesdrop without even decrypting. The vulnerability was fixed in TLSv1.1.
- *SLOTH attack*: The Security Losses from Obsolete and Truncated Transcript Hashes (SLOTH) attack is on the hashing or integrity algorithms. It forces the downgrade of the weaker/older hashing algorithms.

- **Inject attacks and modify attacks**
  Inject attacks and modify attacks are two different forms of the man-in-the-middle attacks. In inject attacks, the attacker or the malicious actor on the network injects a new message or new information to the message. Modify attacks, on the other hand, modify the existing message without adding any new information.

Most of the man-in-the-middle attacks can be avoided by using the latest version of software/protocols and using the authenticated encryption with associated data (AEAD) schemes, which we introduced in Chapter 7. In authenticated encryption, the message is hashed, and the message authentication code (MAC) is generated. This MAC is appended to the encrypted message. The MAC can be generated from the plain text or the cipher text of the message. To provide security to the message, authenticated encryption provides authenticity and confidentiality using the MAC and encryption. To provide integrity, associated data (AD) is added. This is nothing more than a sequence number, a clock, or something simple that differentiates one message from another. If the attacker intercepts one message, then the receiver knows that one message is missing as the sequence number for the next message will be out of order. The AD can be encrypted but it's not necessary.

12.1.3 Other Cryptography Attacks

There are other cryptographic attacks that can't be categorized as cryptanalytic attacks or man-in-the-middle attacks. Let's walk through them:

- **Birthday attack**

  We've already discussed the birthday paradox in Chapter 4, Section 4.1.1, and won't repeat the full explanation here. There is about a 50% probability that out of 23 people in a room 2 people have the same birthday. The same concept is used with the hash function collision. The mathematics behind the birthday paradox is used in finding the hash collision for two different inputs. The birthday attack promises to find the two inputs that generate the same hash value faster than the brute-force attack.

  The attack manipulates the integrity of the content. For example, the sender generates the message's hash value and attaches it to the message. The attacker intercepts and finds the hash value. The attacker's goal is to find other input that has the same hash value and benefits him. Once that is done, the attacker can replace the original message with his message. The receiver will verify the hash value and trust the content.

  If the attacker can find the same hash value for the legitimate message and the forged message, then the receiver won't be able to tell if the message received is from the correct sender or not.

  Another use case of the birthday paradox is password cracking. All passwords are hashed, and the hash value is stored and compared every time a user tries to log in. If the attacker finds the same hash value for a password and for another word, he doesn't need to know the password. He just logs in using the collided hash value!

- **Brute-force attack**

  This is the simplest attack to understand. In this attack, the attacker tries all possible combinations—plain and simple. For example, if the attacker is trying to crack a six-digit password, he can try all the possible combinations—letters, symbols, numbers, and so on.

  If the attacker is using a set of words from a particular database, such as a dictionary, this attack is called a dictionary attack. Although, it's called a dictionary attack, for all practical purposes, it's nothing but a subset of the brute-force attack.

  The attacker can also try to scan from a database called one-upped-constructed passwords. These are just one digit or letter changes in the original dictionary word or even previously used password, such as "We!come."

- **Rainbow attack**

  The rainbow attack can also be considered an extension of the brute-force attack. In the rainbow attack, the attacker scans against the database of the password hash value. It takes a long time to scan huge databases, hash each entry, and then compare. Instead, to save time, attackers maintain a database of the hash values.

- **Side-channel attack**

    The side-channel attack uses external physical parameters to find the plain text or encryption key. These parameters include noise, vibration, electromagnetic interference, power consumption, and heat dissipation. The attacker measures these properties and deduces the bit pattern and, ultimately, the plain text message or the key. The famous TEMPEST (Transient ElectroMagnetic Pulse Emanation STandard) attack is an example of a side-channel attack. In TEMPEST, the attacker studies the electromagnetic radiation from the devices to determine the pattern and tries to deduce the key or the message.

- **Statistical attack**

    The statistical attack focuses on the statistical or mathematical errors in the cryptosystem, such as floating-point errors, linearity errors, the lack of randomness, and so on. The attacker exploits these errors to find the correlation between the plain text and cipher text or find the encryption key.

- **Implementation attack**

    Most of the time, attackers get their foot in the door because of some human error. The implementation attack is nothing but exploiting the implementation errors made by the designers and developers. The WEP and RC4 example that we discussed earlier is an implementation error, and the attacks on WEP were results of the implementation attacks.

12.1.4 Cryptography Attacks in a Nutshell

Table 12.1 summarizes the cryptanalytic attacks, including what information the attacker has and how to protect against such attacks. Figure 12.1 shows all the attacks discussed in this section.

Attack	What the Attacker Knows	Protective Measures
Cipher text only attack (COA)	- Encryption algorithm - Cipher text	- Indistinguishable encryption - Authenticated encryption - Probabilistic cipher - Other considerations
Known plain text attack (KPA)	- Encryption algorithm - Cipher text - Some known plain texts with a corresponding cipher text	
Chosen plain text attack (CPA)	- Encryption algorithm - Cipher text - Chosen plain text	

Table 12.1 Summary of Cryptanalytic Attacks

Attack	What the Attacker Knows	Protective Measures
Chosen cipher text attack (CCA)	■ Encryption algorithm ■ Cipher text ■ Chosen cipher text	
Linear and differential attacks	■ Encryption algorithm ■ Cipher text ■ Chosen plain text	

Table 12.1 Summary of Cryptanalytic Attacks (Cont.)

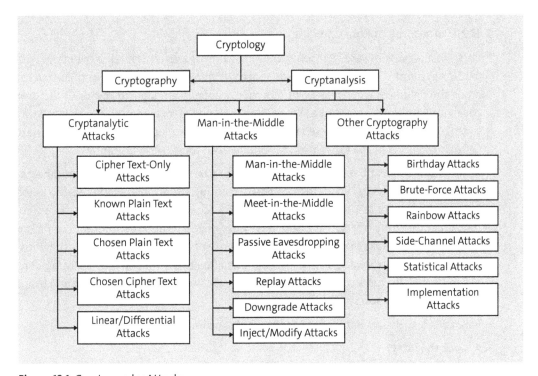

Figure 12.1 Cryptography Attacks

12.2 Ransomware Attacks

In Chapter 1, we compared the development of cryptography with the aviation industry. In this section, we're comparing cryptography with a gun. Yes, cryptography is very similar to a weapon. People invented weapons to protect themselves, but it turned out that these weapons could also harm us. Unfortunately, we haven't learned from experience. We invented cryptography to protect our data and information from hackers and other malicious actors. We designed an extremely robust and secure encryption algorithm so that no one but the rightful owner can access the data. But we forgot the fact that the same enemy can take our gun and fire at us. Now the attackers are using these robust algorithms to extort money from businesses, and—guess what—they use the same robust algorithm that we've designed.

In Chapter 4, we saw that cryptography can be used for confidentiality, integrity, authentication, and nonrepudiation. Sadly, we can add one more service to that lis: ransomware.

In this section, you'll learn about the anatomy of ransomware, which includes the steps to execute it, the types of ransomware, and the concept of ransomware as a service (RaaS).

12.2.1 Anatomy of Ransomware

First, what is ransomware? *Ransomware* is malware or a virus that takes over your data (and/or system). The data isn't usable or accessible by the legitimate user. The hacker holds the data hostage, courtesy of our robust encryption algorithms, until the user pays a ransom. Ransomware is a skilled attack and no doubt a smart (but unethical) way to make money. We're interested in discussing the end-to-end process of ransomware because the heart of the ransomware is our beloved topic—cryptography.

Although we've heard about ransomware in the past 10 years, the history of ransomware is almost 40 years old. In the late 1980s, computer files were encrypted and asked to send money to a P.O. Box in exchange for the decryption key. However, these were rare. It gained some traction after the heavy use of emails and the internet but picked up real steam because of cryptocurrencies. The use of cryptocurrency makes the transactions almost untraceable and that really added fuel to the fire. Conceptually, modern day ransomware works the same way as in the past, but the implementation and usage are very high-tech and sophisticated.

Following are the five parts of the anatomy of ransomware:

- **Lock the target**
  The first and very important part of the ransomware anatomy is locking the target or launching the attack. In this phase, the hacker tries to enter the system. Like most cyberattacks, the attacker needs to go through the information gathering and other reconnaissance stages in ransomware. The hacker can use one of the following attack vectors to launch the attack and lock the target:
  - *Phishing*: Phishing (or social engineering in general) is one of the easiest ways for the hacker to implement, and for the most part, it's easy for the user to click the link. This is typically done by sending a phishing email or inviting users to join a community discussion or comment section. Once the user clicks the link, the malware is executed. Social engineering or email phishing is one of the most popular vectors to launch a ransomware attack.

- *Downloads*: Everyone wants to download free software, books, music, and movies. However, when a user downloads from a site that isn't secure, they get infected with the virus. The hacker takes advantage of the insecure site and uploads or embeds malware in the files that people tend to download, such as popular music, books, and so on. As soon as the user downloads the file, the malware also makes its way to the user's system. And that's what the hacker is waiting for!
- *Remote login*: Remote login is easier for the hacker if they get their hands on the credentials—even better if the credentials belong to the administrator. The hackers siphon the credentials with one of the traditional methods: phishing or purchasing them from the dark web. For the Windows server, the hacker can scan port 3389 and see if there is any open port that can be exploited.
- *Vulnerabilities*: Another way for hackers to enter the target system is to exploit unpatched or open vulnerabilities. Hackers are always looking for zero-day vulnerabilities to exploit. Sometimes, vulnerabilities can give hackers root access to the server. It's like sailing with the full wind behind!

The hacker uses one of these vectors to gain access to the system and launch the attack.

- **Establish the control**

Once the hacker gains access, the first thing they check is the privileges associated with the access. If needed, the next step is to escalate the privilege and gain enough control to establish a strong foothold in the system. The goal is to have enough control to establish the key management and execute the encryption of the system/application data.

During this phase, the hacker also starts scanning and mapping the files, drives, and storage locations. Before the hacker starts encrypting files, they want to have a complete view of the system. You can think of this as a doctor taking X-rays or MRIs before performing surgery.

- **Executing encryption**

This is the heart and soul behind ransomware's success. Once the attacker has enough control and a defined map of the network and system, the encryption process begins. Like security professionals, hackers want to use the speed of symmetric encryption and the security of asymmetric encryption. As a result, hackers also use a combination of symmetric and asymmetric encryption schemes, just like what we do in envelope encryption and in the TLS protocol.

The hacker generates the symmetric key locally on the system. Most ransomware attacks use the AES symmetric cipher—of course, the hackers want the best, most efficient, and most secure cipher! Once the symmetric key is generated, the encryption begins. The hacker encrypts the data.

The next phase is to secure this symmetric key. The hacker generates the asymmetric key pair and sends the public key to the targeted system. The public key encrypts the targeted system's AES or symmetric encryption key. The hacker keeps the private key of the key pair and releases only after the ransom is received. The victim then can use the private key to decrypt the symmetric key and decrypt the rest of the data using the symmetric key. The system is solid and fool-proof. The first and second stages, phishing or social engineering and escalating privileges, are very common attacks. People encounter these types of attacks all the time (sometimes users don't even realize that they have been impacted and the hacker is in the system). However, this stage is critical and needs a skilled hacker to execute this step, which includes creating a symmetric key, encrypting data, generating an asymmetric key pair, and encrypting the symmetric key.

In Chapter 2, you learned how to implement AES symmetric ciphers, and we discussed symmetric or public key encryption in Chapter 3. Refer to these chapters to recap the concepts if needed. The process used here is very similar to envelope encryption that we discussed in Chapter 5. In envelope encryption, the encrypting key or the symmetric key is known as the data encryption key (DEK). The AES symmetric key in this step is no different from the DEK of envelope encryption. The asymmetric public key can be compared with the key encryption key (KEK) of the envelope encryption key.

The hacker also needs to export the encrypted symmetric key (and sometimes even the encrypted data) from the targeted system to the hacker's server. This is another reason why ransomware attack needs hybrid encryption schemes. The process of transmitting the encrypted symmetric key and the encrypted data back to the hacker's server is similar to the TLS protocol functionality. This part of the process (exporting symmetric key and/or encrypted data) is usually executed at the end if needed because the hacker needs the encrypted data and the encryption key to decrypt the data only when the victim doesn't pay the ransom or doesn't pay on time. When the victim doesn't pay, the hacker can threaten the victim with leaking/exposing data. To do that, the hacker needs the data and the decryption key.

- **Warning the victims**

  In this stage, the hacker informs the victim that the system is under attack and locked by the hacker. The hacker also informs the impacted business about the demand and how the ransom should be paid. If you think that the CEO of the impacted business is informed via phone call and asked to bring a bag full of money to some deserted location with a threat that doesn't involve police, then you're dead wrong. It doesn't work that way.

  In ransomware attacks, the hacker locks the screen, and on the blank or plain screen, a warning is displayed with instructions about how much and how to pay the ransom.

- **Recovering from the attack**
  Once the ransom is paid, the hacker (fingers crossed!) releases the private key or the decryption key to decrypt the symmetric encryption key. Ultimately, this decrypted symmetric key is used to decrypt the entire organization's database. The entire process of decrypting everything and entering back into the normal operations mode can take a few days.

  The organizations should also perform a root-cause analysis (RCA) and document the lessons learned. It's also a good idea, at this stage, to rotate all the encryption keys used within the organization as well as change all the passwords. All users must reset their password.

  Obviously, the amount of recovery time varies, depending on the size and footprint of the organization. However, it takes a while to completely sanitize the system, regain full control, and ensure that the hacker isn't in the system.

The anatomy of the ransomware attack consists of these five parts, but the most crucial part is executing the encryption. If the attacker makes any mistake in this step, the entire operation fails. There are several variations of the ransomware attack that involve slight adjustments to the steps. Following are the most common variants of the ransomware attack:

- **Crypto ransomware**
  This is the most common type and widely known ransomware attack. When we talk about ransomware or hear about it in the news, most likely, it's crypto ransomware. In this type of attack, the attacker encrypts the data and demands a ransom.

- **Locker ransomware**
  People have learned from others and their experiences. Many companies now keep backups of their data. In case of a ransomware attack, the data is recovered from the backup, and there's no need to pay any ransom. But guess what? The hackers are a step ahead. They can launch a locker ransomware attack. In this attack, the system is locked out, and the user can't even access the system (and the backup data).

- **Double extortion**
  In this type of attack, the data is not only encrypted and locked but also exported. This imposes another threat of data being exposed if the ransom isn't paid or not paid on time. Like locker ransomware, this threat technique is also used if the victim organization plans to use the backup data and not pay the ransom.

- **Triple extortion**
  In this technique, the hacker adds another dimension to demand the ransom. The hacker asks for ransom from customers and partners also because their data is also locked and unusable.

- **Scareware ransomware**
  Scareware is, as the name suggests, a scare. It generates many warnings and notices on the users' screen but doesn't actually encrypt the data. This is the best-case

scenario—no harm done (assuming nothing is exported!) to the data or system. The victim can disinfect the system and remove the malware that is displaying the warnings. For the most part, this is an easy fix.

- **Doxware ransomware**

    This type of ransomware attack usually targets businesses with personal or sensitive information. The Doxware attack encrypts the sensitive information and threatens the business to release this information if the ransom isn't paid. Releasing the personal and sensitive information can not only create a revenue loss for the victim but could also be a huge reputation loss, which may take years to recover from.

- **Wiper ransomware**

    Wiper ransomware wipes out everything. This type of attack is usually not for ransom or financial gain but for some particular reasons, like taking revenge or deleting the enemy's military documents. You may wonder, as the attacker gains control and deletes all the documents, where is the cryptography in this scenario? This is a valid question. The wiper ransomware uses cryptography to encrypt the data. Once the data is encrypted, the hacker deletes the key and makes the data unusable. The hacker doesn't really delete the entire data physically, but the data is practically useless.

12.2.2 Ransomware as a Service

Over the past quarter century, cybercrimes involving phishing emails, credit card fraud, or hacking into a server or a bank account have been very common. But how do these average hackers get the skills to encrypt data using the most sophisticated symmetric encryption algorithm? How on earth did they figure out the successful implementation of envelope encryption? After all, a team of engineers is required at major organizations to successfully implement these algorithms, but an average hacker with limited resources can execute this flawlessly.

There must be something more to it. This is a business model called *ransomware as a service* (*RaaS*). RaaS is a service offering that's very similar to software as a service (SaaS), where skilled developers write software (malware) and tools necessary to carry out a ransomware attack. Then, hackers purchase or subscribe to this service to launch an attack. The developers of malware and tools are known as RaaS operators, and hackers or subscribers to the service are called RaaS affiliates.

The interaction between the affiliate, the operator, and the victim with the business model of RaaS is shown in Figure 12.2. Apart from the software and tools necessary to execute and manage the attack, the RaaS kit comes with a variety of frills, such as technical support, discussion forums, instructions, and a control panel/dashboard. Premium RaaS services also offer data mining tools that allow attackers to extract or mine data from the victim if the ransom isn't paid on time or not paid at all—this is the data exfiltrate scenario we discussed in the previous section. This feature increases the threat of the attack as hackers can threaten the company and put pressure on them to pay sooner.

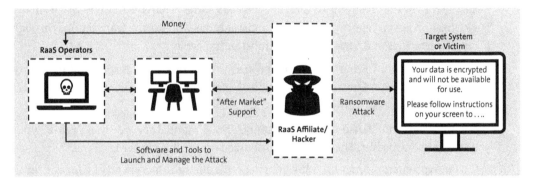

Figure 12.2 RaaS Business Model: Interaction among the Operator, Affiliate, and Victim

The RaaS operators offer various options to the RaaS affiliates to purchase or subscribe to the RaaS kit, such as for a flat fee or a profit-sharing basis. A combination of both purchasing models is also possible, where affiliates pay a percentage of the money made through the attack plus a standard up-front fee to the operator. Because ransomware attacks have been lucrative, some operators have added monthly subscriptions on top of the profit sharing. In this model, the affiliates get regular software updates, the latest tools, and a wide range of products and technologies offered by the operators—all for one monthly fee. How convenient!

RaaS operators are also known as RaaS gangs or groups. Some RaaS gangs are incredibly focused on certain industries. For example, Conti [FBI Fash, 2021] is primarily focused on the healthcare industry. They sell RaaS kits to attack hospitals and healthcare facilities. Conti, for an additional fee, will guide hackers on how to negotiate with hospitals after a successful ransomware attack—a very customer-focused approach. REvil targets high profile personalities and celebrities, while Venus group focuses on C suite executives of major companies. Lockbit, another RaaS group, is running a bounty program on their newly released Version 3 of their RaaS program. If you find any vulnerability in their code, you'll be rewarded just like any other bug bounty program. Aren't they very professional and advanced in their business practices? No wonder they are so successful!

No matter which service model is used, affiliates are responsible for gaining access and escalating the privileges. Thanks to RaaS, ransomware operators have successfully bridged the skill gap, meaning a lack of technical ability isn't stopping anyone from launching a ransomware attack—it only takes a small initial investment and basic IT skills to launch one.

12.2.3 Protecting Against Ransomware Attacks

Everyone understands that ransomware attacks and RaaS businesses are illegal, and there are laws to control or stop them. However, it's very difficult or unlikely (and unrealistic) for law enforcement to catch all the RaaS gangs. To make matters worse, many victims don't even report the incident, and, for those who do, the transactions through cryptocurrencies make the attackers almost untraceable.

In this situation, information security professionals have all the responsibility for protecting against ransomware attacks. Following are some of the helpful tips to protect your data against ransomware and most other cyberattacks:

1. Implement a defense-in-depth strategy. This is mandatory; you can't put all your eggs in one basket. The strategy is discussed briefly in Chapter 1.
2. Encrypt, encrypt, encrypt. We can't stress this enough. Encrypt all of your data in all stages. That way, even if you're attacked and the data is exfiltrated, you're not worried. Your data is encrypted, and the hacker can't threaten you with exposing the sensitive information.
3. Back up your data regularly and create multiple backups if possible. It's also a good idea to keep one backup copy on the public cloud. (Hopefully, the public cloud won't be attacked at the same time as you.)
4. Deploy a centrally managed and integrated endpoint management tool or system. Monitor all the traffic carefully, and set the appropriate alerts.
5. Use a state-of-the-art logging and monitoring system. Detect any intruder or unauthorized access early on. Set the appropriate alerts.
6. In the case of an active ransomware attack, use the monitoring and logging tools to detect and delete the symmetric encryption key before it encrypts your data. If the data is already encrypted, capture the data encryption key before it's encrypted with the attacker's public key.
7. Monitor your network traffic, and intercept the hacker's public key if possible.
8. Some ransomware attackers use the same public-private key pair for multiple victims (to minimize the cost). It can help to ask other victims if they have a private key. It's worth a try—especially if the attacker is the same.
9. Perform periodic checks on all your tools and backups, including retrieving data from the backup. It's also a good idea to set up a dedicated team and perform a table-top exercise at a regular interval.
10. All admin access must use multifactor authentication and follow basic hygiene, such as keeping up with all software updates and segregating the network based on functions or smaller segments.

Unfortunately, ransomware and RaaS are here to stay—at least for now. Cybersecurity professionals can't find or stop all RaaS operators. All they can do is work diligently to protect data and information from such attacks. Don't let your guard down!

12.3 Summary

To protect against a hacker, think like one. It's important for a security professional to learn how cryptographic attacks work and how they can be defended. Many attacks are completely analytical and need almost no or very little technical or mathematical skills. By understating these attacks, designers can easily avoid some of them. Implementation errors can lead to severe implementation attacks, which can be avoided by careful texting and validation methods.

In this chapter, we surveyed various attack techniques on the cryptography algorithms. The attack techniques are divided into three categories: cryptanalytic attacks, man-in-the-middle attacks, and other attacks. You also learned about ransomware and RaaS. Lastly, we covered some of the important measures to protect your data from ransomware attacks.

> **Key Takeaways from This Chapter**
> - Reviewed various cryptographic attacks—cryptanalytic attacks, man-in-the-middle attacks, and various other attacks
> - Discussed ransomware attacks and RaaS
> - Learned ways to protect against ransomware and cyberattacks

In the next chapter, we'll survey various cryptography standards and resources. This will come in handy when you're ready to implement or work on a cryptography algorithm.

Chapter 13
Cryptography Standards and Resources

In this chapter, you'll learn about various standards developed by the government, industries, and organizations for the use and enhancement of cryptography. The chapter will also review resources and best practices you can use to build on the knowledge you've gained throughout this book.

The design and implementation of cryptography algorithms aren't very helpful if they aren't usable by the wider community or society. To achieve this, we need standards and regulations. Fortunately, various government agencies, industries, and organizations have established a set of cryptography standards and best practices for the use and benefit of cryptography.

> **Chapter Highlights**
> - Review government, industry, and organization standards
> - Learn about best practices and resources for use with cryptography

This chapter is divided into four sections. All sections are almost independent of each other, and you can choose to read in any order you prefer. Section 13.1 covers the set of cryptography standards developed by various government agencies around the world. Section 13.2 presents various standards and guidelines developed and proposed by various industry requirements and not-for-profit organizations. Section 13.3 lists various resources and best practices for using cryptography. Section 13.4 provides suggestions for further reading.

The goal of this chapter isn't to list every standard document available on cryptography, but rather to provide a big picture and point you to what is available so you can research/study further and decide what you need to consider while designing or implementing cryptography algorithms.

13.1 Government Standards for Cryptography

Cryptography can deliver confidentiality, integrity, and authenticity which protects the privacy and security of people's data. This piques the interest of government agencies in cryptography around the world. In this section, we'll discuss cryptography standards, guidelines, and requirements developed and enforced by US federal agencies and European agencies.

13.1.1 National Institute of Standards and Technology

The National Institute of Standards and Technology (NIST) publishes various standards under the Federal Information Processing Standards (FIPS) series. These standards are meant to provide quality and interoperability. NIST also releases a special publication (usually under the NIST 800 SP series) to share research, guidelines, and so on. The US federal departments and their vendors are required to adhere to the NIST standards. Nongovernment agencies are free to adopt the standards as well.

> **National Bureau of Standards**
>
> At the end of the 19th century, US industries were lagging behind the top economic powers of the world. To compete against the top industrial developments, in 1901, congress established an agency to meet the challenges faced by the US industries: the National Bureau of Standards (NBS). In 1988, NBS was renamed as the National Institute of Standards and Technology (NIST). NIST has contributions in all industries, but NIST's contribution to cryptography is unmatched. Cryptography wouldn't have matured so much without the leadership of NIST in developing various cryptography standards—from the Data Encryption Standard (DES) to the latest lightweight cryptography!

The use of cryptography in nonmilitary applications was expanded only after the Second World War. Interestingly, from the beginning, NIST was involved in the development and standardization of cryptography algorithms.

In the respective chapters, we've already discussed how NIST was involved in developing DES, Advanced Encryption Standards (AES), post-quantum cryptography, lightweight cryptography (LWC), and so on. The selection process for the AES was conducted by NIST in the late 1990s, but even today, it stands tall above all. It's still considered very robust and secure. At the time of writing, AES is expected to be secured against post-quantum cryptography, although Grover's algorithm can pose some challenges. NIST has a dedicated page for cryptography at *www.nist.gov/cryptography*.

The following is a selected list of cryptography standards and guidelines published by NIST:

- Block cipher techniques
- Crypto publications review
- Digital signatures
- Hash functions

- Interoperable randomness beacons
- Key management
- Lightweight cryptography (LWC)
- Message authentication codes (MACs)
- Multi-party threshold cryptography
- Post-quantum cryptography
- Privacy-enhancing cryptography
- Random bit generation

The complete list of cryptography standards and guidelines is available here: *https://csrc.nist.gov/projects/cryptographic-standards-and-guidelines*.

Additionally, the Computer Security Division of NIST maintains the Computer Security Resource Center (CSRC). The CSRC website is one of the most visited sites of NIST. You can learn more about what CSRC has to offer at *www.nist.gov/itl/csd/computer-security-resource-center*.

13.1.2 European Standards and Regulations for Cryptography

When we discuss cybersecurity, we must mention the European Union Agency for Cybersecurity (ENISA) and the General Data Protection Regulation (GDPR). ENISA is an agency that works with the EU member nations to provide and support the cybersecurity framework. The GDPR is a law to protect the privacy of consumer data within the European Union.

On the cryptography web page, ENISA affirms the following:

> *In cooperation with the European Commission, Member States, and other EU bodies, ENISA engages with expert groups to address emerging challenges and promote good practices on the implementation of cryptographic solutions, with a major focus on post-quantum computing.*

ENISA's technical guidelines for minimum security requirements [ENISA, 2014] mandate the following:

- Software and data in the network and information systems is protected using input controls, firewalls, encryption, and signing.
- Security critical data is protected using protection mechanisms like separate storage, encryption, hashing, and so on.

ENISA also focuses heavily on post-quantum cryptography and has published many white papers, documents, and blogs. ENISA has also published a Recommended Cryptography Measures for Securing Personal Data [ENISA, 2013]. More information about cryptography related standards, guidelines, and requirements is available at *www.enisa.europa.eu/topics/cryptography*.

GDPR is one of the toughest privacy laws in the world. GDPR strongly advocates consumer data privacy and severely punishes businesses that don't meet the minimum requirements. In the event of a personal data breach, GDPR requires that the data controller notify the incident to the data owner within a specified time. However, in Article 34, Section 3(a), GDPR exempts the data controller from this requirement if the data is encrypted (unintelligible to any unauthorized user)! Even GDPR, one of the toughest privacy laws in the world, recognizes the importance and the power of cryptography.

> **A Word of Caution**
> This section (and the book) doesn't intend to provide legal advice about GDPR or any other laws, regulations, and standards. You should do your own homework and research the applicable laws and regulations. The author, publisher, or anyone related to this work isn't responsible or liable.

13.2 Other Standards for Cryptography

In this section, we'll outline the standards and guidelines proposed or managed by nongovernmental organizations, such as ISO, PCI-DSS, HIPAA, and so on. The International Organization for Standards (ISO) publishes ISO 27001 – Information Security Management Standards, which outlines the minimum security requirements. ISO 27001 provides "guidance for establishing, implementing, maintaining and continually improving an information security management system." ISO 27001:2022, ISMS, Annex A describes the policy for using cryptography to meet various security requirements.

The Payment Card Industry-Data Security Standards (PCI-DSS) requirements state to "encrypt cardholder data across open, public networks." PCI-DSS requires point-to-point encryption (P2PE) with the use of a hardware security module (HSM) to store the encryption key. The European counterpart of PCI-DSS, the European Payment Council, also has similar requirements for cryptography protection in their standards.

The US government's Health Insurance Portability and Accountability Act (HIPAA) protects personal health information (PHI). HIPAA standards require that data be transmitted using end-to-end encryption (E2EE) so that only the sender and the authorized recipient can access it.

13.3 Best Practices and Other Resources

In this section, we'll discuss the best practices to use and implement cryptography and other useful resources.

13.3.1 Best Practices for Cryptography

The following are the recommended best practices to consider for implementing cryptography in your environment:

- **Don't invent.**
  The first rule of implementing cryptography is to use a vetted algorithm. Don't invent your own algorithm. If you must, it's always a good idea to get it validated from multiple sources. Use algorithms that have been in use for a while, such as AES, RSA, and so on.

- **Develop a cryptography policy.**
  Develop a policy that outlines the goal, minimum security level, validation parameters, and other criteria. This creates a cryptography implementation blueprint for the organization.

- **Prioritize the encryption.**
  Encrypting data comes with a cost, and you don't want to encrypt everything unless there is a reason. Based on the data classification, decide the security and level, and prioritize the implementation.

- **Randomize the initialization vector (IV).**
  Make sure to randomize the IV. Add salt if required. If the IV is repeated, the chances are the key (or part of it) will eventually repeat. This will give an attacker some clues and predictability.

- **Key size matters, so make it long.**
  The key size is very crucial. Keep the key as long as possible. For example, for the AES algorithm, the key size can be up to 256 bits, while for RSA, 3,072 bits is preferred, if your implementation allows it.

- **Perform effective key management.**
  Key management is one of the most important parts of cryptography. We've already discussed it extensively and won't repeat it here, but to recap, the key should be rotated at a regular interval and deleted when it's no longer used. Once rotated out, the key must be deactivated first and then deleted after the cooling period. The key must be stored, depending on the requirement, in a key management system or in the dedicated HSM.

  Access to the key must be closely guarded and permitted only on a need-to-know basis. No one, absolutely no one, should have access to the key unless there is a business case for it. Access management is one of the very important aspects of key management.

- **Monitor and audit logs.**
  Monitor and audit logs for the key management service (KMS) and key management. Set up the system to generate alerts when a login is suspicious. Perform regular automatic and manual audits of the logs [Dholakia, 2022].

- **Use the latest versions.**
  Use the latest versions where possible, and avoid/discontinue using the older versions such as MD5, SSL3.0, TLS.1.1, TLS1.2, and so on.
- **Manage vulnerabilities and unused ports.**
  Patch software and infrastructure regularly. Unpatched vulnerabilities and open ports give easy access to attackers.

13.3.2 Other Resources

In this section, we'll discuss other useful resources to design, implement and use cryptography:

- **International Association for Cryptography Research (IACR)**
  The IACR homepage (*https://IACR.org*) states, "The International Association for Cryptologic Research (IACR) is a nonprofit scientific organization whose purpose is to further research in cryptology and related fields." IACR was first organized by the initiatives of David Chaum, (digital cash fame; see Chapter 8 for more information).

  IACR is one of the best and most comprehensive resources for cryptography practitioners. It offers several valuable publications, including the *Journal of Cryptography*, *Cryptography ePrint Archive*, Proceedings from IACR Conferences and Workshops, and so on. One interesting publication is *The Museum of Historic Papers in Cryptography*. The Claude Shannon paper, "A Mathematical Theory of Cryptography," from 1945 has made the list! Although it has only two papers so far (and the organization is requesting nominations), we found the concept very interesting.

  IACR offers the following conferences on various topics around the world throughout the year:
  - **General conferences**
    - Crypto
    - Eurocrypt
    - Asiacrypt
  - **Topic-centric conferences**
    - Cryptography Hardware and Embedded Systems
    - Fast Software Encryption
    - Public Key Cryptography
    - Theoretical Cryptography Conferences
  - **Symposia**
    - Real World Cryptography

  We highly recommend that every cryptography practitioner join the IACR. It costs only $50/year and has a wealth of information to offer.

- **Homomorphic Encryption Standardization**
  Homomorphic Encryption Standardization, per the website, is "an open consortium of industry, government, and academia to standardize homomorphic encryption." This is a homomorphic encryption–specific organization.

  Since 2017, Homomorphic Encryption Standardization has organized regular standardization workshops and meetings, which have resulted in whitepapers. Based on these whitepapers, homomorphic standards are now being developed. For more information on the homomorphic standardization, please visit *https://homomorphicencryption.org*.

- **Post-Quantum Cryptography Standardization Project**
  Unlike homomorphic encryption, the Post-Quantum Cryptography Standardization Project is organized by NIST. We discussed the project and the standardization process in Chapter 10 and won't repeat them here.

- **Python Library**
  Python is an easy-to-use language for cybersecurity. The Python project offers a package called "cryptography." Python programmers and developers can use recipes and algorithms from the cryptography project.

  The library can easily be installed by using the following command:

  ```
  $pip install cryptography
  ```

  More information about the cryptography project and download files are available at *https://pypi.org/project/cryptography/*. The documentation is available at *https://cryptography.io/en/latest/*.

- **OpenSSL**
  OpenSSL is a cryptography and SSL/TLS toolkit. More information is available at *www.openssl.org*.

- **Let's Encrypt**
  Let's Encrypt (*https://letsencrypt.org*) provides the digital certificates in order to enable HTTPS (SSL/TLS) for websites, for free, in a user-friendly format.

- **Computer Security and Industrial Cryptography (COSIC)**
  COSIC (*https://www.esat.kuleuven.be/cosic/*) is another good resource for cryptography. It provides free courses and seminars for people who are interested in keeping in touch with the latest trends and developments in cryptography.

- **Applied crypto hardening**
  Applied crypto hardening (*https://bettercrypto.org/*) provides guidelines on implementing SSL, PGP, SSH, and other cryptography protocols. This is a very useful resource for developers.

13.4 Further Reading

Although we have provided references throughout the book, I recommend these wonderful textbooks and/or reference books for further reading (see Table 13.1).

Book Title	Author(s)
The CodeBreakers	Kahn, David
The Code Book	Singh, Simon
Understanding Cryptography	Paar, Christoph et. al.
Introduction to Modern Cryptography	Katz, Jonatha et. al
Serious Cryptography: A Practical Introduction to Modern Encryption	Aumasson, Jean-Philippe
Applied Cryptography: Protocols, Algorithms, and Source Code in C	Schneier, Bruce
Encyclopedia of Cryptography	Van Tilborg, Henk
Data Privacy and Security	Salomon, David
Cryptography Engineering: Design Principles and Practical Applications	Ferguson, Niels et. al.
Handbook of Applied Cryptography	Menezes, Alfred et. al.
Real-World Cryptography	Wong, David
A Graduate Course in Applied Cryptography	Boneh, Dan et. al.
The Joy of Cryptography	Rosulek, Mike et. al.
Post-Quantum Cryptography	Berstein, Daniel et.al
Security and Artificial Intelligence	Batina, Lejla et. al.
Protecting Privacy through Homomorphic Encryption	Lauter, Kristin et. al.
Homomorphic Encryption and Applications	Yi, Xun et. al.
The Basics of Bitcoins and Blockchains: An Introduction to Cryptocurrencies and the Technology that Powers Them	Lewis, Antony
Handbook of Research on Blockchain Technology	Krishnan, Sarvanan
Implementing SSL/TLS Using Cryptography and PKI	Davies, Jashua

Table 13.1 Recommended Further Reading (in No Particular Order)

13.5 Summary

Technology is advancing at a rapid speed, and cryptography is no exception. New developments in cryptography and new attacks on cryptography are on the horizon every day. The only way for a security professional to keep pace with this continuously evolving field is to learn every day. This chapter is intended to help you do just that.

In this chapter, we learned about various governmental and industry standards for cryptography. We also reviewed various resources and best practices for implementing and using cryptography.

> **Key Takeaways from This Chapter**
> - Learned about NIST, CSRC, FIPS, ENISA, and GDPR
> - Reviewed various industry standards for cryptography
> - Recapped the best practices and other resources

In the next chapter, you'll learn about the future trends in cryptography and concluding remarks.

Chapter 14
Future Trends and Concluding Remarks

If we had a crystal ball, we could tell who would prevail in the end—a cryptographer or a cryptanalyst. Instead, we must rely on the current state of technology to predict the future trends in cryptography. In this chapter, we'll briefly look at the future trends in cryptography. We'll wrap up our discussion with concluding remarks.

Cryptography has a very colorful history and has been discussed and refined over centuries. Today, cryptography is living up to the hype and protecting society from various attacks and vicious hackers. If trends are any indicator, the future of cryptography is very bright, and it will be the backbone of technology and our daily lives.

Section 14.1 will briefly examine future trends in cryptography and review various technological advancements that can impact this field. We'll conclude the chapter and the book with short remarks in Section 14.2.

14.1 Future Trends in Cryptography

Cryptography was invented to provide confidentiality, but over the past 50 years, it has slowly made its way into providing integrity, authenticity, and nonrepudiation. Today, people have a lot more expectations from cryptography compared to Caeser's day, and this is only going to grow. Will cryptography meet these expectations?

Before we answer the question, let's first look at what's in store for cryptography. The following are the major trends in cryptography that could impact the future of cryptography:

- **Lightweight cryptography (LWC)**
  It wouldn't be an overstatement to say that we use Internet of Things (IoT) devices every day. IoT devices are everywhere—at work, at home, in public transportation, in cars, at the airport, at coffee shops, and so on. The list can go on and on. All these devices need to be secure. The powerful and robust cryptography algorithms we have, such as Advanced Encryption Standard (AES), can consume lots of power and throughput. Our tiny IoT devices aren't capable of handling conventional ciphers.

14 Future Trends and Concluding Remarks

LWC is the solution. As you learned earlier in Chapter 7, NIST finished the standardization process to select the LWC algorithm and crowned the winner, Ascon. A large number of IoT devices are already in use, and this number is only going to grow. If this trend continues, then eventually, we'll have more use of lightweight cryptography.

Issues such as power consumption/battery life, computation capacity, and so on are imposing some limitations on the technology, but researchers are determined to improve LWC.

- **Homomorphic encryption**

    Homomorphic encryption could be a game changer for privacy requirements and cloud computing. Fully homomorphic encryption (FHE) has lots of potential for improving privacy and security. The technology is still evolving, and researchers are working to improve its efficiency and narrow the targeted malleability. The community is very active, and research on the subject has gained momentum over the past six to eight years.

    This is one of the most promising trends in cryptography and can change the way we do business today. Refer to Chapter 11 for further details on homomorphic encryption.

- **Quantum computers**

    Quantum computers, when realized, can potentially crack asymmetric cryptography and weaken the symmetric cryptography. Development in the following three areas of quantum computers and cryptography are of interest to research:

    - *Quantum cryptography*: You learned earlier about quantum cryptography and how it works based on photons. This technology is very fragile, and it has yet to take off. Once it does, it could be a game changer. However, it's very expensive and very sensitive to the environment at this stage, and it will take some time for this technology to mature.

    - *Quantum key distribution*: This is an extension of quantum cryptography. Quantum encryption key distribution, which we also studied earlier, is based on the principles of physics and photons, the same as quantum cryptography. Once the technology matures, quantum key distribution will be a very robust and almost foolproof way to share the key.

    - *Post-quantum cryptography*: This is a more realistic goal in the short term, since cryptographers are already on top of the issue. Post-quantum cryptography works on mathematical concepts only, not on quantum physics. NIST has already gone through the standardization process. This is almost here (and not much of a future trend.).

    Refer to Chapter 10 for more information on this topic.

- **Zero-knowledge proof (ZKP)**
  The principle behind zero-knowledge proof (ZKP) is to prove to the other party or the system that you're the authorized user without revealing any other information. There is a lot of interest in this topic, as it helps a lot with privacy-preserving applications. We briefly touched on this topic in Chapter 1.

- **Blockchain and cryptocurrency**
  There is a lot of buzz and media attention on this topic. The technology uses encryption and other cryptography services, such as hashing and digital signatures. Any improvement in cryptography in the future will tremendously help blockchain technology and cryptocurrency. We've discussed the technology in detail in Chapter 8. Future trends in cryptography could impact/improve this technology.

- **Privacy-enhancing cryptography**
  Cryptography used to improve a user's privacy is grouped under this category. Some examples are ZKP, FHE, and multiparty communication.

- **AI and cryptography**
  Apart from cryptocurrency, AI is one of the most talked-about technologies right now. AI-based algorithms can improve the efficiency and security of encryption algorithms. There is a lot of interest in this research area, and the technology is developing quickly. On the flip side, cryptography can help AI by using homomorphic encryption to secure the training data. AI and cryptography together result in a win-win situation!

- **DNA cryptography**
  The concept of DNA cryptography isn't totally new but hasn't taken off yet. The idea is to encode information or data into the DNA sequence and then store the data in DNA. Compared to conventional storage, DNA can store a lot of data, has a long shelf-life of data, and consumes very little power. The technology is still in a very early stage.

These are some of the major trends in cryptography. The list isn't exhaustive by any means. New methods and algorithms are constantly appearing on the horizon. Keep yourself up to date by learning new technologies, algorithms, and advancements in the field.

14.2 Concluding Remarks

I'm equally fascinated by the way Egyptians used cryptography 4,000 years ago and the way it will be used with DNA cryptography. In this book, we tried to cover everything from classical cryptography to modern cryptography and beyond.

14   Future Trends and Concluding Remarks

Part I of the book covered the fundamental concepts of cryptography, including symmetric and asymmetric cryptography, hashing algorithms, digital signature schemes, and more.

Part II of the book explained how you can apply the concepts learned to protect our data and information. This part of the book is all about the practical applications in the field of cryptography. It warms up with some routine applications, such as storage security and web security, and then takes you deep into the latest and greatest applications of cryptography—from cloud computing to blockchain technology and from AI to homomorphic encryption. And we explained everything with a minimum use of mathematics!

Cryptography has a greater impact on people's lives now than ever before. Its use isn't restricted to protecting data-at-rest and data-in-transit but extends far into our modern lives. Cryptography has reached our smartphones, bank accounts, health records, doorbells, automobiles, and almost everything we do today.

We live in an era where news of data breaches can tumble the stock market, and attackers often hold data hostage rather than people. Protecting data and information is of paramount concern to every business and individual. How can you ensure that your organization isn't the next victim of ransomware or a data breach? Hopefully, reading this book motivated you to take the next step in protecting your data and information. Let's review what we've covered:

- We provided instructions on how to design and implement some of the most commonly used cryptography algorithms. We also explained how the basic concepts are tied into modern applications.
- If you're aiming to develop a quantum-resistant strategy for your organization, we've provided a wealth of information to help you understand the concept, review and compare various options, and finally select and implement the best algorithm.
- How can I have a third-party vendor process my customer data without accessing it? If this question is keeping you up at night, then we've got you covered. You learned about FHE and determined whether and how it can help your use case.
- Everyone wants to use AI, so why shouldn't you join the frenzy? This book got your feet wet with AI and cryptography. You explored how AI-based algorithms can be used in cryptography as well as how cryptography can be used with AI.
- And yes, even after applying cryptography, things can still go wrong. There are hackers out in the wild and malicious insiders or disgruntled employees in your organization. We showed you how hackers think and discussed some of the possible cyberattacks on cryptography.
- Cryptography is an evolving field, and you must keep learning. We've summarized some useful standards, resources, and best practices for you to learn more as you embark on the journey.

The concepts and applications you learned in this book go well beyond the classroom. You can and should apply the concepts learned in this book to protect data, information, and applications. No matter what your role is in the organization, you can create a profound impact by using the techniques and principles outlined in this book.

14.3 Summary

The journey from transferring cryptic messages on horses to computing encrypted messages without decrypting is truly amazing. The possibility to learn more about cryptography and apply the techniques at your work is limitless. Don't stop after this book; rather, continue to grow in the field of cryptography. I encourage you to read, research, and react!

Appendix A
Bibliography

Armknecht, F., Boyd, C., Carr, C., Gjøsteen, K., Jäschke, A., Reuter, C., and Strand, M. (2015). *A Guide to Fully Homomorphic Encryption*. Cryptology ePrint Archive, Paper 2015/1192. *https://eprint.iacr.org/2015/1192*.

Bakas, A., Frimpong, E., and Michalas, A. (2022). Symmetrical Disguise: Realizing Homomorphic Encryption Services from Symmetric Primitives. In *International Conference on Security and Privacy in Communication Systems* (pp. 353-370). Cham: Springer Nature Switzerland.

Batina, L., Bäck, T., Buhan, I., and Picek, S. (2022). *Security and Artificial Intelligence: A Crossdisciplinary Approach*. Springer Nature Switzerland AG.

Bellare, M., Canetti, R., and Krawczyk, H. (1996). Keying Hash Functions for Message Authentication. In: Koblitz, N. (eds) *Advances in Cryptology — CRYPTO '96*. CRYPTO 1996. *Lecture Notes in Computer Science*, vol 1109. Springer, Berlin, Heidelberg. *https://doi.org/10.1007/3-540-68697-5_1*.

Bellare, M., Rogaway, P. (1995). Optimal asymmetric encryption. In: De Santis, A. (eds) *Advances in Cryptology — EUROCRYPT'94*. EUROCRYPT 1994. *Lecture Notes in Computer Science*, vol 950. Springer, Berlin, Heidelberg. *https://doi.org/10.1007/BFb0053428*.

Benaloh, J., Lampson, B., Simon, D., Spies, T., and Yee, B. (1995). The Private Communication Technology (PCT) Protocol. Microsoft Corporation. *http://graphcomp.com/info/specs/ms/pct.htm \l ref1*.

Bennett, C., Brassard, G. (1984). Quantum Cryptography: Public Key Distribution and Coin Tossing. *Proceedings of IEEE International Conference on Computers, Systems and Signal Processing*, 175-179.

Bernstein, D., Buchmann, J., and Dahmen, E. (Editors). (2008). *Post-Quantum Cryptography*. Springer Nature Switzerland AG.

Bernstein, D., and Lange, T. (2019). eBACS: ECRYPT Benchmarking of Cryptographic Systems. *http://bench.cr.yp.to*.

Bertoni, G., Daemen, J., Peeters, M., Van Assche G., and Van Keer, R. (2012). Keccak Implementation Overview. *http://keccak.noekeon.org/Keccak- implementation- 3.2.pdf*.

Biham, E., Anderson, R., and Knudsen, L. (1998). Serpent: A New Block Cipher Proposal. In: Vaudenay, S. (eds) *Fast Software Encryption*. FSE 1998. *Lecture Notes in Computer Science*, vol 1372. Springer, Berlin, Heidelberg. *https://doi.org/10.1007/3-540-69710-1_15*.

Biham, E., Dunkelman, O., Keller, N. et al. (2015). New Attacks on IDEA with at Least 6 Rounds. *Journal of Cryptology* 28, 209–239. https://doi.org/10.1007/s00145-013-9162-9.

Biham, E., Shamir, A. (1991). Differential cryptanalysis of DES-like cryptosystems. *Journal of Cryptology* 4, 3–72. https://doi.org/10.1007/BF00630563.

Biryukov, A., Kushilevitz, E. (1998). From differential cryptanalysis to ciphertext-only attacks. In: Krawczyk, H. (eds) *Advances in Cryptology—CRYPTO '98*. CRYPTO 1998. *Lecture Notes in Computer Science*, vol 1462. Springer, Berlin, Heidelberg. https://doi.org/10.1007/BFb0055721.

Blatt, M., Guseva, A., Polyakova,Y., and Goldwassera, S. (2020). Secure Large-Scale Genome Wide Association Studies Using Homomorphic Encryption. Cryptology ePrint Archive, Paper 2020/563. https://doi.org/10.1073/pnas.1918257117.

Brakerski, Z., Gentry, C., and Vaikuntanathan, V. (2012). Fully Homomorphic Encryption without Bootstrapping. In *Proceedings of the 3rd Innovations in Theoretical Computer Science Conference (ITCS '12)*. Association for Computing Machinery, New York, NY, USA, 309–325. https://doi.org/10.1145/2090236.2090262.

Britannica, Encyclopedia. (n.d.). Disquisitiones Arithmeticae. www.britannica.com/topic/Disquisitiones-Arithmeticae.

Burkacky, O., Deichmann, J., Klein, B., Pototzky, K., and Gundbert Scherf, G. (2020). Cybersecurity in automotive: Mastering the challenge. A report by McKinsey & Company on Advanced Industries – Automotive & Assembly.

Burton, R. (1991). *The Kama Sutra of Vatsayana* (English translation). Inner Traditions and Company.

Burwick, C.G., Coppersmith, D., D'Avignon, E., Gennaro, R., Halevi, S., Jutla, C.S., Matyas, S., O'Connor, L.J., Peyravian, M., Safford, D.R., and Zunic, N. (1999). MARS - a candidate cipher for AES. https://shaih.github.io/pubs/mars/mars.pdf.

CEMC. Center for Education in Mathematics and Computing, University of Waterloo. (n.d.). Breaking the Vatsyayana Encryption Scheme. www.cemc.uwaterloo.ca/resources/real-world/images/BreakingTheVatsyayanaEncryptionScheme.pdf.

CENC. (2023). Common Encryption in ISO Based Media Files Format. https://www.iso.org/standard/84637.html.

Chaum, D. (1979). *Computer Systems Established, Maintained, and Trusted by Mutually Suspicious Groups*. Electronics Research Laboratory, UC Berkeley.

Cheon, J. H., Kim, A., Kim, M., and Song, Y. (2017). Homomorphic Encryption for Arithmetic of Approximate Numbers. In: Takagi, T., Peyrin, T. (eds.), *Advances in Cryptology – ASIACRYPT 2017. Lecture Notes in Computer Science*, vol 10624. Springer, Cham. https://doi.org/10.1007/978-3-319-70694-8_15.

Chillotti, I., Gama, N., Georgieva, M., and Izabachène, M. (2016). Faster Fully Homomorphic Encryption: Bootstrapping in Less Than 0.1 Seconds. In: Cheon, J., Takagi, T. (eds)

Advances in Cryptology – ASIACRYPT 2016. ASIACRYPT 2016. *Lecture Notes in Computer Science*, vol 10031. Springer, Berlin, Heidelberg. *https://doi.org/10.1007/978-3-662-53887-6_1*.

Cypherpunk. (1994). *http://cypherpunks.venona.com/date/1994/09/msg00304.html*. Retrieved January 2024.

Damgård, I. (1989). A Design Principle for Hash Functions. CRYPTO 1989. Springer LNCS, vol 435.

Dastin, J. (2018). Insight - Amazon Scraps Secret AI Recruiting Tool That Showed Bias Against Women. *https://www.reuters.com/article/world/insight-amazon-scraps-secret-ai-recruiting-tool-that-showed-bias-against-women-idUSKCN1MK0AG/*.

Davies, J. (2011). *Implementing SSL/TLS Using Cryptography and PKI*. Wiley Publishing, Inc.

De Cannière, C. (2006). TRIVIUM: A Stream Cipher Construction Inspired by Block Cipher Design Principles. In: Katsikas, S.K., López, J., Backes, M., Gritzalis, S., Preneel, B. (eds) *Information Security*. ISC 2006. *Lecture Notes in Computer Science*, vol 4176. Springer, Berlin, Heidelberg. *https://doi.org/10.1007/11836810_13*.

De Cannière, C., Preneel, B. (2008). TRIVIUM. In: Robshaw, M., Billet, O. (eds) *New Stream Cipher Designs. Lecture Notes in Computer Science*, vol 4986. Springer, Berlin, Heidelberg. *https://doi.org/10.1007/978-3-540-68351-3_18*.

De Feo, L., Jao, D., and Plût, J. (2011). Towards Quantum-Resistant Cryptosystems from Supersingular Elliptic Curve Isogenies. *https://eprint.iacr.org/2011/506.pdf*.

den Boer, B., and Bosselaers, A. (1992). An Attack on the Last Two Rounds of MD4. In: Feigenbaum, J. (eds) *Advances in Cryptology — CRYPTO '91*. CRYPTO 1991. *Lecture Notes in Computer Science*, vol 576. Springer, Berlin, Heidelberg. *https://doi.org/10.1007/3-540-46766-1_14*.

Dhall, H., Dhall, D., Batra, S., and Rani, P. Implementation of IPSec Protocol. (2012). *Second International Conference on Advanced Computing & Communication Technologies*, Rohtak, India, 2012, pp. 176-181. doi: 10.1109/ACCT.2012.64.

Dholakia, S. (2022). *Logging for SAP S/4HANA Security*. SAP PRESS.

Diffie, W., Hellman, M., (1976). New directions in cryptography. IEEE Transactions on Information Theory, vol 22, no. 6, pp. 644-654, November 1976. doi: 10.1109/TIT.1976.1055638.

DigiCert. (2015). FREAK Attack: What You Need to Know. *https://www.digicert.com/blog/freak-attack-need-know*.

Dobbertin, H. (1998). Cryptanalysis of MD4. Journal of Cryptology 11, 253–271. *https://doi.org/10.1007/s001459900047*.

Dobraunig, C., Eichlseder, M., Mendel, F., and Schlaffer, M. (2019). Ascon Version 1.2. Submission to NIST. Lightweight Cryptography Competition. *https://csrc.nist.gov/CSRC/*

media/Projects/lightweight-cryptography/documents/finalist-round/updated-spec-doc/ascon-spec-final.pdf.

Dobraunig, C., Grassi, L., Helminger, L., Rechberger, C., Schofnegger, M., and Walch, R. (2021). Pasta: A Case for Hybrid Homomorphic Encryption. IACR Cryptology ePrint Arch. Paper 2021/731. https://eprint.iacr.org/2021/731.

Dougherty. (2008). Vulnerability Note VU#836068 MD5 Vulnerable to Collision Attacks. Carnegie Mellon University Software Engineering Institute. https://www.kb.cert.org/vuls/id/836068.

Ducas, L., Micciancio, D. (2015). FHEW: Bootstrapping Homomorphic Encryption in Less Than a Second. In: Oswald, E., Fischlin, M. (eds) *Advances in Cryptology -- EUROCRYPT 2015*. EUROCRYPT 2015. *Lecture Notes in Computer Science*, vol 9056. Springer, Berlin, Heidelberg. https://doi.org/10.1007/978-3-662-46800-5_24.

EFF. (1998). Cracking DES. https://w2.eff.org/Privacy/Crypto/Crypto_misc/DESCracker/.

Elgamal, T. (1985). A Public Key Cryptosystem and a Signature Scheme Based on Discrete Logarithms. In: Blakley, G.R., Chaum, D. (eds) *Advances in Cryptology*. CRYPTO 1984. *Lecture Notes in Computer Science*, vol 196. Springer, Berlin, Heidelberg. https://doi.org/10.1007/3-540-39568-7_2.

EME. (2024). Encrypted Median Encryption. W3C. https://w3c.github.io/encrypted-media.

ENISA (n.d.). Cryptography. www.enisa.europa.eu/topics/cryptography.

ENISA. (2013). Recommended Cryptographic Measures. https://www.enisa.europa.eu/publications/recommended-cryptographic-measures-securing-personal-data.

ENISA. (2014). Technical Guideline on Security Measures. https://www.enisa.europa.eu/publications/technical-guideline-on-minimum-security-measures.

FBI Flash. (2021). Conti Ransomware Attacks Impact Healthcare and First Responder Networks. https://www.ic3.gov/Media/News/2021/210521.pdf.

Feistel, H. (1973). Cryptography and Computer Privacy. *Scientific American*. https://www.scientificamerican.com/article/cryptography-and-computer-privacy/.

Ferguson, N., Schneier, B., and Kohno, T. (2010). *Cryptography Engineering: Design, Principles and Practical Application*. Wiley Publications.

FIPS (Federal Information Processing Standard) Publication 180-4. (2015). *Secure Hash Standard (SHS)*. US Department of Commerce, Technology Administration, National Institute of Standards & Technology.

FIPS (Federal Information Processing Standard) Publication 180-5. (2023). *Secure Hash Standard (SHS)*. US Department of Commerce, Technology Administration, National Institute of Standards & Technology.

FIPS (Federal Information Processing Standard) Publication 186-5. (2023). *Digital Signature Standard (DSS)*. US Department of Commerce, Technology Administration, National Institute of Standards & Technology.

FIPS (Federal Information Processing Standard) Publication 197. (2001). *Advance Encryption Standard (AES)*. US Department of Commerce, Technology Administration, National Institute of Standards & Technology.

FIPS (Federal Information Processing Standard) Publication 203. (2023). *Module-Lattice-Based Key-Encapsulation Mechanism Standard*. US Department of Commerce, Technology Administration, National Institute of Standards & Technology.

FIPS (Federal Information Processing Standard) Publication 204. (2023). *Stateless Hash-Based Digital Signature Standard*. US Department of Commerce, Technology Administration, National Institute of Standards & Technology.

FIPS (Federal Information Processing Standard) Publication 205. (2023). *Module-Lattice-Based Digital Signature Standard*. US Department of Commerce, Technology Administration, National Institute of Standards & Technology.

FIPS (Federal Information Processing Standard) Publication 46-2. (1993). *Data Encryption Standard (DES)*. US Department of Commerce, Technology Administration, National Institute of Standards & Technology.

Fluhrer, S., Mantin, I., and Shamir, A. (2001). Weaknesses in the Key Scheduling Algorithm of RC4. In: Vaudenay, S., Youssef, A.M. (eds) *Selected Areas in Cryptography*. SAC 2001. *Lecture Notes in Computer Science*, vol 2259. Springer, Berlin, Heidelberg. *https://doi.org/10.1007/3-540-45537-X_1*.

Gentry, C. (2009). A Fully Homomorphic Encryption Scheme (Ph.D. thesis). *https://crypto.stanford.edu/craig/*.

Gentry, C., Sahai, A., and Waters, B. (2013). Homomorphic Encryption from Learning with Errors: Conceptually-Simpler, Asymptotically-Faster, Attribute-Based. In: Canetti, R., Garay, J.A. (eds) *Advances in Cryptology – CRYPTO 2013*. CRYPTO 2013. *Lecture Notes in Computer Science*, vol 8042. Springer, Berlin, Heidelberg. *https://doi.org/10.1007/978-3-642-40041-4_5*.

GIMPS. (2018). GIMPS Discovers: Largest Known Prime Number. *https://www.mersenne.org/primes/press/M82589933.html*.

Goldreich, O., Goldwasser, S., and Halevi, S. (1997). Public-key cryptosystems from lattice reduction problems. In: Kaliski, B.S. (eds) *Advances in Cryptology — CRYPTO '97*. CRYPTO 1997. *Lecture Notes in Computer Science*, vol 1294. Springer, Berlin, Heidelberg. *https://doi.org/10.1007/BFb0052231*.

Grover, L. (1996). A Fast Quantum Mechanical Algorithm for Database Search. ACM Symposium on Theory of Computing. Proceedings of the Twenty-Eighth Annual ACM Symposium on Theory of Computing Pages: 212-219. *https://doi.org/10.1145/237814.237866*.

Halevi, S. (2017). *Homomorphic Encryption*. IBM Research. https://shaih.github.io/pubs/he-chapter.pdf.

Hankerson, D., Menezes, A., and Vanstone, S. (2004). *Guide to Elliptic Curve Cryptography*. Springer Professional Computing.

Hellman, M. (2002). An overview of public key cryptography. *IEEE Communications Magazine*, vol 40, pp 42-49.

Hernández-Castro, C., Liu, Z., Serban, A. Tsingenopoulos, I., and Joosen W. (2022). Adversarial Machine Learning. Security and Artificial Intelligence: A Crossdisciplinary Approach. *Lecture Notes in Computer Science*. Springer Nature Switzerland AG.

Heys, H. (2002). A Tutorial on Linear and Differential Cryptanalysis. *Cryptologia*, 26(3), 189-221. https://doi.org/10.1080/0161-110291890885.

Hoffmann, T. (2015). *Bitcoin: The End of Money as We Know It!* Documentary.

Hoffstein, J., Pipher, J., and Silverman, J. (2008). *An Introduction to Mathematical Cryptography*. Springer.

Holevo, A. (1973). Bounds for the Quantity of Information Transmitted by a Quantum Communication Channel. Probl. Peredachi Inf., 9:3 (1973), 3–11; Problems Inform. Transmission, 9:3 (1973), 177–183.

Internet Archive. (n.d.). *Disquisitiones arithmeticae*. https://archive.org/details/disquisitionesa00gaus.

Ireland, C. (2012). Alan Turing at 100. *The Harvard Gazette*. https://news.harvard.edu/gazette/story/2012/09/alan-turing-at-100/.

Kahn, D. (1973). *The Codebreakers*. Macmillan Company.

Kaliski, B. (1992). The MD2 Message-Digest Algorithm. *IETF*. RFC1319. https://datatracker.ietf.org/doc/html/rfc1319 \l page-3.

Keccak. (2023). Figures. https://keccak.team/figures.html.

Koblitz, N. (1987). Elliptic Curve Cryptosystems. *Mathematics of Computation*, vol 48, no. 117, pp 203-209.

Lai, X., Massey J. (1991). A Proposal for a New Block Encryption Standard. Advances in Cryptology. *EUROCRYPT '90 Proceedings*, Springer–Verlag.

Lewis, A. (2018). *The Basics of Bitcoins and Blockchains: An Introduction to Cryptocurrencies and the Technology that Powers Them*. Mango Publishing.

Linn, J., Rivest, R. (1989). RSA-MD2 Message Digest Algorithm. *IETF*. RFC1115. https://datatracker.ietf.org/doc/html/rfc1115 \l section-4.2.

Mariot, L. (2023). AI and Cryptography. Recording of a Ph.D. Course. Semantics, Cybersecurity and Services Group, University of Twente. https://lucamariot.org/teaching/ai-crypto/.

Mariot, L., Jakobovic, D., Bäck, T., and Hernandez-Castro, J. (2022). Artificial Intelligence for the Design of Symmetric Cryptographic Primitives. In: *Security and Artificial Intelligence - A Crossdisciplinary Approach. Lecture Notes in Computer Science.* Springer Nature Switzerland AG.

Matsui, M. (1993). Linear Cryptanalysis Method for DES Cipher. In: *Advances in Cryptology - EUROCRYPT'93*. Springer-Verlag LNCS, vol 765, pp. 386–397.

Matthewson, N., Laurie, B. (2013). Deprecating gmt_unix_time in TLS. *https://datatracker.ietf.org/doc/html/draft-mathewson-no-gmtunixtime-00 \l page-2.*

Merkle, R. (1978). Secure Communications Over Insecure Channels. *Communications of the ACM. Association for Computing Machinery*, vol 21, no. 4. *https://doi.org/10.1145/359460.359473.*

Merkle, R.C. (1990). One Way Hash Functions and DES. In: Brassard, G. (eds) *Advances in Cryptology — CRYPTO' 89 Proceedings*. CRYPTO 1989. *Lecture Notes in Computer Science*, vol 435. Springer, New York, NY. *https://doi.org/10.1007/0-387-34805-0_40.*

Merkle, R. (1988). A Digital Signature Based on a Conventional Encryption Function. CRYPTO '87.

Merkle, R.C. (1988). A Digital Signature Based on a Conventional Encryption Function. In: Pomerance, C. (eds) *Advances in Cryptology — CRYPTO '87*. CRYPTO 1987. *Lecture Notes in Computer Science*, vol 293. Springer, Berlin, Heidelberg. *https://doi.org/10.1007/3-540-48184-2_32.*

Messmer, E. (2012). Father of SSL, Dr. Taher ElGamal, Finds Fast-Moving IT Projects in the Middle East. *Network World. https://www.networkworld.com/article/2161851/father-of-ssl--dr--taher-elgamal--finds-fast-moving-it-projects-in-the-middle-east.html.*

Miller, V.S. (1986). Use of Elliptic Curves in Cryptography. In: Williams, H.C. (eds) *Advances in Cryptology — CRYPTO '85 Proceedings*. CRYPTO 1985. *Lecture Notes in Computer Science*, vol 218. Springer, Berlin, Heidelberg. *https://doi.org/10.1007/3-540-39799-X_31.*

Nakamoto, S. (2008). Bitcoin: A Peer-to-Peer Electronic Cash System. *http://www.bitcoin.org.*

Nguyen, P. (1999). Cryptanalysis of the Goldreich-Goldwasser-Halevi Cryptosystem from Crypto '97. In: Wiener, M. (eds) *Advances in Cryptology — CRYPTO' 99*. CRYPTO 1999. *Lecture Notes in Computer Science*, vol 1666. Springer, Berlin, Heidelberg. *https://doi.org/10.1007/3-540-48405-1_18.*

NIST 800-45. (2007). Guidance on Email Security. NIST special publication Revision 2. *https://csrc.nist.gov/pubs/sp/800/45/ver2/final.*

NIST 800-77. (2020). Guide to IPsec VPN. NIST special publication Revision 1. *https://csrc.nist.gov/pubs/sp/800/77/r1/final.*

OWASP. (2021). OWASP Top 10 2021. *https://owasp.org/www-project-top-ten/.*

A Bibliography

Paar, C., Pelzl J. (2010). *Understanding Cryptography: A Textbook for Students and Practitioners*. 1st Edition. Springer.

Paar, C., Pelzl J., and Guneysu, T. (2024). *Understanding Cryptography: From Established Symmetric and Asymmetric Ciphers to Post-Quantum Algorithms*. 2nd Edition. Springer.

Paar, C., Pelzl J. (2017). SHA-3 and The Hash Function Keccak. A supplement to *Understanding Cryptography: A Textbook for Students and Practitioners*. 1st Edition. Springer.

Paillier, P. (2020). Introduction to FHE. FHE.org Meetup. *https://youtu.be/umqz7kK-Wxyw?si=MjmWpjOLIJpa5P3a*.

Pattanayak, A. (2012). Revisiting Dedicated and Block Cipher Based Hash Functions. Cryptology ePrint Archive, Paper 2012/322. *https://eprint.iacr.org/2012/322*.

Picek, S., Mariot, L., Yang, B., Jakobovic, D., and Mentens, N. (2017). Design of S-Boxes Defined with Cellular Automata Rules. Association for Computing Machinery Proceedings of the Computing Frontiers Conference, CF'17, pp 409-414. *http://dx.doi.org/10.1145/3075564.3079069*.

Rabin, M. (1978). *Digitalized Signatures*. Foundations of Secure Computation, pp 155-168. 1978 Academic Press.

RFC 821. (1982). Simple Mail Transfer Protocol. IETF. *https://datatracker.ietf.org/doc/html/rfc821*.

RFC 1991. (1996). PGP Message Exchange Format. IETF. *https://datatracker.ietf.org/doc/html/rfc1991*.

RFC 2246. (1999). TLS Protocol Version 1.0 IETF. *https://datatracker.ietf.org/doc/html/rfc2246*.

RFC 4301. (2005). Security Architecture for Internet Protocol. IETF. *https://datatracker.ietf.org/doc/html/rfc4301*.

RFC 4880. (2007). OpenPGP Message Exchange Format. IETF. *https://datatracker.ietf.org/doc/html/rfc4880*.

RFC 5246. (2008). Transport Layer Protocol Version 1.2. *https://www.ietf.org/rfc/rfc5246.txt*.

Rivest, R. (1990). The MD4 Message Digest Algorithm. Network Working Group. *https://datatracker.ietf.org/doc/html/rfc1186*.

Rivest, R.L. (1993). Cryptography and machine learning. In: Imai, H., Rivest, R.L., Matsumoto, T. (eds) *Advances in Cryptology — ASIACRYPT '91*. ASIACRYPT 1991. *Lecture Notes in Computer Science*, vol 739. Springer, Berlin, Heidelberg. *https://doi.org/10.1007/3-540-57332-1_36*.

Rivest, R., Adleman, L., and Dertouzos. M. (1978). *On Data Banks and Homomorphisms*. Foundations of secure computation, Academic Press, pp 169–177.

Rivest, R., Robshaw, M., Sidney, R., and Yin, Y. (1998). The RC6 Block Cipher. M.I.T. Laboratory for Computer Science. *https://people.csail.mit.edu/rivest/pubs/RRSY98.pdf*.

Rivest, R., Shamir, A., and Adleman, L. (1977). Cryptographic Communications Method and Systems. U.S. patent 4,405,829 (filed in 1977 and granted in 1983). *https://patents.google.com/patent/US4405829*.

Rivest, R., Shamir, A., and Adleman, L. (1978). A Method for Obtaining Digital Signatures and Public-Key Cryptosystems. *Communications of the ACM*, 21(2): 120–126.

Salomon, D. (2003). *Data Privacy and Security*. Springer Professional Series.

Sangster, P. (2011). *Virtualized Trusted Platform Architecture Specification*. TCG Publication.

Schneier, B. (1994). *Applied Cryptography*. John Wiley & Sons.

Schneier, B. (1998). The Twofish Encryption Algorithm. *www.schneier.com/academic/archives/1998/12/the_twofish_encrypti.html*.

SEC. (2010). SEC 2: Recommended Elliptic Curve Domain Parameters. Certicom Research. *www.secg.org/sec2-v2.pdf*.

Shannon, C. (1949). Communication Theory of Secrecy Systems. *Bell Systems Technical Journal*, 28, 656 – 715.

Shor, P. (1994). Algorithms for Quantum Computation: Discrete Logarithms and Factoring. *Proceedings 35th Annual Symposium on Foundations of Computer Science*, Santa Fe, NM, 1994, pp. 124-134. doi: 10.1109/SFCS.1994.365700.

Singh, S. (2000). *The Code Book: The Science of Secrecy from Ancient Egypt to Quantum Cryptography*. Anchor Books.

Smid, M. (2021). Development of the Advanced Encryption Standards. *Journal of Research of the National Institute of Standards and Technology*, vol 126, article no. 126024. *https://nvlpubs.nist.gov/nistpubs/jres/126/jres.126.024.pdf*.

Sorkin, A. (1984). Lucifer, A Cryptographic Algorithm. *Cryptologia*, 8(1), 22–42. *https://doi.org/10.1080/0161-118491858746*.

Stevens, H. (2018). Hans Peter Luhn and the Birth of the Hashing Algorithms. *IEEE Spectrum. https://spectrum.ieee.org/hans-peter-luhn-and-the-birth-of-the-hashing-algorithm*.

Teicher, J. (2018). The Little Known Story of the First IoT Device. *https://www.ibm.com/blog/little-known-story-first-iot-device/*.

The Register. (2001). RSA Poses $200,000 Crypto Challenge. *https://www.theregister.com/2001/07/25/rsa_poses_200_000_crypto/*.

Turing, A. (1950). Computing Machinery and Intelligence. *Oxford Academic Journal. https://academic.oup.com/mind/article/LIX/236/433/986238?login=false*.

A Bibliography

van Tilborg, H.C.A., Jajodia, S. (eds) Encyclopedia of Cryptography and Security. Springer, Boston, MA. *https://doi.org/10.1007/978-1-4419-5906-5_870* (Original version: Kerckhoffs, 1883. LaCryptographie militaire. *Journal des Sciences Militaires*, 9, 5–38).

Wang, X., Yu, H., and Yin, Y.L. (2005). Efficient Collision Search Attacks on SHA-0. In: Shoup, V. (eds) *Advances in Cryptology – CRYPTO 2005*. CRYPTO 2005. *Lecture Notes in Computer Science*, vol 3621. Springer, Berlin, Heidelberg. *https://doi.org/10.1007/11535218_1*.

White, J. (1971). Network Specifications for Remote Job Entry and Remote Job Output Retrieval at UCSB. *https://datatracker.ietf.org/doc/html/rfc105*.

Yuan, K., Sang, P., Zhang, S., Chen, X., Yang, W., and Jia, C. (2023). An Electronic Voting Scheme Based on Homomorphic Encryption and Decentralization. *PeerJ Computer Science*, 9:e1649. *https://doi.org/10.7717/peerj-cs.1649*.

Appendix B
The Author

Sandip Dholakia is an author, cryptographer, and information security professional. He works as a principal security architect within the Global Security and Compliance group at SAP America. Sandip is a co-chair of the Cryptography Center of Excellence (CoE) and a core member of the AI for Security task force at SAP. As a co-chair of Cryptography CoE, he is actively engaged in developing an enterprise-wide cryptoagility program, focusing on post-quantum cryptography strategies. Before joining SAP, Sandip was an information security architect at Cisco and a security architect and compliance leader at General Motors.

Sandip holds a US patent and defensive publication for designing and developing a cryptography-based antitheft algorithm. He has also published a book and papers about various security topics, including log management, zero-trust architecture, encryption, and security configuration. Sandip graduated from Carnegie Mellon University with dual master's degrees in software engineering and information systems management. He holds CISSP, CCSP, CCSK, GWAPT, and AWS security certifications. Please visit Google Scholar (*https://scholar.google.com/citations?user=mMOjuGcAAAAJ&hl=en*) for the complete list of his publications.

Index

A

A5/1 69
Access control 181, 250
Advanced Encryption Standard (AES) ... 95, 184
   architecture 98
   background 95
   decryption 104
   finalists 107
   implementation 96
   key expansion schedule 101
   security 106
Adversarial attack 310
Algorithm 38
   optimization 304
Amazon Web Services (AWS) 237, 255
   cryptography services 255
American Standard Code for
   Information Interchange (ASCII) 39
Anonymity 279
Application layer 205, 208
Application-level encryption (ALE) 194
   cloud 247
   pros and cons 195
Application-specific integrated
   circuits (ASIC) 240
Arbitrary length input 148
Artificial general intelligence (AGI) 302
Artificial intelligence (AI) 297, 298, 401
   algorithms 305
   best practices 315
   cryptography 303, 309
   defense techniques 313
   domains 300
   ethics 315
   homomorphic encryption 360
   security risks 310
   types 302
Artificial narrow intelligence (ANI) 302
Artificial super intelligence (ASI) 302
Ascon algorithm 264
Associated data (AD) 376
Asymmetric cryptography 50, 113
   algorithms 118
   digital signatures 170
   mathematics 119
   properties 115

Atom 322, 324
   neutral 324
Authenticated encrypted mode 77
Authenticated encryption with
   associated data (AEAD) 264
Authentication Header (AH) 225
Authenticity 47, 115
Authorization 232
Availability 42, 182, 239

B

Babington Plot 32
Backdoor attack 312
Barter system 272
BEAST attack 376
Biased output 315
Binding 241
Birthday attack 377
Birthday paradox 150
Bitcoin 276
   mining 277
BitLocker 185
Block 154
Blockchain 278, 401
   advantages 286
   characteristics 278
   networks 280
   nodes 283
   protocols 281
   transactions 283
Block cipher 70, 85
   CBC mode 72
   CFB mode 75
   CTR mode 76
   ECB mode 71
   GCM 76
   hash functions 153
   OFB mode 74
Bootstrapping 347
Botnet attack 263
Bring your own key (BYOK) 253
Brute-force attack 91, 377
Byte substitution 97, 99
   lookup table 100

417

Index

C

Caesar cipher 29, 57, 80, 89
Candidate block 285
Cellular automata 308
Centralization 272
Certificate 216, 233
   formats 235
   pinning 235
   registration 234
   revoke 235
Certificate authority (CA) 216, 234
   hierarchy 235
Certificate revocation list (CRL) 235
Certificate signing request (CSR) 234
ChangeCipherSpec message 219
ChatGPT 298
Checksum 150
Chi (state) 167
Chinese remainder theorem 122
Chosen cipher text attack (CCA) 372
Chosen plain text attack (CPA) 371
Chromosomes 306
Cipher 35, 57
   block 70
   deterministic versus probabilistic .. 369
   stream 62
   substitution 57
   suite 213, 215
   transposition 60
Cipher block chaining (CBC) mode 72
Cipher feedback (CFB) mode 64, 75
Cipher text 35, 49, 50
   refreshed 347
Cipher text only attack (COA) 369
Circuit privacy 351
ClientHello message 211, 213
ClientKeyExchange message 219
Client random 212
Client-side encryption (CSE) 245, 350
Closed-loop system 261
Closest vector problem (CVP) 333
Cloud applications 247
Cloud computing 237
Cloud cryptography 238
   encryption key management 248
Cloud service provider (CSP) 238, 254
   AWS 255
   Google Cloud Platform 258
   IBM 259
   managed key 251
   Microsoft Azure 257

Cloud service provider (CSP) (Cont.)
   Oracle 259
   SAP 259
Code-based cryptography 331
Code book 30, 72
Coherence time 322
Coinbase 285
Cold backup 292
Cold wallet 292
Collision resistance 149
Columnar cipher 62
Column-level encryption 191, 192
Commodity money 272
Common Encryption (CENC) scheme 231
Communication security 314
Compression method 213, 215
Computer Security and Industrial
   Cryptography (COSIC) 395
Confidentiality 41, 46, 115, 147, 239, 297
Confusion 55, 83, 304
Congruence 56
Consensus 279, 286
Content Decryption Module (CDM) 230
Coprime 123
Corda 282
Counter (CTR) mode 76
Credential storage 241
CRIME attack 221
Cryptanalysis 35
   attacks 368
   categories 48
Cryptanalytic attack 368
Cryptocurrency 117, 271, 275, 401
   Bitcoin 276
   blockchain 279
   cryptographic transactions 282
   ECC 287
   ECDSA 289
   history of money 272
   Merkle root 289
   outlook 293
   SHA256 287
   wallets 290
Cryptographically secure pseudorandom
   number generator (CSPRNG) 66
Cryptography 27, 34
   AI 297, 303, 305, 309
   as a service 254
   asymmetric 50, 113
   attacks 367, 378
   best practices 393
   ciphers 57

Cryptography (Cont.)
   classical 36
   cloud 237
   commercial success 31
   connected devices 237
   cryptocurrency 271, 282
   definitions 34
   early 28
   future trends 399
   government standards 389
   history 27
   homomorphic 345
   keyless 51
   mathematics 56
   modern 37, 48
   necessity 33
   post-quantum computing 341
   quantum 328
   quantum computing 325
   quantum-resistant algorithms 331
   ransomware 379
   resources 394
   rise of hackers 32
   security concepts 40
   security services 46
   services 147
   symmetric 48, 79
   vehicles 266
   vulnerabilities 178
   world wars 30
Cryptology 34
Crypto ransomware 383
Crypto-shredding 179
CSP-managed key 251
Customer-managed key (CMK) 252
Cyclic redundancy checksum (CRC) 207
Cypherpunks 69

D

Data archiving 179
Database encryption 190
Data classification 180
Data controller 244
Data custodian 244
Data encryption at rest 177
   CSE 245
   methods 183
   reasons 182
Data encryption in transit 199
Data encryption key (DEK) 184

Data Encryption Standard (DES) 37, 83, 114
   background 84
   decryption 90
   function f 87
   implementation 85
   key whitening 91
   security 91
   subkey scheduling 89
   triple 91, 93
   variants 91
Data integrity 47
Data lifecycle 178
Data link layer 207
Data owner 244
Data poisoning attack 311
Data processing 179
Data processor 244
Data security 178, 181, 239
   cloud 243
   training 313
Data stages 179
Data storage 178
Decapsulation 204
Decentralization 278
Decentralized application (DApp) 279
Decryption 35
   asymmetric 50
   symmetric 49
Deep learning 300
Defense-in-depth 44, 183
Denial of service (DoS) 263
DES-X algorithm 92
Deterministic cipher 369
DevSecOps 43
Differential cryptanalysis 373
Differential privacy (DP) 314
Diffie–Hellman key exchange
   algorithm 37, 113, 130, 201, 217
Diffie–Hellman–Merkle key
   exchange algorithm 131, 135
   process 134
Diffusion 55, 83, 304
Digital rights management (DRM) 230
Digital signature 117, 152, 169
   Merkle tree 173
   process 170
Digital Signature Algorithm (DSA) 170
Digital Signature Standard (DSS) 170
Discrete logarithms 132
Disk encryption 184
   client-side devices 185
   pros and cons 185

Index

Division algorithm 120
DNA cryptography 401
Double encryption 92
Double extortion 383
Double-slit experiment 320
Downgrade attack 221, 375
Doxware ransomware 384
Duplex mode 206
Dynamic application security
    testing (DAST) 43

E

Ekert's algorithm 330
Electron 322, 324
Electronic code book (ECB) mode 71
Elgamal cryptosystem 136
    process 138
Elliptic curve algorithm 118
Elliptic curve cryptography (ECC) 139, 295
    considerations 143
    cryptocurrency 287
    elliptic curves 140
    key exchange 142
    process 140
Elliptic Curve Digital Signature
    Algorithm (ECDSA) 171
    cryptocurrency 289
Email communication 227
Encapsulating Security Payload
    (ESP) 225, 226
Encapsulation 204
Encoding 39, 42
Encrypted Media Extension (EME)
    specification 231
Encryption 35, 40, 42
Encryption key 35, 38, 248
    length 116
    symmetric 49
    types 48
Encryption key management 248
    decentralized versus centralized 249
    public cloud options 251
    symmetric 81
Enigma 31
Entanglement 319, 322, 330
Envelope encryption 184
Ethereum 281
Euclidean algorithm 119
Euler's totient 123
European Union Agency for
    Cybersecurity (ENISA) 391

Evaluation key 349
Evasion attack 312
Exclusive OR (XOR) 64, 65
Expansion box 88
Extended Euclidean algorithm 120
Extension field 213, 215, 234

F

Factorization algorithm 118
Fastest Homomorphic Encryption in
    the West (FHEW) 347
Federal Information Processing
    Standards (FIPS) 390
Feistel network 83, 107, 110
Fiat money 275, 276
Field 191, 192
Field-level encryption 192
    pros and cons 193
File-based encryption (FBE) 188
    pros and cons 189
File encryption 188
File encryption key (FEK) 188
File storage 188
Finished message 220
Forward secrecy 221
FREAK attack 376
Frequency analysis 32, 369
Full-disk encryption (FDE) 184
Fully homomorphic encryption
    (FHE) 346, 352
    open-source libraries 364

G

Galois/Counter mode (GCM) 76
Galois field 77, 100, 101
General Data Protection Regulation
    (GDPR) 391
Genesis block 285
Genetic algorithm 305
    process 305
Genetic programming 307
Genome-wide association studies
    (GWAS) 356
Gold 274
Google Cloud Platform 258
Google-managed key 258
Goppa code 332
Greatest common divisor (GCD) 119
Grover's algorithm 326

Index

H

Hackers	32, 368, 380
Half-duplex mode	206
Hardware security module (HSM)	250
Hash-based cryptography	334
Hash function	51, 148, 150
algorithms	152
applications	151
block cipher-based	153
dedicated	153
properties	148
Hashing	40, 42, 51, 148, 287
Health Insurance Portability and Accountability Act (HIPAA)	355, 392
Hex	148
Holevo's theorem	319
Homomorphic encryption	345, 400
advantages	361
AI	360
challenges	362
electronic voting system	357
healthcare	354
history	346
hybrid	353
outlook	364
outsourcing	361
PIR	360
process	348
properties	350
types	351
with symmetric key	353
Homomorphic Encryption Standardization	395
Homophonic cipher	60
Host-to-network layer	208
Hot backup	292
Hot wallet	291
Hyperledger	281

I

IBM	31, 259
Identity and access management (IAM)	55
Immobilizer	267
Immutability	279
Implementation attack	378
Indistinguishable encryption	373
Initialization vector (IV)	70, 73, 76, 393
Inject attack	376
Inner padding (ipad)	168
Integrity	42, 47, 169, 239, 279
International Association for Cryptography Research (IACR)	394
International Data Encryption Algorithm (IDEA)	93
International Organization for Standards (ISO)	392
Internet Engineering Task Force (IETF)	203
Internet Key Exchange (IKE)	224
Internet layer	208
Internet of Things (IoT)	260, 399
open versus closed loop	261
risks and attacks	261
Internet Protocol (IP)	207
Internet Protocol Security (IPsec)	223
ESP	226
modes	224
versus TLS	227
Internet Security Association Key Management Protocol (ISAKMP)	224
Iota (state)	167
IP Payload Compression Protocol (IPComp)	227
Isogeny-based cryptography	335
Iterative cipher	85

K

Keccak	162, 163
implementation	165
message processing	163
Kerchoff's principle	38, 248
Key addition layer	97, 98, 101
Key distribution	115
Key encryption key (KEK)	184
Key expansion	102
Key expansion schedule	101
Keyless cryptography	51
Key management policy	249
Key sharing	116
Keystream	35, 65
Key vault	250
Key whitening	91
Keyword	30
Knowledge representation	300
Known-plain text attack (KPA)	370

L

Lamport-Diffie One-Time Signature (LD-OTS)	335
Lattice	333
Lattice-based cryptography	332

Index

Layered security 44
Let's Encrypt! 234, 395
Leveled fully homomorphic encryption 352
Lightweight cryptography (LWC) 263, 399
   implementation 264
Linear cryptanalysis attack 373
Linear feedback shift register (LFSR) 67
Locker ransomware 383
Logical link control (LLC) layer 207
Logjam attack 376
Lucifer 31, 83, 84

M

Machine learning 299, 300
   algorithms 302
MacOS 185
Malleability 351, 372
Man-in-the-middle attack 81, 201, 262, 330, 374
MARS 110
Master encryption key (MEK) 184, 192
McEliece's algorithm 332
MD2 algorithm 153
MD4 algorithm 153
MD5 algorithm 153
   functions 156
   message processing 154
Media access control (MAC) 207
Meet-in-the-middle attack 93, 374
Merkle-Damgård construction 153, 157
Merkle root 172
   cryptocurrency 289
Merkle signature scheme (MSS) 335
Merkle tree 172, 289
Message authentication code (MAC) 152, 168, 376
   hash-based 168
Message digest (MD) algorithm 153
Microdot technique 52
Microsoft Azure 257
MixColumn layer 97, 100, 104
Model security 314
Model stealing 312
Modify attack 376
Modular arithmetic 65
Modulo arithmetic 56
Monoalphabetic cipher 58
Morse code 39
Multichain 281
Multiplicative inverse 121, 127
Multivariate cryptography 334

N

National Bureau of Standards (NBS) 390
National Institute of Standards and
   Technology (NIST) 84, 95, 157, 162, 336, 390
   resources 390
National Institute of Standards
   Technology (NIST) 32
Natural language programming (NLP) 300
Netscape 202
Network layer 207, 208
Node 279, 283
Nonce 284
Nonrepudiation 36, 47, 116
Number theory 119

O

Oakley protocol 224
Object storage 188
Old Quantum Theory 318
One-time pad 53
One-way function 40, 51, 125
Online Certificate Status Protocol (OSCP) ... 235
Open-loop system 261
OpenPGP 229
Open Systems Interconnect (OSI) model 203
Open Worldwide Application Security
   Project (OWASP) 178
Oracle 259
Outer padding (opad) 168
Output feedback (OFB) mode 64, 74
Outsourcing 361, 362

P

Padding 71, 129, 158
Partially homomorphic encryption 352
Partition 187
Passive eavesdropping 375
PassLock system 266
Password verification 151
Payment Card Industry-Data Security
   Standards (PCI-DSS) 392
Penetration testing 44
Peppering 152
Performance optimization 304
Period analysis 60
Permutation 83
Permutation box 89
Phishing 380

Index

Photon 322, 324
    polarization 328
Photonic 324
Physical layer 208
Pi (state) 166
Plain text 35, 49, 50
Platform-managed key 257
Polyalphabetic cipher 58
Polybius Square 29
Polygraphic cipher 60
POODLE attack 203, 375
Post-quantum cryptography 317, 331, 336
    algorithms 337
    outlook 341
    preparation 338
    standardization 336
Post-quantum resistance 80
Pre-fully homomorphic
    encryption (Pre-FHE) 346
Preimage resistance 149
Presentation layer 205
Pretty Good Privacy (PGP) 228
Prime number 122
Privacy 315
Privacy attack 312
Private information retrieval (PIR) 360
Private key 50, 80, 114, 170, 201, 234
    forgotten 293
Privilege creep 263
Probabilistic cipher 369, 374
Product cipher 85
Proof-of-stake 284
Proof of transaction 169
Proof-of-work 284
Proton Mail 230
Pseudorandom number generator (PRNG) ... 66
Public key 50, 114, 170, 201, 232, 348
    subject 233
Public Key Cryptography
    Standards (PKCS) 235
Public key infrastructure (PKI) 118, 233
    AWS 256
PyCrypto 106
Python 106, 308, 395

Q

Quanta of energy 318
Quantum 318
Quantum computer 294, 323, 400
Quantum computing 317
    cryptography 325

Quantum computing (Cont.)
    entanglement 322
    quibits 322
    risks 325
    superposition 321
    technologies 323
    wave-particle duality 320
Quantum cryptography 328
Quantum dot 324
Quantum key distribution (QKD) 330
Quantum leap 318
Quantum supremacy 341
Qubit 319, 322
Quorum 282

R

Rail fence cipher 61
Rainbow attack 377
Random number generator (RNG) 65, 304
Ransomware 367, 379
    anatomy 380
    as a service 384
    attacks 182, 383
    defense 386
RC4 .... 68, 370
RC6 109
Record header 218
Red team testing 44
Registration authority (RA) 234
Reinforcement learning 303
Remainder 56
Remote login 381
Replay attack 375
Representation money 273
Rho (state) 166
Rijndael 96
Robotics 301
Role-based access control (RBAC) 193
Round coefficient (RC) 103
Round key 102
Route cipher 62
RSA 122, 124, 229, 295
    challenge 130
    considerations 129
    encryption-decryption 127
    key generation 125

S

Salting 152
SAP 259

423

Index

Scareware ransomware … 383
Schrodinger's cat … 321
Scytale cipher … 61
Sealing … 241
Second preimage resistance … 149
Secure development and operational lifecycle (SDOL) … 43
Secure Hash Algorithm (SHA) … 157
Secure Multipurpose Internet Mail Extension (S/MIME) … 228, 229
Secure Socket Layer (SSL) … 202
Security … 33, 40
   AI … 310
   applications in the cloud … 247
   cloud … 238
   communication … 314
   cryptanalytic attacks … 373
   cryptography attacks … 368
   data … 177, 313
   data at rest … 182
   data in the cloud … 243
   infrastructure in the cloud … 240
   model … 314
   post-quantum … 336
   ransomware … 386
   services … 46
   storage … 177
   streaming and downloading … 230
   triad … 41, 239
   web … 200, 209
   web-based applications … 227
Security association (SA) … 224
Security Key Exchange Mechanism (SKEME) … 224
Seed value … 63
Sensor … 260
Sequence control … 206
Serpent … 107
ServerCertificate message … 216
ServerHelloDone message … 218
ServerHello message … 214, 215
ServerKeyExchange message … 217, 218
Server random … 215
Server-side encryption (SSE) … 246
Session ID … 212, 215
Session key … 228
Session layer … 206
Session resumption … 221
SHA-1 algorithm … 157
SHA-256 algorithm … 285
   cryptocurrency … 287
SHA-2 algorithm … 157

SHA-3 algorithm … 161
   implementation … 164
   message processing … 163
   states … 166
SHA-512 algorithm … 157
   message processing … 158
Shared key … 49, 80
Shift-left approach … 43
Shift register … 67
ShiftRow layer … 97, 100, 104
Shor's algorithm … 295, 325
Shortest vector problem (SVP) … 333
Side-channel attack … 264, 378
Simple Mail Transfer Protocol (SMTP) … 227
Simplex mode … 206
SLOTH attack … 376
Smart contract … 279
Software-defined network (SDN) … 223
Software licensing … 232
Somewhat homomorphic encryption … 352
SP-network … 265, 304
Sponge construction … 162
SSLv3.0 … 202
State … 97
Static application security testing (SAST) … 43
Statistical attack … 378
Steganography … 51
Storage security … 177
Stream cipher … 62
   asynchronous … 64
   examples … 68
   implementation … 65
   pros and cons … 67
   security … 65
   synchronous … 63
Streaming and downloading … 230
Subkey scheduling … 97
Substitution … 83
Substitution box (S-box) … 88
Substitution cipher … 57
Substitution-permutation network … 97
Supercomputer … 323
Superconducting … 324
Superposition … 319, 321
Supersingular isogeny Diffie-Hellman (SIDH) algorithm … 335
Supervised learning … 302
Symmetric cryptography … 48, 79
   algorithms … 83
   encryption keys … 81
   pros and cons … 80

Index

T

T2 chip	185
Table	190
Tablespace	190
Targeted malleability	351
TCP/IP networking model	208
Telegram	30, 39
Theta (state)	166
TLSv1.2	210, 215
TLSv1.3	215, 220
Transducer	260
Transparent data encryption (TDE)	190
pros and cons	191
Transport Control Protocol (TCP)	207
Transport layer	206
Transport Layer Security (TLS)	203
handshake	210
implementation	209
IPsec	227
record	220
versions	203, 214
Transport mode	224
Transposition cipher	60
Trapdoor function	125
Trapped ion	324
Triple DES	91, 93
Triple encryption	93
Triple extortion	383
Trivium	69
True random number generator (TRNG)	66
Trusted Platform Module (TPM)	185, 240
architecture	241
cryptographic functions	240
virtual	242
Tunnel mode	224
Turing test	298
Twofish	108

U

Unicode	39
Unsupervised learning	303
User Datagram Protocol (UDP)	207, 222

V

Vatsyayana cipher	28
Vehicle Anti-Theft System (VATS)	266
Vernam cipher	53, 63
Vigenère cipher	32, 58
Virtual machine (VM)	242
manage vTPMs	243
Virtual private network (VPN)	222
set up with IPsec	224
transport versus tunnel	224
Virus detection	152
Volume	186, 187
Volume encryption	186
pros and cons	187

W

Wallet	290
hot versus cold	291
types	292
Wave-particle duality	319, 320
Web-based application	227
Web of trust	228
Web security	200
email communication	227
protocols	209
Windows	185
Winternitz One-Time Signature (WOTS)	335
Wiper ransomware	384
Wired Equivalent Privacy (WEP)	370
Wolfram's rule 30	308

X

X.509 certificate	233

Z

Zero-knowledge proof (ZKP)	54, 401
Zimmermann Telegram	31

- Understand IT system vulnerabilities and identify attack vectors
- Learn to secure multiple infrastructures, including Linux, Microsoft Windows, cloud, and mobile
- Master pen testing with tools like Metaspolit, Kali Linux, hydra, OpenVAS, Empire, Pwnagotchi, and more

Kofler, Gebeshuber, Kloep, Neugebauer, Zingsheim, Hackner, Widl, Aigner, Kania, Scheible, Wübbeling

Hacking and Security

The Comprehensive Guide to Penetration Testing and Cybersecurity

Uncover security vulnerabilities and harden your system against attacks! With this guide you'll learn to set up a virtual learning environment where you can test out hacking tools, from Kali Linux to hydra and Wireshark. Then expand your understanding of offline hacking, external safety checks, penetration testing in networks, and other essential security techniques, with step-by-step instructions. With information on mobile, cloud, and IoT security you can fortify your system against any threat!

1141 pages, pub. 07/2023
E-Book: $54.99 | **Print:** $59.95 | **Bundle:** $69.99

www.rheinwerk-computing.com/5696

- Master blockchain fundamentals and implement applications on the Ethereum network
- Develop smart contracts and decentralized applications (dApps) with Solidity
- Test, debug, and secure your blockchain applications

Tobias Fertig, Andreas Schütz

Blockchain

The Comprehensive Guide to Blockchain Development, Ethereum, Solidity, and Smart Contracts

Demystify the blockchain—and learn how to use it—with this practical guide. Start from the ground up: What is Ethereum? What is Solidity? And how are they used to create smart contracts? Then see how to implement your own blockchain, including configuring a peer-to-peer network, managing miner accounts, and more. Follow step-by-step instructions and detailed code examples to develop smart contracts and dApps. Work with cutting-edge technologies such as Bitcoin, DeFi, NFTs, and more. Welcome to the world of blockchain!

654 pages, pub. 09/2024
E-Book: $54.99 | **Print:** $59.95 | **Bundle:** $69.99

www.rheinwerk-computing.com/5800

- A practical guide to Python for nonprogrammers
- Work with NumPy, SymPy, SciPy, Matplotlib, and VPython
- Automate numerical calculations, create simulations and visualizations, perform statistical analysis, and more

Veit Steinkamp

Python for Engineering and Scientific Computing

It's finally here—your guide to Python for engineers and scientists, by an engineer and scientist! Get to know your development environments and the key Python modules you'll need: NumPy, SymPy, SciPy, Matplotlib, and VPython. Understand basic Python program structures and walk through practical exercises that start simple and increase in complexity as you work your way through the book. With information on statistical calculations, Boolean algebra, and interactive programming with Tkinter, this Python guide belongs on every scientist's shelf!

511 pages, pub. 03/2024
E-Book: $54.99 | **Print:** $59.95 | **Bundle:** $69.99

www.rheinwerk-computing.com/5852

- Learn to work with scripting languages such as Bash, PowerShell, and Python
- Get to know your scripting toolbox: cmdlets, regular expressions, filters, pipes, and REST APIs
- Automate key tasks, including backups, database updates, image processing, and web scraping

Michael Kofler

Scripting

Automation with Bash, PowerShell, and Python

Developers and admins, it's time to simplify your workday. With this practical guide, use scripting to solve tedious IT problems with less effort and fewer lines of code! Learn about popular scripting languages: Bash, PowerShell, and Python. Master important techniques such as working with Linux, cmdlets, regular expressions, JSON, SSH, Git, and more. Use scripts to automate different scenarios, from backups and image processing to virtual machine management. Discover what's possible with only 10 lines of code!

470 pages, pub. 02/2024
E-Book: $44.99 | **Print:** $49.95 | **Bundle:** $59.99

www.rheinwerk-computing.com/5851

- Your complete, cross-distribution, professional guide to Linux, for beginners and advanced users

- Get detailed instructions for installation, configuration, and administration, on both desktop and server

- Set up security, virtualization, and more

Michael Kofler

Linux

The Comprehensive Guide

Beginner or expert, professional or hobbyist, this is the Linux guide you need! Install Linux and walk through the basics: working in the terminal, handling files and directories, using Bash, and more. Then get into the nitty-gritty details of configuring your system and server, from compiling kernel modules to using tools like Apache, Postfix, and Samba. With information on backups, firewalls, virtualization, and more, you'll learn everything there is to know about Linux!

1178 pages, pub. 05/2024
E-Book: $54.99 | **Print:** $59.95 | **Bundle:** $69.99

www.rheinwerk-computing.com/5779

- The complete Python 3 handbook
- Learn basic Python principles and work with functions, methods, data types, and more
- Walk through GUIs, network programming, debugging, optimization, and other advanced topics

Johannes Ernesti, Peter Kaiser

Python 3

The Comprehensive Guide

Ready to master Python? Learn to write effective code, whether you're a beginner or a professional programmer. Review core Python concepts, including functions, modularization, and object orientation and walk through the available data types. Then dive into more advanced topics, such as using Django and working with GUIs. With plenty of code examples throughout, this hands-on reference guide has everything you need to become proficient in Python!

1036 pages, pub. 09/2022
E-Book: $54.99 | **Print:** $59.95 | **Bundle:** $69.99

www.rheinwerk-computing.com/5566